Left; Underground Mine Model, World Museum of Mining, Butte, Montana. **Below;** Montana's largest coal strip mine, Western Energy Co., Colstrip, Montana. **Right;** Montana's largest open-pit copper mine, Anaconda Company Mineral Resources Group.

MINING Methods & Equipment

Koehler S. Stout

Associate Dean Engineering Division
MONTANA COLLEGE OF MINERAL SCIENCE AND TECHNOLOGY
Butte, Montana

ACKNOWLEDGMENTS

Special acknowledgment is given to the mining companies and mine equipment suppliers for the excellent photographs supplied for this book.

As often happens, a book turns into a family affair. My wife, Phyllis, patiently read, corrected and typed the text. My oldest daughter Karen, an environmental engineer, took many of the photographs used. Carla, our second daughter and a mining engineer, drafted most of the new sketches. Our youngest daughter, Janet, finished drafting the last five plates.

Many fine photographs of coal stripping operations were taken and supplied to us by Bob Webster of the Western Energy Co. Excellent underground and surface photos were taken by Bob Stewart of the Anaconda Company. Photos supplied by others were professionally done, and although the names of the photographers are unknown, their works are gratefully acknowledged.

Selected parts of the introduction and acknowledgments for *Mining Methods and Equipment Illustrated* printed in 1967.

In 1959, Richard W. Banghart, a senior student in mining engineering, wrote a report entitled *Mining Methods in Montana.* Much information was taken from that excellent report and incorporated in this work. Special appreciation is expressed to Mr. Banghart for his original work. Mine operators throughout the state were most cooperative in permitting inspection of their properties and answering numerous questions about their operations. Many pictures in the report were provided by Anaconda Company Research Department, Stewart Hurlbut, director, and the Communication Department, Tom Wigal, manager. The Gardner-Denver Co., Ingersoll-Rand Co., Archibald Co., and Joy Manufacturing Co. were most generous in supplying pictures and providing data.

The excellent drafting of sketches was done by Roger Holmes. The late Professor William A. Vine, former head of the Mining Department at the Montana College of Mineral Science and Technology, was most cooperative in the photographing of mine models in his department.

MINING INFORMATIONAL SERVICES
McGraw-Hill, Inc.
1221 Avenue of the Americas, New York, N.Y. 10020

Library of Congress. Catalog Card Number 80-81937
0-07-039794-5

CONTENTS

ILLUSTRATIONS

MINING METHODS AND EQUIPMENT

by Koehler Stout

In December 1967 the Montana Bureau of Mines and Geology published *Mining Methods and Equipment Illustrated.* That booklet proved to be very popular and had several printings. However, in the intervening years there have been a great many developments in the mining industry. This new publication includes a number of the improvements and innovations in mining since the first booklet was written. Much of the material from the previous edition is just as applicable today, and has been retained in this book. The new edition also includes sections on finding, evaluating, and opening a mineral deposit. Land reclamation now an essential part of mining also has been included.

Mining is one of our basic industries. Without mining, civilization would still be in the dark ages. We wouldn't have modern conveniences such as television, refrigerators, automobiles, airplanes or essential services such as electric power, telephones, water and gas. Even though mining is a basic industry, the public has little knowledge of it except that it takes place at or below the surface of the ground. The problems peculiar to mining and the innovative ways that miners produce ore under hostile conditions are for the most part unknown to the public.

While most books on mining are written for people who have some knowledge of the subject, this book is intended as an introduction to the fundamental concepts used, and to the application of those concepts in producing ore and coal. Therefore, it does not probe the details of any one system.

The author of this book believes the proverb that one picture is worth a thousand words. For ease of understanding, illustrations have been placed on one page with pertinent explanations on the facing page.

It is almost impossible to describe a mining method unless at least a sketch of the system is shown. A sketch, however, has only two dimensions. Therefore, special techniques are used to illustrate three-dimensional effects. Because not everyone has had experience in reading sketches and maps, the first part of this publication describes those functions. The second part illustrates and describes the shape and type of ore bodies which influence the mining methods chosen. Photographs of mine models are used in conjunction with sketches throughout the publication because the three dimensional effects of the photos are more easily understood. The third part of the book is new and discusses the techniques of finding and evaluating ore bodies. Once an ore body has been found, it must be opened or developed for mining. Part four treats ways to accomplish this. Part five discusses and describes the major operations in mining and some of the techniques used in these operations. The sixth part covers development, or the work that prepares the ore for mining. Mining is logically divided into surface and underground methods, and the seventh part of the book describes surface methods, including placer mining, open-pit, and stripping. The eighth part is devoted to underground mining methods. Part nine also is new and describes reclamation procedures used in surface mining. Finally, an explanation of common mining terms, and suggestions for further reading are contained in part ten.

PART 1

PICTURES, MODELS, SKETCHES AND MAPS

The story of mining can best be told with models, pictures, sketches, and maps. Certain parts of underground mines can be photographed, but it is impossible to illustrate all of an underground mine except by pictures of models, by sketches, or by maps. The operation of an open-pit mine can be easily seen, but sketches and maps illustrate more clearly the mining operation or sequence of operations.

For those who may not have had experience in reading and interpreting sketches or maps, this part of the bulletin will explain what sketches and maps can show and how to interpret them.

TOP, BOTTOM, SIDE, AND END VIEWS

(PLATE 1-1)

Picture A shows a box lying on a table; the picture was taken at an angle.

C is a space or block diagram, sometimes called a space sketch, of the box showing the same general view as shown in the picture. The purpose of this sketch is to show the similarity between the picture and the sketch. The directions of various other views are indicated by arrows. The top view looks directly down on the box; this view is illustrated in D. Sketch E shows the front view and F the side, or end view.

SECTION VIEWS

To illustrate a mine by sketches and maps, it may be necessary to use section views, hidden lines, and other devices to illustrate certain details. A mine is an operation in three dimensions, length, depth, and width, but our sketches and maps show only two dimensions, length and width. Therefore, to illustrate certain features, it is necessary to use certain tricks or conventions of drafting.

In E, a front view of the box, the dashed lines indicate the thickness of the sides, which cannot be seen from the front. Drafting convention has us illustrate the hidden thickness by a dashed line where the edge would be if we could see it.

If the structure is complicated, there may be so many hidden lines in a view that their meaning is impossible to interpret. To show complicated details, an imaginary cut is made through the object, but in picture B the box was actually cut. The cut portion of the object is shown with slanted lines or "hatched lines." G is a sketch of the right portion in picture B. Note the hatches slant in different directions on the bottom and the side pieces, a common practice where two different pieces of material come together.

H and I are space sketches and resulting views of the box, had it been cut in different directions.

In mining, a section as shown in G is called a longitudinal section, or simply a long section. The top view is called the plan view H, and the side view is a cross section I.

PLATE 1-1 SECTION VIEWS

A. Picture of a box

B. Picture of a box cut in two

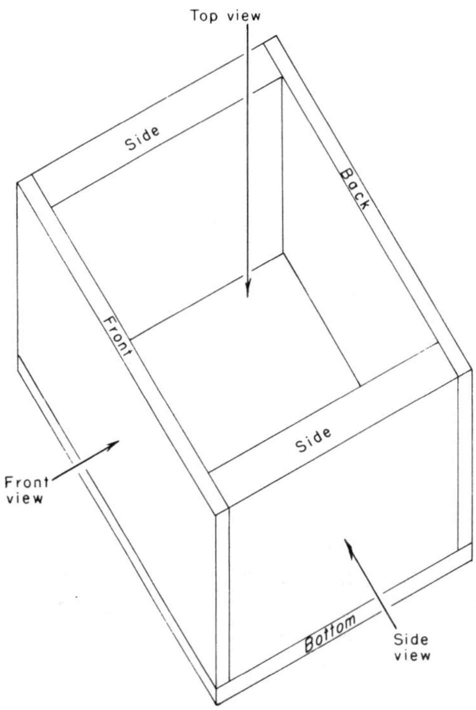

C. Block diagram of box

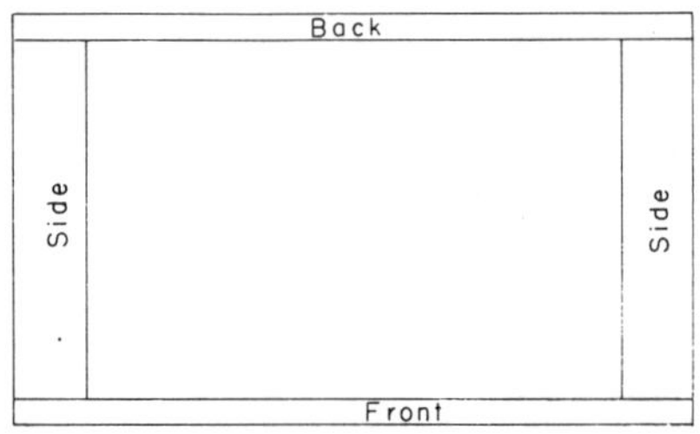

D. Top view of a box

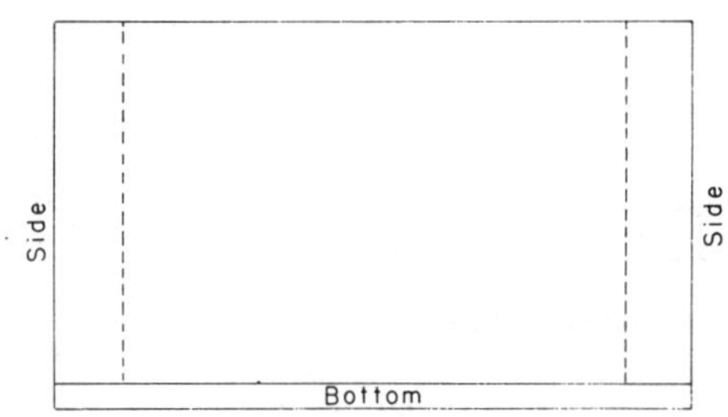

E. Front view of a box

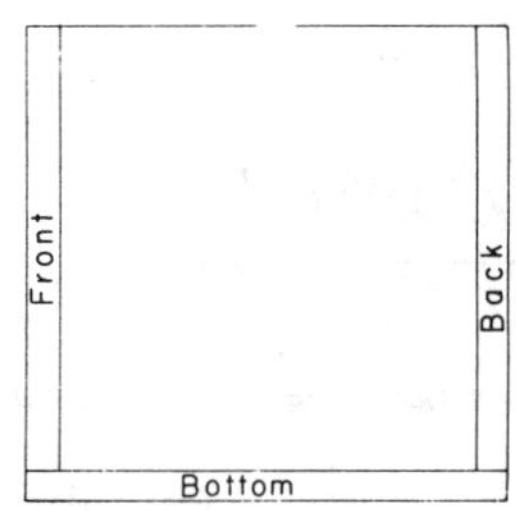

F. Side view of a box

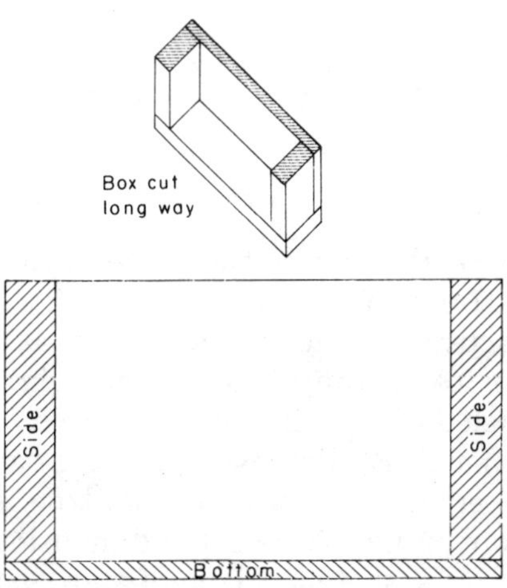

G. Front or longitudinal section

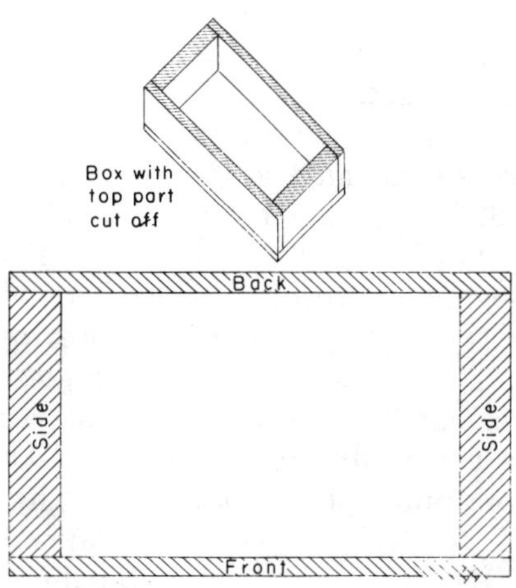

H. Top or plan section

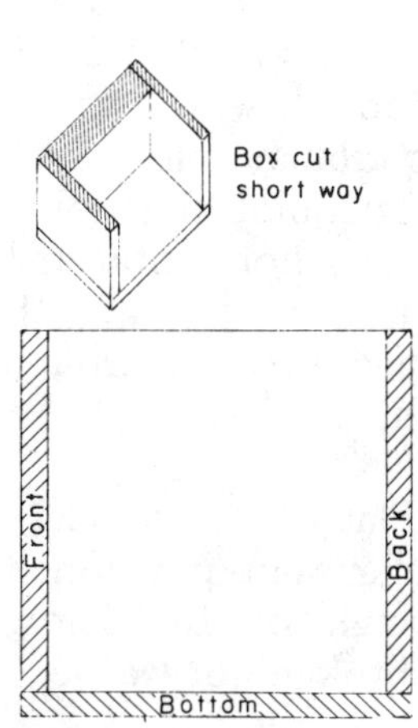

I. Side or cross-section

COMMON VIEWS OF A MINE

A mine is contained in three-dimension space—length, width, and depth—but to present a picture on two-dimension maps, as many as four different views may be required. Many ore bodies are in tabular veins, that is, similar in shape to a piece of cardboard, which has width, length, and thickness. This tabular vein may be positioned in the earth in a horizontal position, vertical, or at any angle between horizontal and vertical. To further complicate the picture, the vein may be curved or warped or even broken into segments within the earth.

Plate 1-2 includes a sketch and views of a tabular vein positioned in the earth (dipping) at an angle of 45° below the horizontal plane, but not broken nor warped.

BLOCK DIAGRAM

A is a sketch of an imaginary block of the earth containing this vein. The lines depicting the outline of the rock overlying the vein are dashed to avoid confusion. The vein is indicated by hatched lines, except where it was cut on the edges of this block; these cuts are indicated by short wavy lines. Notice the dimensions on this block and on all the other views.

The plan view is shown in C, the side or cross-section view is shown in D, and the front or longitudinal (long) view is shown in E. These views are comparable to the same views in Plate 1-1.

PLANE-OF-VEIN VIEW

B is a view of the vein in the plane of the vein. This view is different from any of the other views previously discussed. The direction is at right angles (90° angle) to the plane of the vein. Note the arrow labeled "view in plane of vein" in A and D. Note that in B, dimension a is the same as in the other sketches, g now shows its true length on the sides, but none of the other dimensions are true length in this sketch. Plane-of-vein sections are often used where all workings are in the plane of the vein, because in this view the workings are shown in true length and width.

PLATE 1-2 VIEWS OF A VEIN

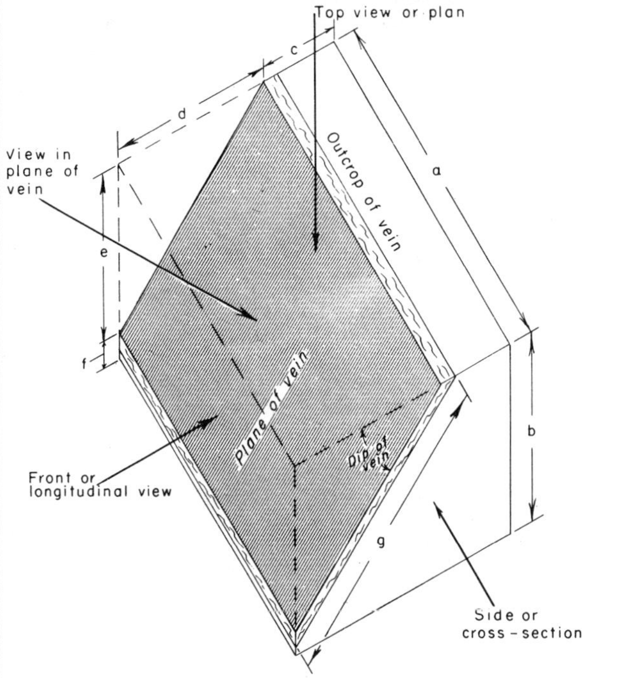

A. Block diagram of a vein contained in a block of earth

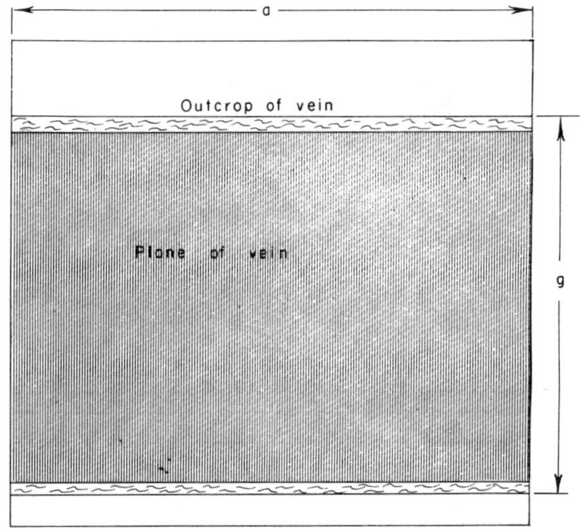

B. Plane-of-vein view

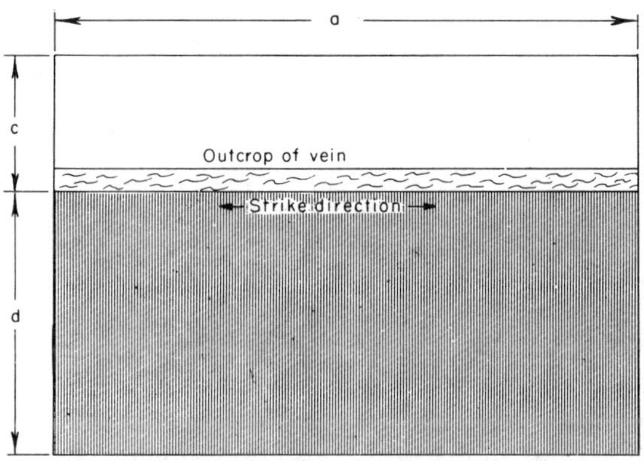

C. Top or plan view

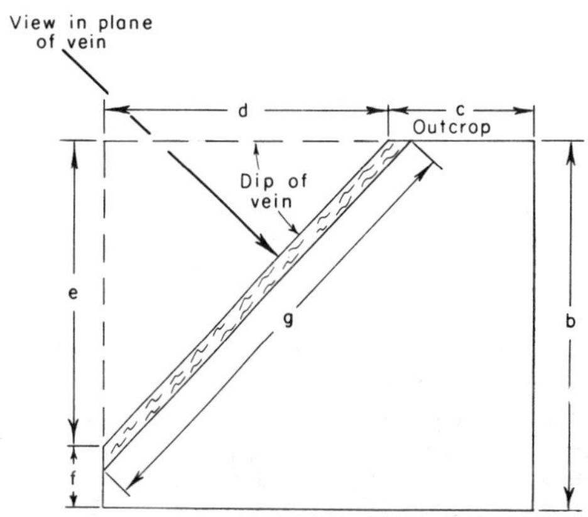

D. Side or cross-section view

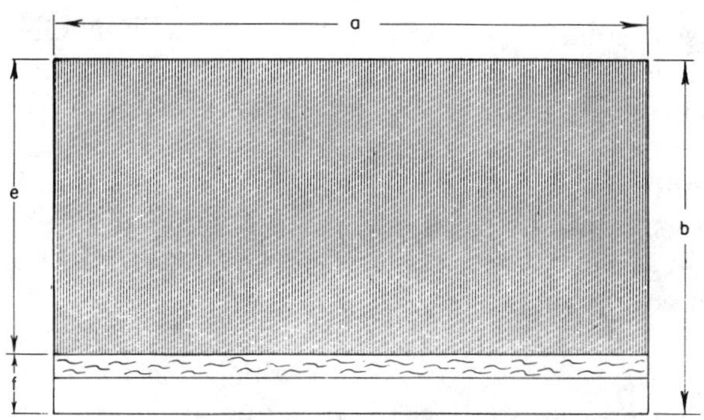

E. Front or longitudinal view

PROJECTION VIEWS OF A MINE

Most drawings or sketches of solid objects are designed to show the solid object as it appears in space. In mining, however, the objects that appear on maps or sketches are spaces or openings in the solid enclosing rock. The model shown in Plate 1-3A depicts the mine openings as solid pieces of construction material set out in space, whereas the solid rock that encloses the openings in the actual mine is depicted as space in the model.

Plate 1-3 is a block diagram of a vein similar to that in Plate 1-2, except that mine workings are included and labeled in these sketches. A is a photograph of the model of the mine, B is the block diagram of the mine. C, D, E, and F show the four most common views of the mine. C is a plane-of-vein view; D the cross-section view; E the long view; and F the plan view of the workings. These views were explained in Plates 1-1 and 1-2.

PROJECTION ON A PLANE

To show various openings that are unequal distances from the point of observation, a technique is used that is called projection on a plane. Objects not in the plane are projected to the plane in their true dimensions and at their true distance from one another, rather than in perspective.

Study views B, C, D, E, and F, and notice which views show the correct lengths or dimensions of the various openings. If the vein were curved, warped, or broken, even more views would be necessary to show the true lengths of certain mine openings.

PLATE 1-3 PROJECTION VIEWS OF A MINE

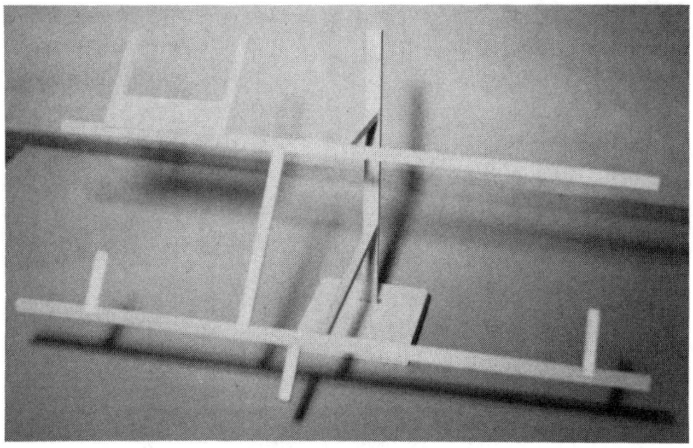

A. Picture of mine model

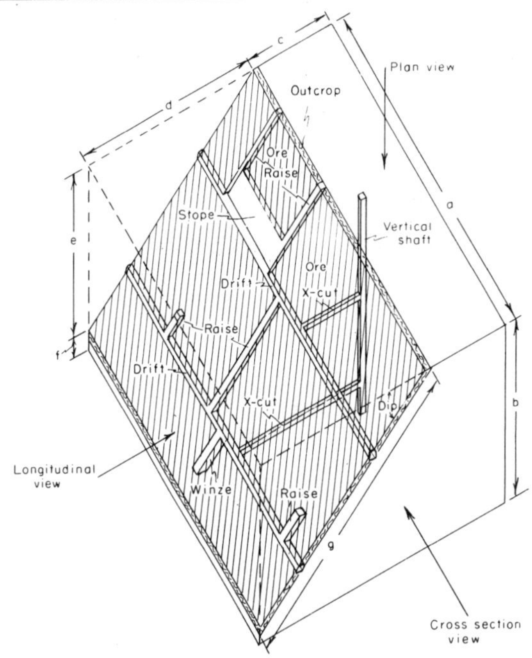

B. Block diagram of mine

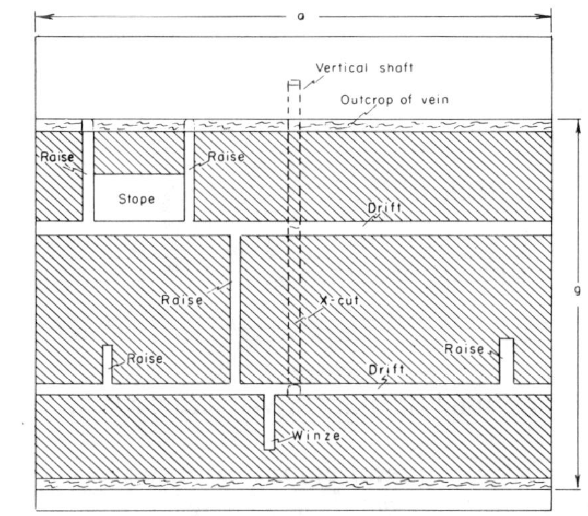

C. Plane-of-vein view

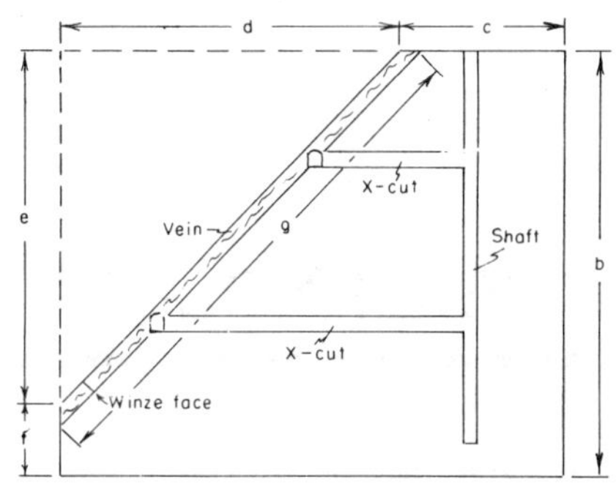

D. Cross-section view

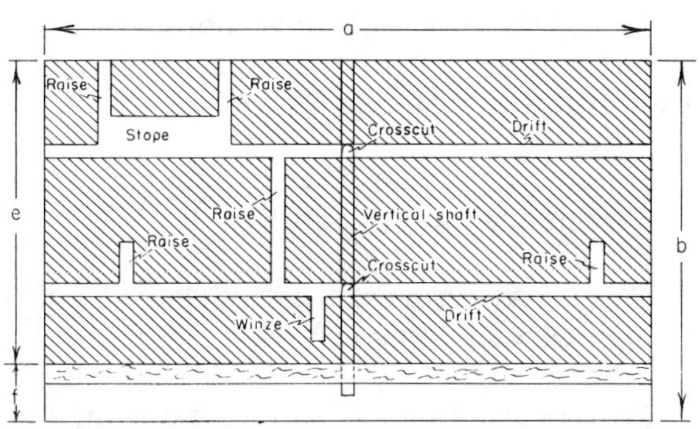

E. Longitudinal view of workings

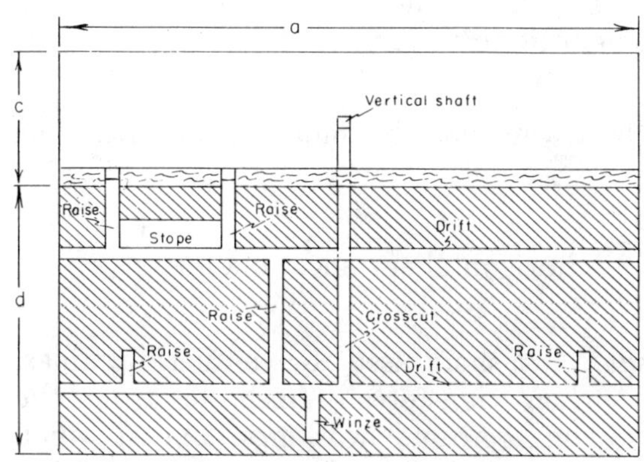

F. Plan view of workings

MINE TERMINOLOGY

To understand any report on mining it is necessary to know what the terms mean. On Plates 1-3, and 1-4 showing models of mines, the different kinds of openings are labeled. The model in Plate 1-4 shows the openings in black, and the ore in white. (This model was built to demonstrate the movement of air underground.)

The following are the most common mining terms:

Adit.—A horizontal entrance to a mine.

Air shaft.—A shaft used for ventilating mines, downcast when transferring fresh air from the surface to underground workings, upcast when discharging exhausted air to the surface.

Crosscut.—A horizontal or nearly horizontal underground opening driven to intersect a vein.

Drift.—A horizontal or nearly horizontal underground opening driven along a vein.

Entry.—A term commonly used in coal mining for an adit or a gently inclined shaft.

Intermediate level.—A system of horizontal workings started from a raise and not connected directly to the main working shaft.

Level.—A horizontal system of underground workings, the basis of operations for excavating the ore above or below. Manner of designation of levels differs in different districts.

Raise.—An underground opening driven upward from one level to another.

Shaft.—A vertical or inclined excavation through which a mine is worked.

Stope.—An underground excavation made by extracting ore.

Winze.—A vertical or inclined underground opening driven downward.

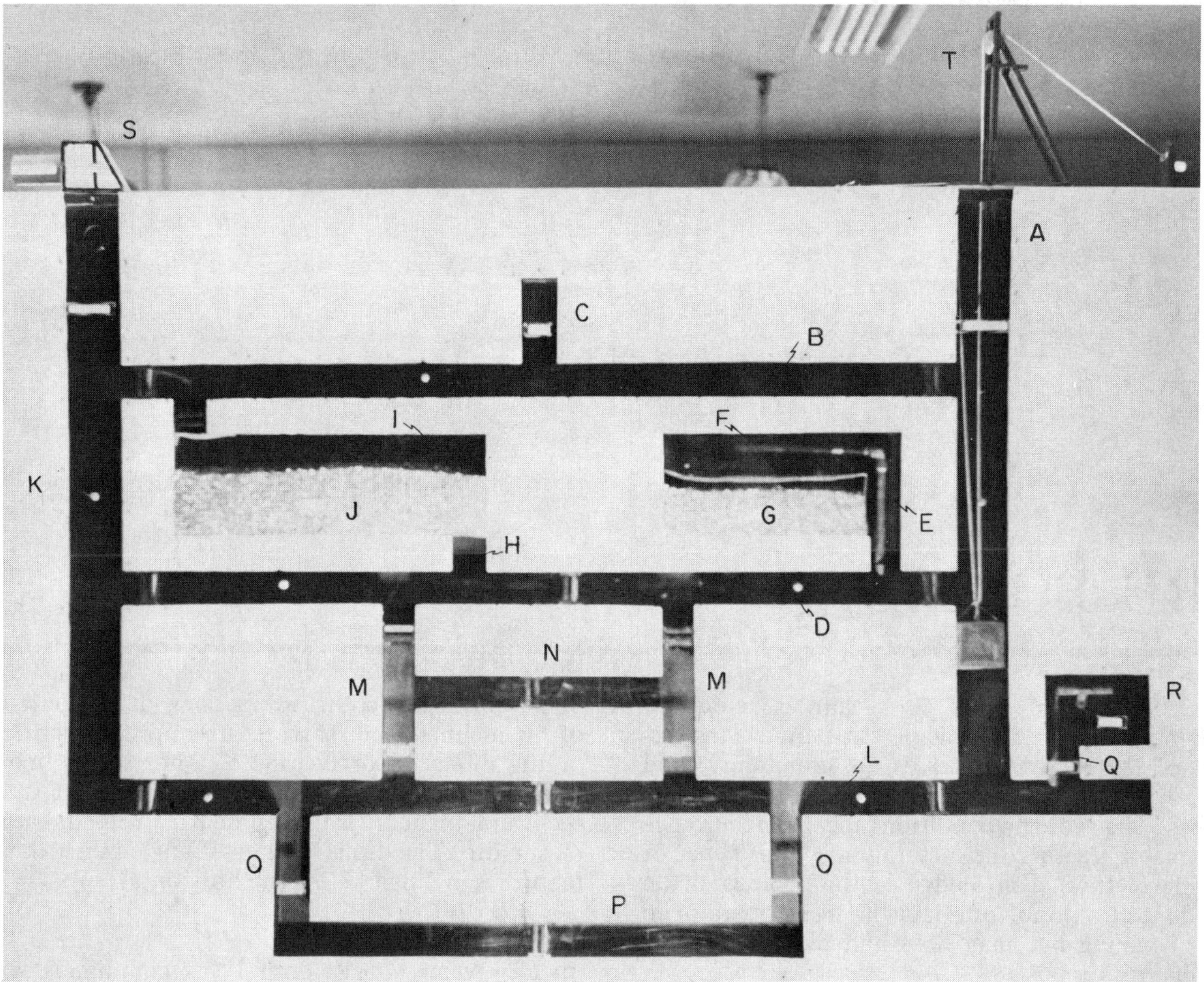

MINE MODEL

Mine openings appear black, ore and waste appear white.

Explanation:

A	- Main hoisting shaft	K	- Air raise or shaft
B	- Level no. 1	L	- Level no. 3
C	- Raise	M	- Raises
D	- Level no. 2	N	- Intermediate drift
E	- Raise	O	- Winzes or internal shafts
F	- Stope	P	- Level no. 4
G	- Waste fill	Q	- Raise
H	- Sealed raise	R	- Blind stope
I	- Stope	S	- Main ventilating fan
J	- Waste fill	T	- Headframe

PART 2

ORE BODIES

Ore occurs in many forms and is associated with many types of geologic structures. This section of the report illustrates, by commonly used methods, a few of the more common types of ore bodies. The geologic conditions and ore occurrence are major factors in determining the type of mining method. Knowledge of the geology of an ore deposit will not only aid the mine operator in actual mining but may help him to discover new ore deposits.

As in the previous section, both sketches and pictures are used to illustrate ore deposits, which are difficult to picture in two-dimension drawings only.

The types of ore bodies discussed in this section are those most commonly observed in Montana.

SOME TYPES OF ORE BODIES

Plates 1-2, and 1-3 show a tabular vein that dips at an angle into the earth; Plate 1-4 shows a vertical vein. Plate 2-1A shows a tabular vein dipping into the earth, but the block is cut at an oblique angle to the strike of the vein. The strike of a vein is the course or bearing of the outcrop of an inclined bed or structure on a level surface, or the direction or bearing of a horizontal line in the stratum, joint, fault, cleavage plane, or other structural plane. It is perpendicular to the direction of the dip. The dip is the largest angle at which the feature is inclined below the horizontal.

ORE SHOOTS

Few veins consist entirely of commercial ore; in most veins the ore is in zones called ore shoots. Some ore shoots do not follow down the dip of the vein but angle off in a different direction. This trend is called the plunge as shown in Plate 2-1A.

Coal, uranium, many nonmetallic minerals, and some metallic minerals may occur in horizontal or nearly horizontal layers or beds. See Plate 2-1B. It is not unusual for the bed or vein to be broken by a fault and displaced in position in either side of the fault. See Plate 2-1C.

ORE PIPES

Not all ore is in veins. It may occur in an ore "pipe" as shown in D. Where two veins intersect, ore may form a shoot in a zone at the intersection as shown in E. There is no apparent reason for some ore bodies, which seem to occur irregularly in the earth. See Plate 2-1F. Irregular ore bodies lacking apparent structural control are difficult to discover.

PLATE 2-1 SOME TYPES OF ORE BODIES

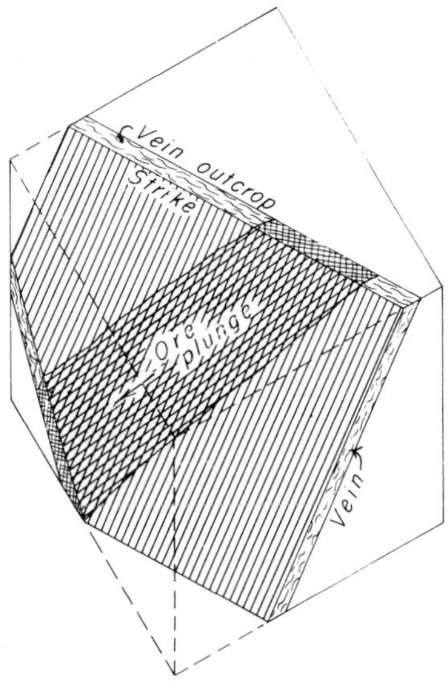

A. Block diagram of ore shoot; plunge in plane of vein

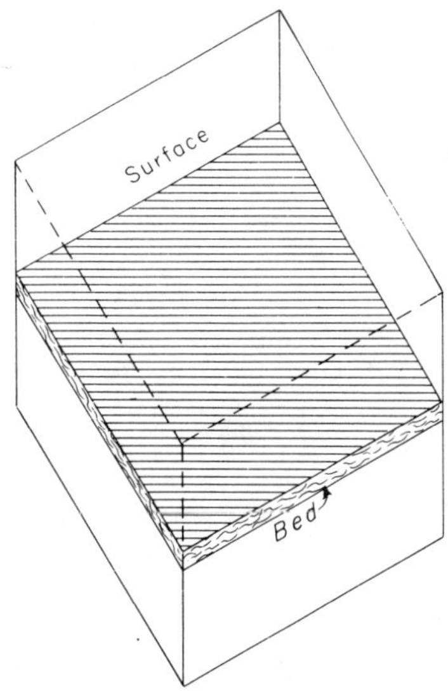

B. Block diagram of a flat, bedded deposit, such as a coal seam

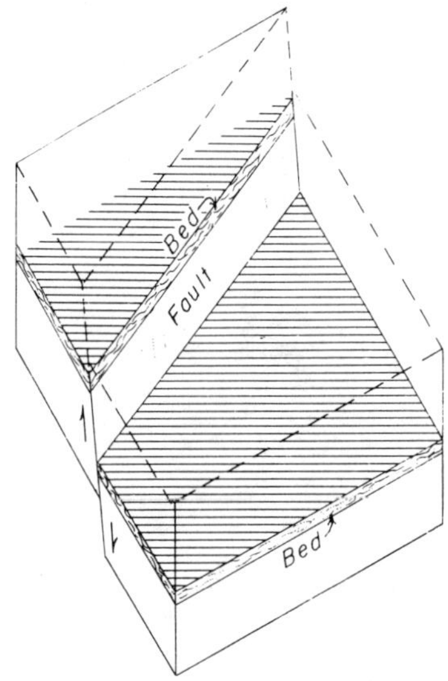

C. Block diagram of a bed or seam broken and displaced by a fault

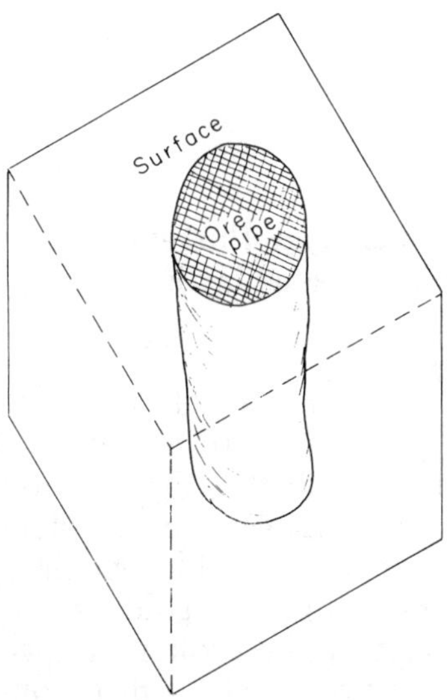

D. Block diagram of an ore pipe or ore chimney

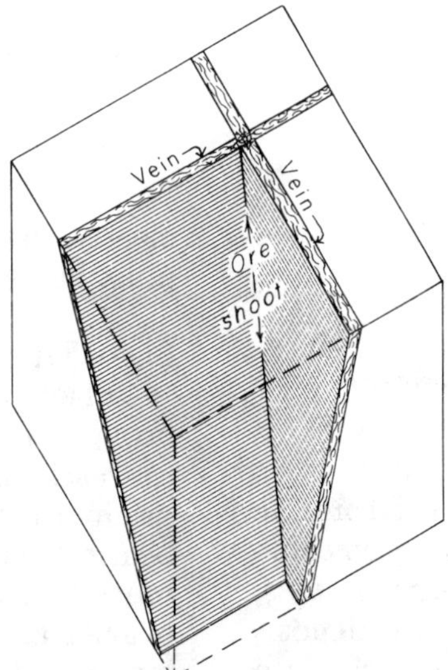

E. Block diagram of an ore shoot at vein intersection

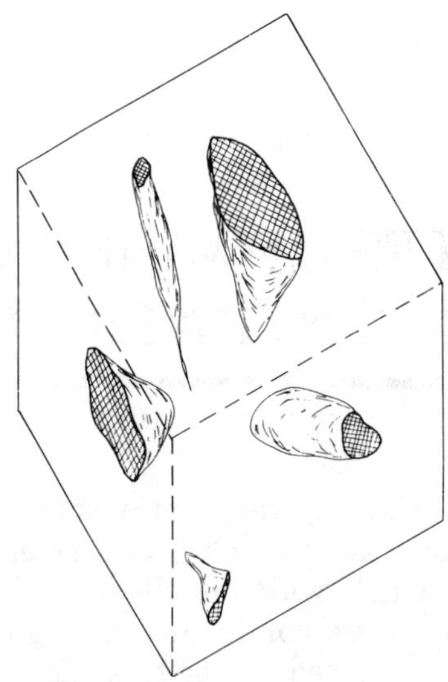

F. Block diagram or irregular ore bodies

TYPES OF MINERAL CONCENTRATIONS

If the upward movement of ore-forming solutions is stopped by a layer of impermeable rock, which will not allow the solutions to pass, and if other conditions are right, an ore deposit may be formed under this cap or cover. See Plate 2-2A. If ground is fractured and broken, numerous small veins or veinlets may fill these fractures B, but stringers and veinlets may be found even in regular veins.

An ore vein may be very distinct within the ground and show definite walls C; it may have only one distinct wall D; or there may be no clearly defined walls, and the extent of the ore must be determined by assaying E.

PLACER MINING

Placer gold was the chief attraction that brought miners into the western United States in the early days. F is a block diagram showing the origin of placer gold. Erosion of a vertical gold-bearing vein liberates the gold, which tends to move downstream. But the gold, being much heavier than the rest of the broken and weathered material, tends to move slowly and to concentrate near bedrock. Bedrock is defined as the solid rock underlying gold-bearing gravel, sand, or clay. Although gold tends to concentrate on or near bedrock, it is not evenly distributed along the length or across the width of the bedrock. It tends to concentrate on the inside downstream side of channel bends G or on the downstream side of boulders or projecting irregularities in bedrock H.

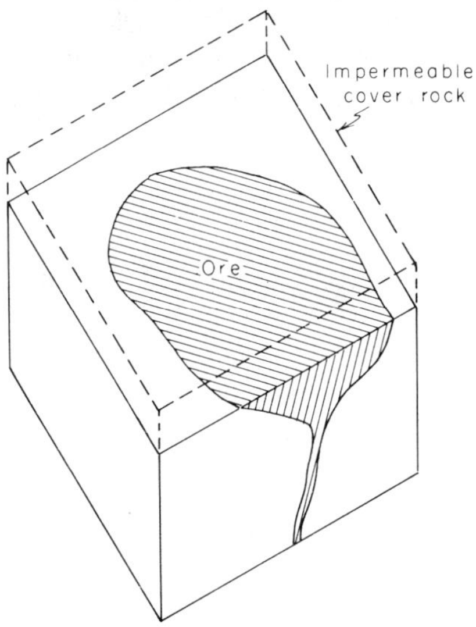

A. Block diagram of ore body localized by impermeable cover rock

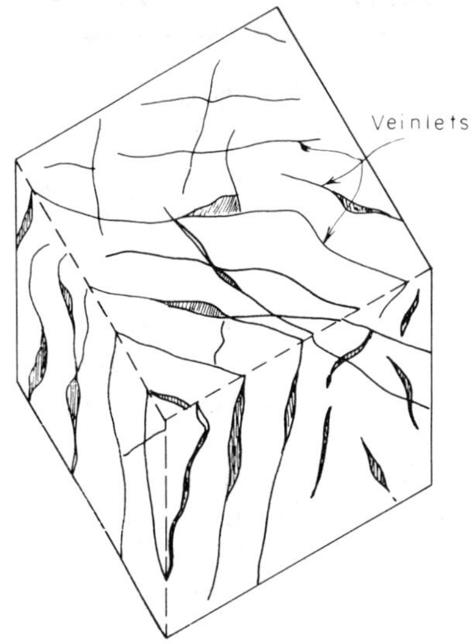

B. Block diagram of numerous mineralized veins and veinlets

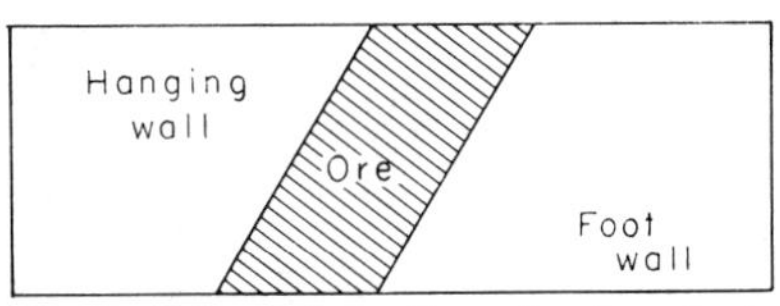

C. Cross-section of a vein showing clear-cut walls between ore and waste

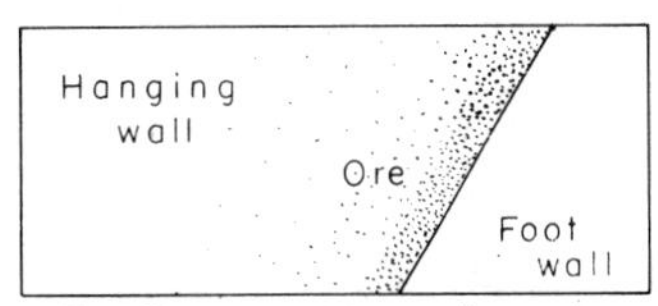

D. Cross-section of a vein showing hard foot wall and assay cutoff hanging wall

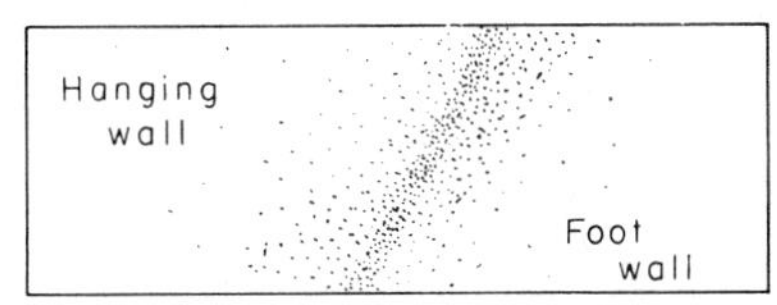

E. Cross-section of a vein showing assay cutoff on both walls

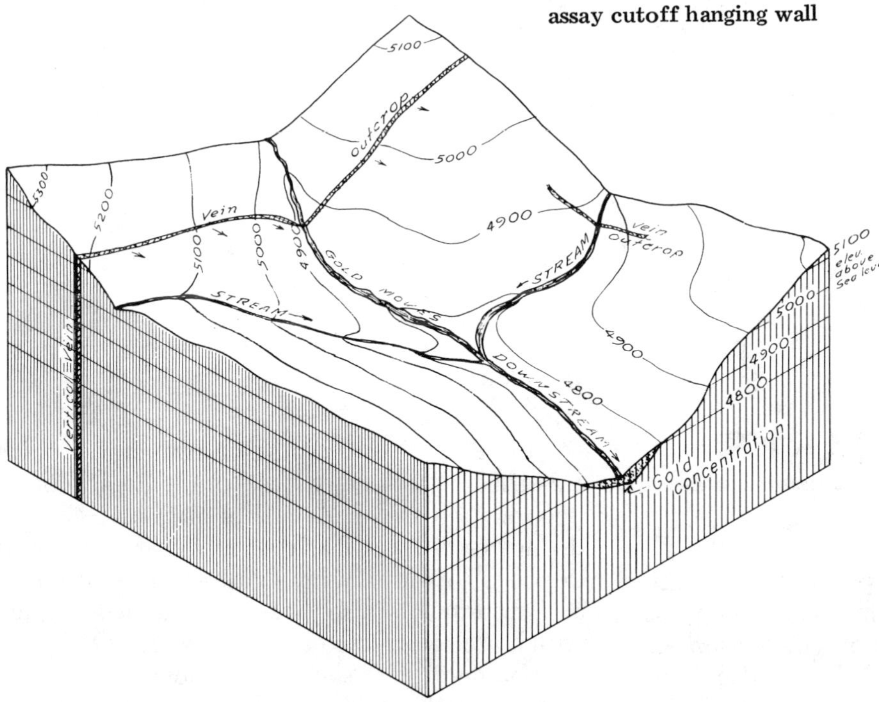

F. Block diagram of a portion of the earth showing erosion of veins and concentration of placer gold

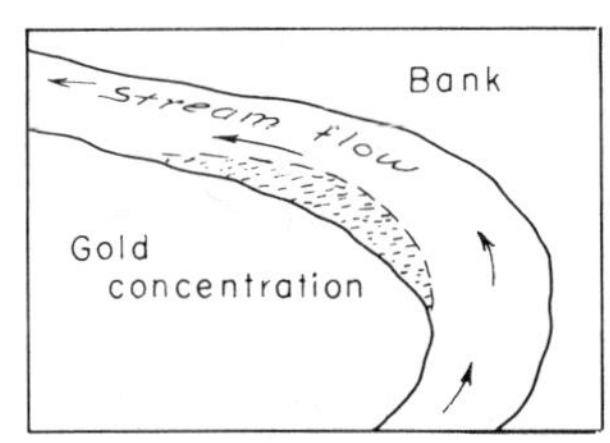

G. Concentration of gold in stream

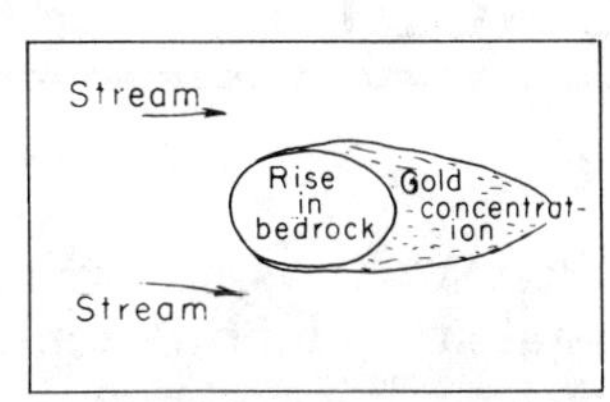

H. Concentration of gold on downstream side of projection in bed rock

MODELS OF ORE BODIES

Most of the discussion so far has dealt with regular ore bodies generally unbroken by faults. To get a picture of a more complicated deposit it may be necessary to build a model of the ore body and mine.

Plate 2-3 shows pictures of various complicated mine models and the different ways of constructing these models. A and B show models constructed of wood. In A, the white members are development openings, shafts, drifts, crosscuts, and raises; the darker members show the ore bodies. The dark and light members in B show different ore bodies or vein systems.

In C the ore body and the development openings are white. The dark lines (bent wire) are contour lines showing the configuration of the surface of the ground. The outcrop of the vein is shown as a gray solid line from upper center toward lower left.

Complicated geological features can be drawn on glass or plastic sheets, which are then stacked or placed in order D. A light placed back of the sheets brings out the feature in three dimensions.

Wood, plastic sheets, and colored cardboard sheets glued in position can be used to depict veins and faults E.

A. Method of depicting ore bodies and development openings by solid material in a mine model. **B.** Different style of depicting ore bodies by solid material. **C.** Ore body near the surface; ore is shown in solid white. Bent wires depict the configuration of the surface. **D.** Ore body depicted by drawing geology and veins on sheets of glass. **E.** Geologic models showing veins and faults.

PART 1 & 2 CONCLUDING REMARKS

PART 1

Part 1 shows how mines or parts of mines are illustrated and how some of the terminology is used in mining. A more complete definition of mining terms is found in Part 10. It is difficult to understand a complicated process unless some common form of understanding is used. Part 1 illustrates the basis used for showing the more complicated mining systems discussed in the following parts of this publication.

PART 2

In Part 2, all valuable mineral bodies are classified generically as ore. Coal commonly occurs in a flat bedded deposit, and coal miners do not refer to coal as ore; however, many valuable deposits of lead, zinc, copper and other metals and non-metals also occur in flat bedded deposits which are referred to as ore deposits. Ore is defined as a mineral aggregate of sufficient value, both in quantity and quality, which can be mined at a profit. Hence, even though certain mineral commodities such as coal are not called ore by the industry, the general term ore may be used in this publication to cover all mineral materials.

PLATE 2-3 MODELS OF MINES AND ORE BODIES

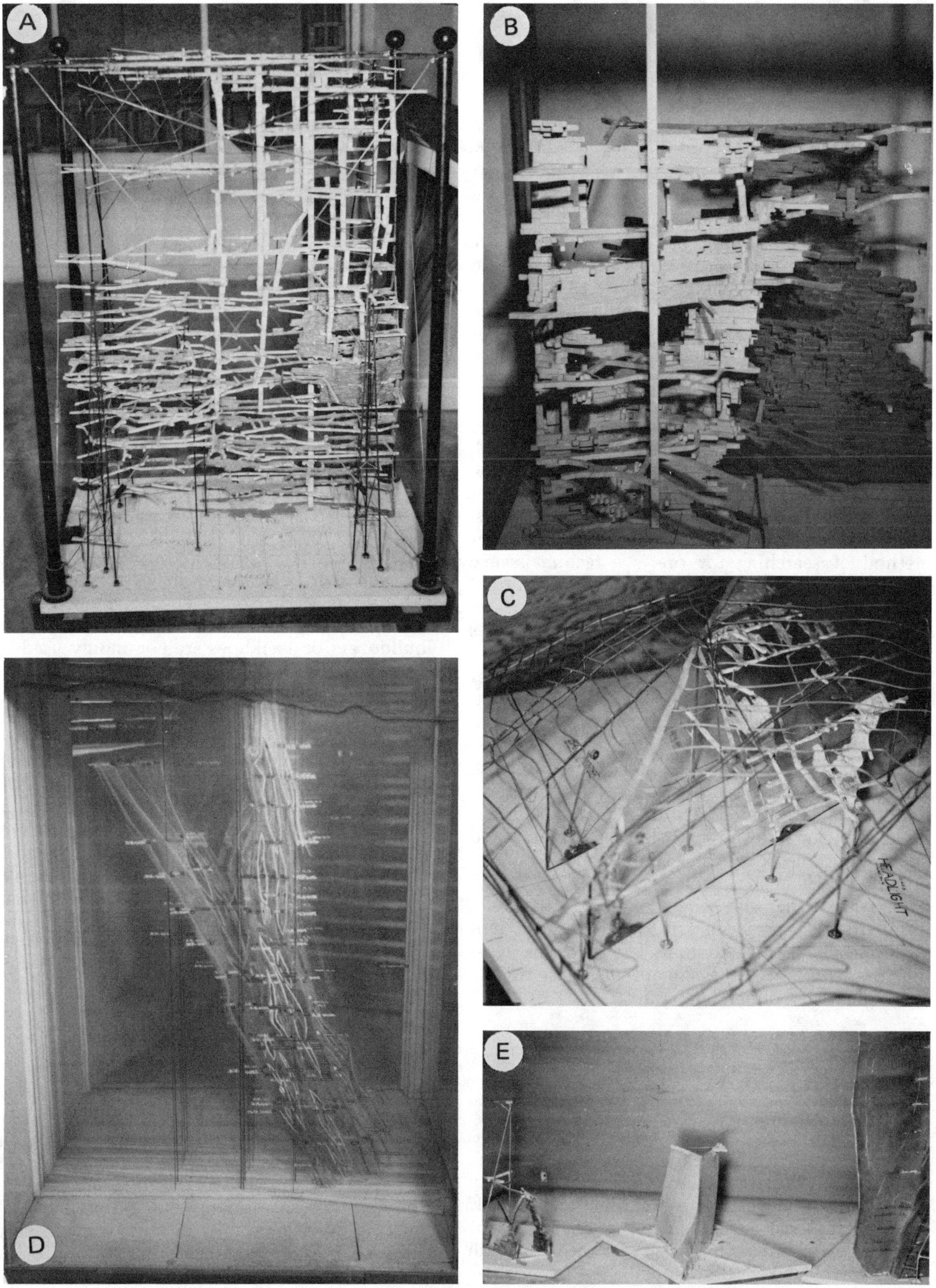

PART 3

FINDING AND EVALUATING ORE BODIES

In the early days of prospecting many outcrops of minerals were found because the veins were exposed on the surface of the earth. They could easily be seen, and followed. Most of those ore deposits have now been found. However, the early prospectors' techniques for tracing float are still used for finding minerals which have become valuable in today's economy.

RECONNAISSANCE SURVEYS

Techniques have also been developed which sometimes indicate ore bodies lying close to the surface but not exposed. Although far from perfect, these techniques are used to search for undiscovered ore bodies. Usually the first phase or method of searching for ore is a reconnaissance survey, or simply, a preliminary examination of an area of interest. Discoloration of the surface, mineral chips or float, or unusual earth structures may be signs of a mineral deposit. Often these zones are investigated in more detail at a later time to see if a mineral deposit actually does exist.

TRENCHING

If valuable minerals are found as float or chips, trenching is often necessary to expose the surface outcrop. If no minerals are found on the surface with trenching techniques, more sophisticated exploration methods using the sciences of geobotany, geochemistry, geology and geophysics may be employed. If, and when, a mineral deposit is found it must be evaluated to determine if it is economically feasible to mine. The following pages illustrate some of the techniques used for finding and evaluating a mineral deposit.

TRACING FLOAT AND TRENCHING

If a vein is exposed near the surface, but is covered with overburden or detrital material, remnants of the vein called float will frequently appear downhill in the overburden material. If the vein material is resistant to weathering, float usually starts appearing some distance downhill from the hidden vein outcrop.

FLOAT MATERIAL

A cross-section of a vertical vein is shown in A, and the broken off pieces of the vein are shown scattered throughout the overburden material. B is a plan view of the same vein showing how the float material can be scattered randomly throughout the overburden.

TRENCHING AT LEVEL GRADE

A trench is usually dug to expose the vein. One way of digging a trench is shown in C. It is common practice to go down hill a short distance below the uppermost showing of float and start a trench at a level grade to intersect bedrock. The reason for going in on a level grade is to provide a drain in the event water is encountered. Also, it is easier to travel in and out of the cut with a level grade. Mechanical equipment such as bulldozers or backhoes are commonly used to dig trenches, but small, shallow hand-dug trenches are still common.

After bedrock is encountered and if float is still found in the overburden, the trench is continued uphill exposing the bedrock until the vein is uncovered. The uncovered vein is shown in D. It is considered good practice to dig the trench uphill beyond the vein exposure. If that isn't done the material will sluff into the vein. If no more float is found in the overburden, and the vein is not found in the bottom of the trench, it may be necessary to start further downhill with the trench. Unfortunately, the trenching method doesn't always uncover a vein.

DETERMINE OUTCROP LIMITS

Once a vein is encountered it is necessary to determine the extent of the outcrop. Therefore, it is necessary to dig a series of trenches along the outcrop as shown in E. The distance between the trenches varies with circumstances. Trenches require considerable effort and are expensive to dig, but the closer the trenches, the better the results for assaying and evaluating the vein outcrop. A compromise is usually made between the cost of digging cuts and the information desired for evaluating the total project.

PLATE 3-1 TRACING FLOAT AND TRENCHING

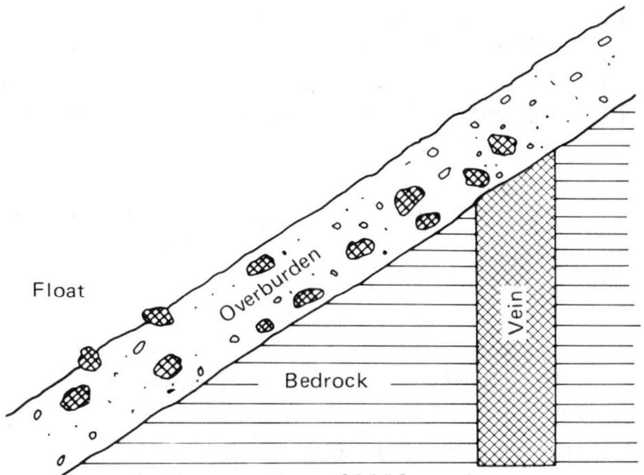

A. Cross-section of hidden vein

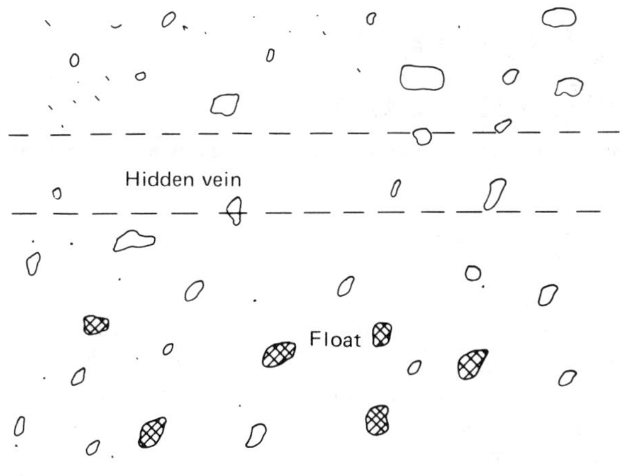

B. Plan view of hidden vein

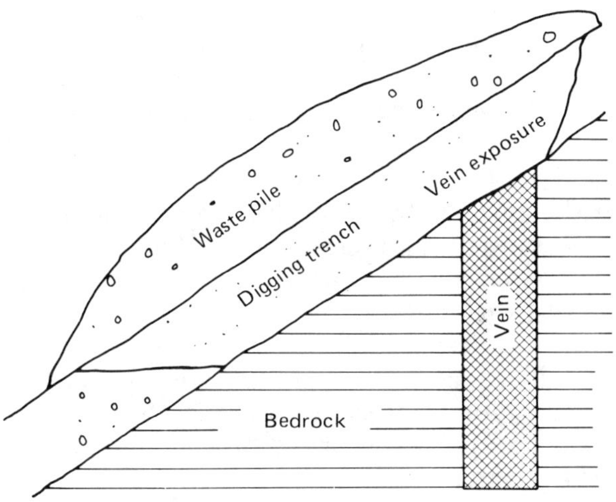

C. Cross-section showing trench

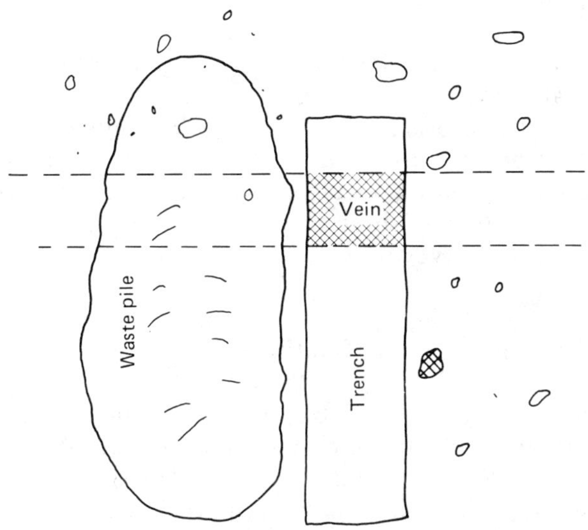

D. Plan view; exposed vein in trench

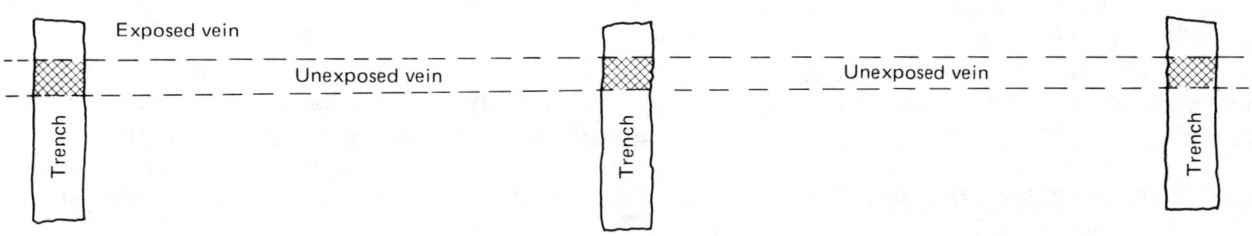

E. Plan view of a series of trenches exposing the vein

RECONNAISSANCE MINERAL SURVEYS

Generally speaking, Mother Nature hides her mineral wealth well, and most of the easily found deposits have been discovered. It is now necessary to use the tools of science to find favorable locations of mineral deposits. This technique is called mineral exploration and involves considerable work and expense.

MINERAL EXPLORATION

A geologic map taken from Bulletin 17, Plate 4 published by the Montana Bureau of Mines and Geology is shown in A. Before starting any detailed mineral reconnaissance survey, the geology of the area, if known and published, is studied in detail to see if there are any favorable host rocks that might contain ore bodies. Much of the United States has been mapped for geologic formations on a broad scale, and these maps are used as a basis for starting the survey.

For example, A is a geologic rock formation in Western Montana. Southwest of this region lies the famous Coeur d' Alene Silver District of Northern Idaho. The Ravalli group, mostly the quartzites, are favorable host rocks for the rich ore deposits in the Coeur d' Alene District. Some production has also come from the Prichard formation in Idaho.

These rocks are of interest because they may also contain ore bodies in the area under investigation. It must, however, be emphasized that because a rock type contains ore bodies in one area, it doesn't insure that it may not be barren in a different area. Conversely, barren rock in an active mining area may contain ore bodies in a different location.

SELECTING TARGET AREAS

Because mineral exploration is time consuming and costly, some target area must be selected. Assume that after careful study and consideration, it is decided to investigate the Ravalli group in more detail in A. This group is shown in the upper right part of the map. This is still a big area of land to investigate since each of the squares on the map measures one square mile or 640 acres. This rock group covers about 24 square miles in the map area in rugged mountainous country. Consequently, it takes considerable time to examine in detail.

Although different organizations plan mineral reconnaissance surveys in different sequences, one often used method is to send geologists into the area to look at the rocks, alteration zones and structure in more detail. Their work will consist of doing more detailed geologic mapping and to collect a great many more rock and soil samples for further analysis. In addition, they will commonly collect water and sediment samples from streams to determine if they contain high amounts of valuable minerals.

SAMPLING

B shows a plan and cross-section of a stream system where water and sediment samples may be collected. In addition, some geologists and geobotanists collect leaves, twigs, blades of grass and roots as shown in C and have them analyzed to determine whether high amounts of minerals are present in these plants. High mineral values indicated by the analysis call for more thorough investigation.

It is also necessary to collect samples from surrounding areas to identify what is called a background level or count of minerals analysis for the area.

Some areas, for example, have a general high background count in mineral content as compared to normal background in other areas. Thus, unless the samples can be compared to surrounding areas, an anamolous indication may not be identified.

When enough samples have been gathered and analyzed, high background areas can be identified, and further investigation in the area can be done. This technique narrows or confines the search to more promising areas of mineral deposits.

Although stream and vegetation samples appear simple to gather, interpreting the results of analysis is difficult. Sometimes these methods give misleading information. They are merely tools in the exploration process and should be used with full realization of their limitations.

PLATE 3-2 RECONNAISSANCE MINERAL SURVEYS

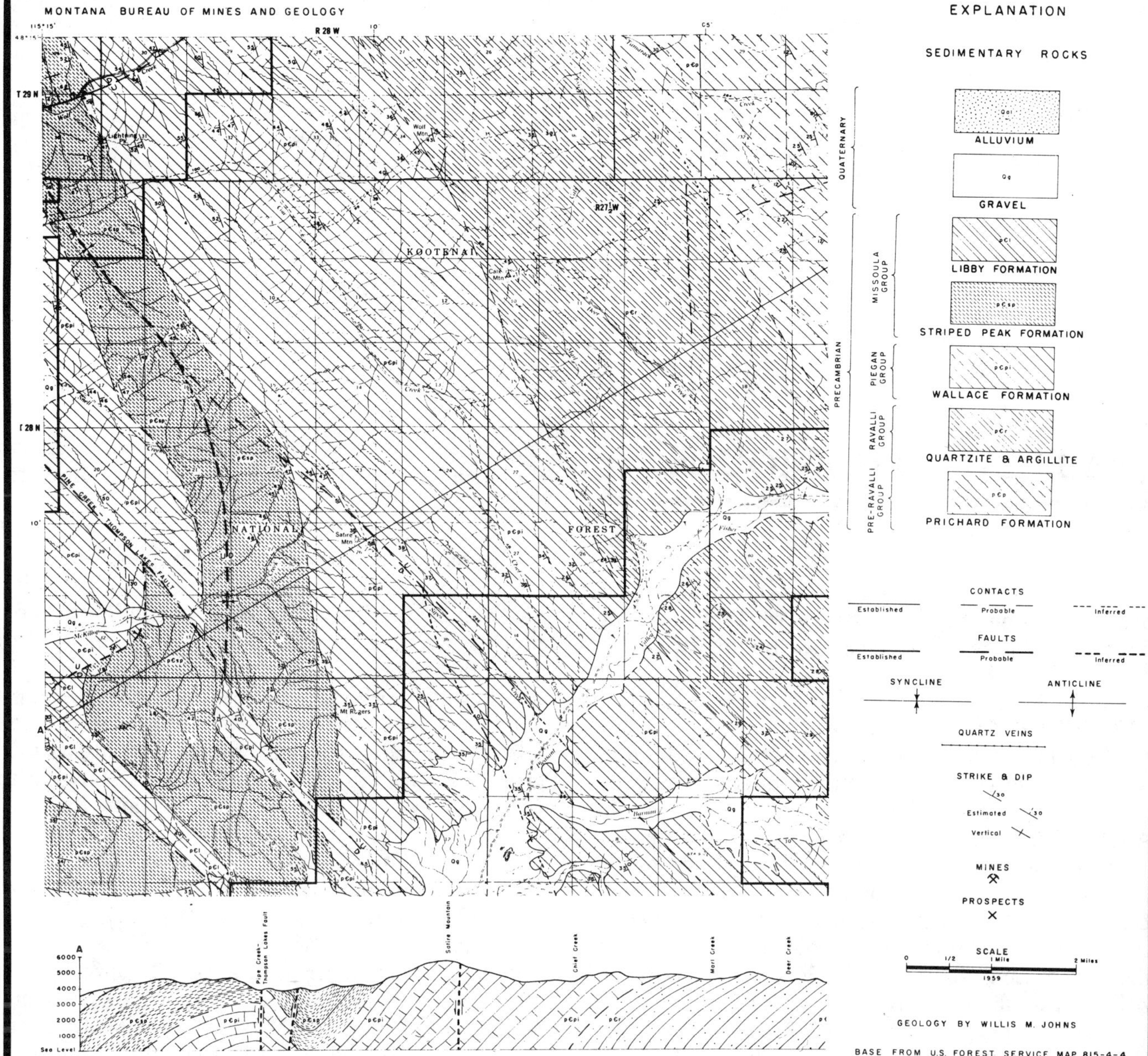

A. GEOLOGIC MAP THOMPSON LAKES S E QUADRANGLE LINCOLN COUNTY, MONTANA (SHOWING CROSS-SECTION) *(Courtesy Mont. Bureau of Mines & Geol.)*

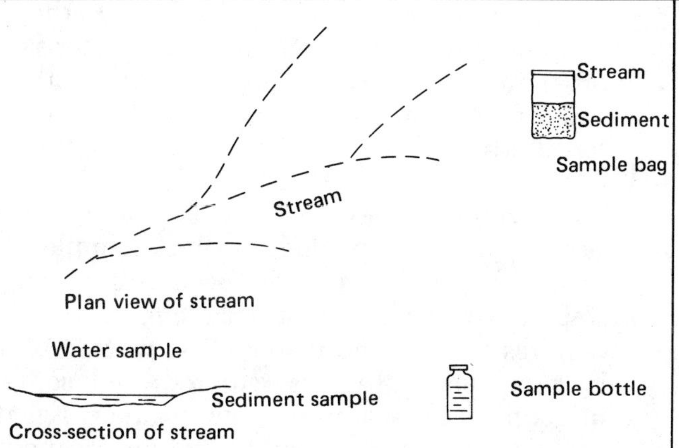

B. Sampling streams

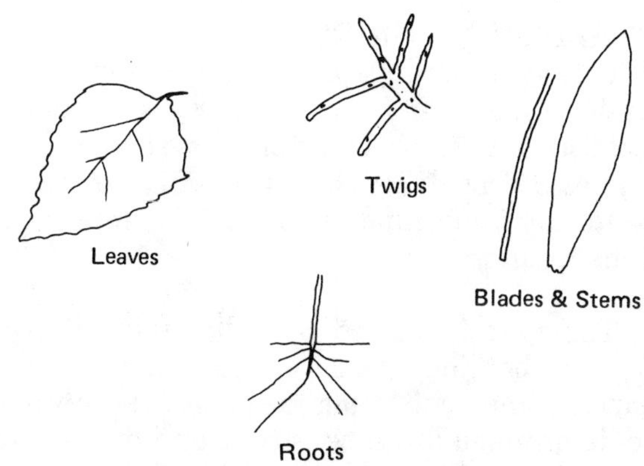

C. Sampling vegetation

SAMPLING TECHNIQUES

After a promising mineral bearing zone has been identified, the next step is to do a sampling program so potential targets can be identified for more detailed work.

GEOCHEMICAL SAMPLING

Geochemical sampling is one rather common way of identifying a potential mineral area. The collection of a geochemical sample is shown in A. A small hole is dug, either with a prospector's pick or a small shovel, and a portion of the pile is placed in a sample sack for further analysis. Usually this hole is dug deep enough to get below the organic bearing surface soil. If the area has much exposed rock, a rock chip sample is often collected as shown in B. The point of the pick can be used to pry off fractured chips as shown, or the hammer end can be used to break off solid chunks of rock.

ESTABLISHING GRIDS

In any sampling program, or even in a geophysical measuring program, it is necessary to establish a grid or coordinate system. When the samples are analyzed it is necessary to record the results on some kind of a map for interpretation of the results.

The grid is started by following a designated line or bearing, usually following a compass bearing, and collecting samples at some equal pre-determined distance. At the end of the distance two right angle turns are made, and another grid line is started in the opposite direction as shown in C. The sampling is continued as shown in D until the initial grid and sampling program is completed.

Analytical techniques have improved greatly in the past few years, and the samples can be analyzed for metals of interest. The grid shown in E has values in silver in parts per million (ppm) plotted at each sample point. When making the grid and collecting the samples some systematic numbering method must be used so the samples and their location can be identified clearly. It is also a good idea to clearly mark and identify the grid lines on the ground so that at some future time the grid can be expanded in any direction.

LINES OF EQUAL VALUE

Lines of equal values of silver are drawn on the grid in E. In doing this the person drawing the lines must use judgment as to just where the lines go. Some exploration firms feel that 2 ppm in silver are anomalous and require a closer look. As can be seen in E, the higher silver values occur in the North center of the area selected.

After examining the results of the initial sampling program it appears that another sampling program should be extended to the north, south and possibly west to more closely identify the target area.

Although the results of the geochemical survey are encouraging in the example shown, an ore body has not yet been found. The next step in the procedure for finding an ore deposit requires much more detailed work. A closer look at the geology is required, and some kind of geophysical survey may be in order. Eventually drilling will be required before an ore body can definitely be identified.

PLATE 3-3 SAMPLING TECHNIQUES

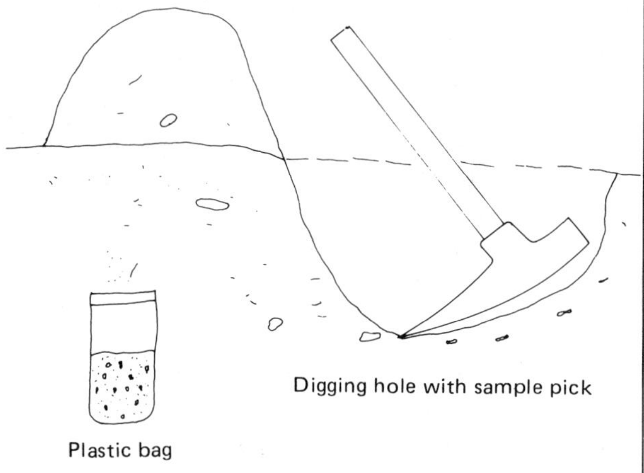

A. Collecting soil samples

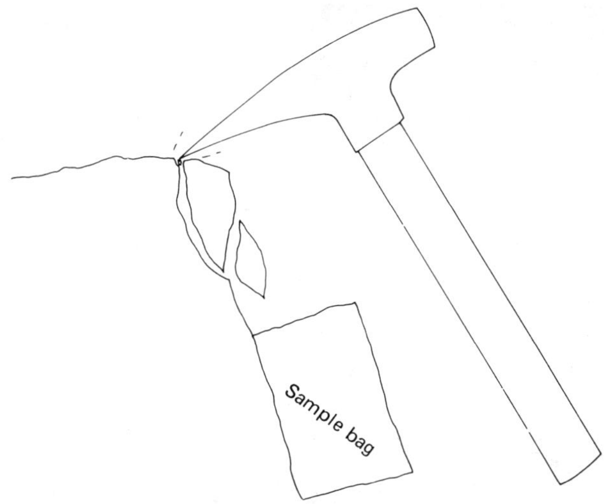

B. Collecting rock chip samples

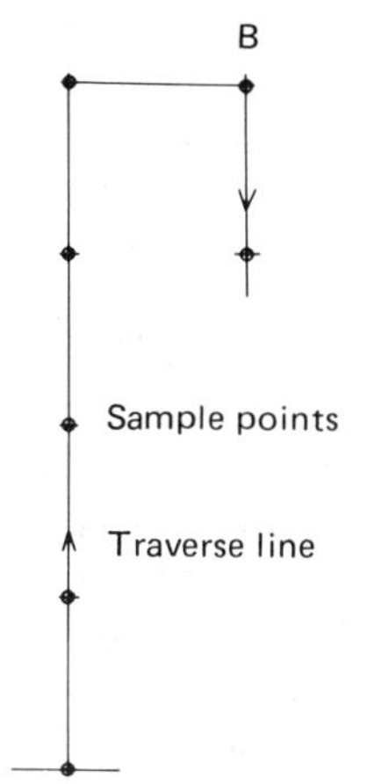

C. Starting grid for sample collection

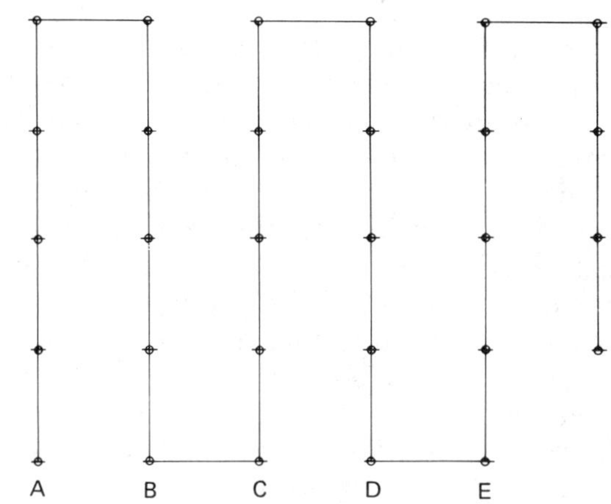

D. Continuation of grid for sample collection

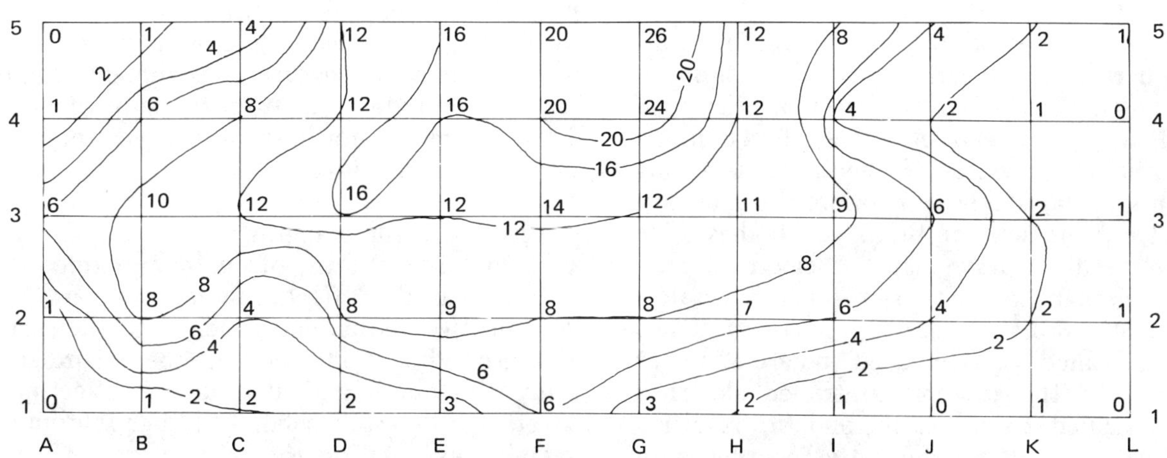

E. Grid with assay values plotted. Anomaly lines plotted

FUNDAMENTALS OF GEOPHYSICAL PROSPECTING

Some mineral deposits have physical characteristics which can be detected by certain types of geophysical measurements. The three most common types of physical phenomena that are usually measured in a mineral search are magnetic, electrical and radiometric.

MAGNETIC FIELDS CREATED BY ORE BODIES

The earth has a natural magnetic field which can be detected by an ordinary magnetic compass. When the compass is held level, the needle points to the magnetic poles of the earth. There is also a vertical component or dip of the magnetic field, and a dip needle points into the earth toward the magnetic pole. A magnetic ore body will set up a magnetic field as shown in A. An instrument called a magnetometer measures the intensity of the magnetic field. These instruments can be mounted in aircraft, land vehicles, or they can be positioned on stations on the ground. A grid is normally established for a magnetic survey similar to the one shown on Plate 3-3 for geochemical surveys. Magnetic intensities are recorded at each station after which a magnetic contour map is prepared similar to the process shown in Plate 3-3. Some ore bodies have high magnetic attraction while others have low attraction as shown in B. The magnetic exploration methods have found some good blind ore bodies which have high magnetic characteristics.

ELECTRICAL FIELDS

Some ore bodies containing metallic minerals act like a battery under certain ground and groundwater conditions. They can project an electrical field similar to the magnetic field shown in A. Self potential (abbreviated SP) electrical measuring instruments are available to measure the intensity of the electrical field of those ore bodies. The procedure used to draw maps of electrical anomalies is similar to that used for geochemical surveys Plate 3-3. A more popular electrical method is called the induced polarization or IP method. In this system, controlled electrical energy is applied to the earth, and the resulting electrical behavior of the ground is recorded with measuring instruments. In IP work, the conductivity of mineralized ground changes with the frequency of the applied current while the conductivity of barren ground stays constant.

AIRCRAFT RECONNAISSANCE SURVEYS

Electromagnetic systems are often used in aircraft. The ground is energized inductively by means of an alternating current flowing in a transmitter coil. The ground response signals are detected by a receiver coil, and again variations or anomalies are noted from the recorded data. Aircraft reconnaissance surveys, which detect areas of interest, must be followed by more precise ground exploration methods.

The previously discussed electrical and electomagnetic methods are the ones most commonly used for metallic ore exploration. Although the methods work well when ore deposits have the proper physical response characteristics, many ore bodies do not respond to these geophysical measuring techniques. Also, certain ground conditions such as water, organic material in the rock, and other factors will give erroneous anomalies over non-ore deposits. The methods are being perfected, and undoubtedly more ore deposits will be found by these methods in the future.

GAMMA RAY RADIATION DETECTORS

Gamma ray radiation detectors have been very useful in finding deposits of uranium, thorium and in some cases potassium. However, these rays penetrate only a few inches of soil and a few hundred feet of air. The deposit usually must outcrop for detection as shown in D. Geiger counters and scintillometers, which are portable, detect this radiation and are often used in aircraft and land vehicles.

Two geophysical methods used in oil and gas exploration with success are gravimeter and seismic. Some gas and oil formations have densities different from that of the surrounding rock. Sensitive gravity measuring instruments can record the different densities from the subsurface, see C. This system has not been very successful in finding ore deposits at this time.

In the seismic method, a heavy shock is impressed on the earth, often by an explosive force as shown in E. Reflected shock waves are recorded on sensitive geophones attached to the surface of the ground. By interpreting the recorded timed data, the depth and attitude of reflecting earth layers can be established. Although seldom used in metallic exploration the seismic system is used to trace coal beds and some other non-metallic ore bodies.

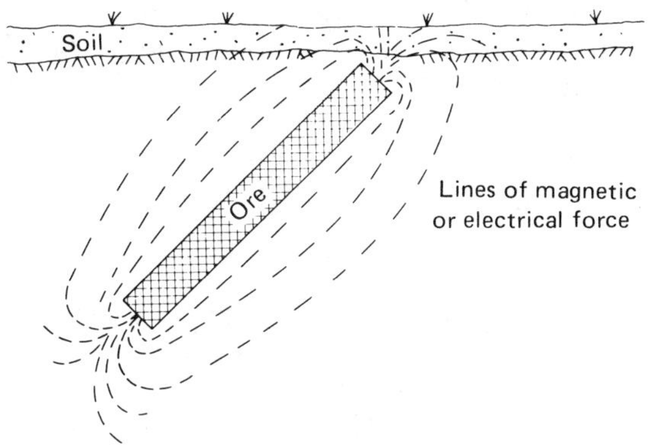

A. Lines of force about some orebodies

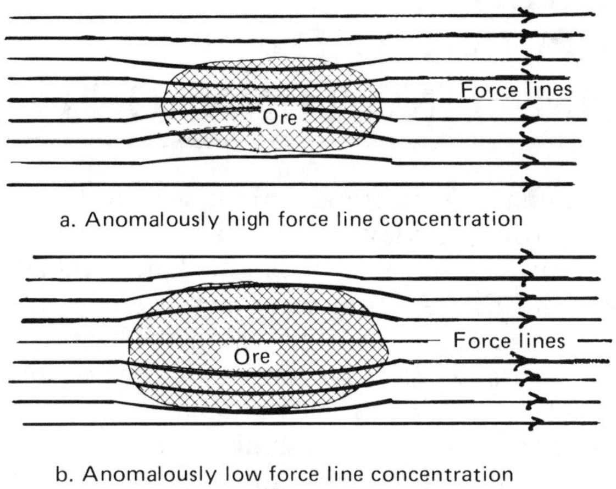

B. Horizontal distortion of force lines

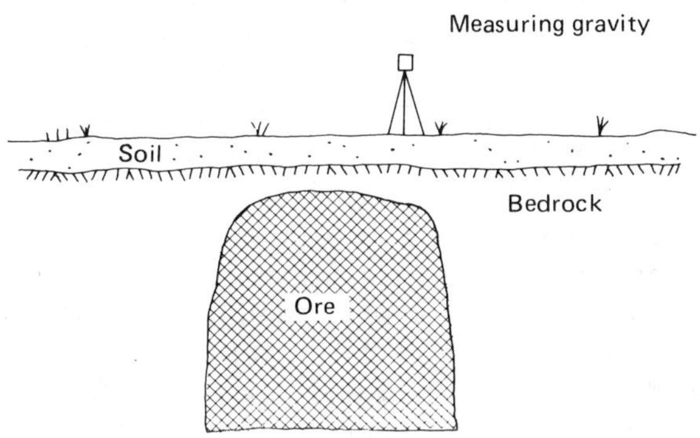

C. Measuring gravity about an orebody

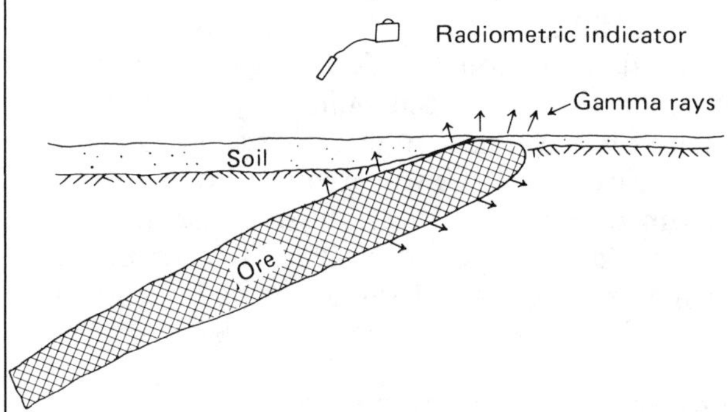

D. Detecting gamma ray radiation

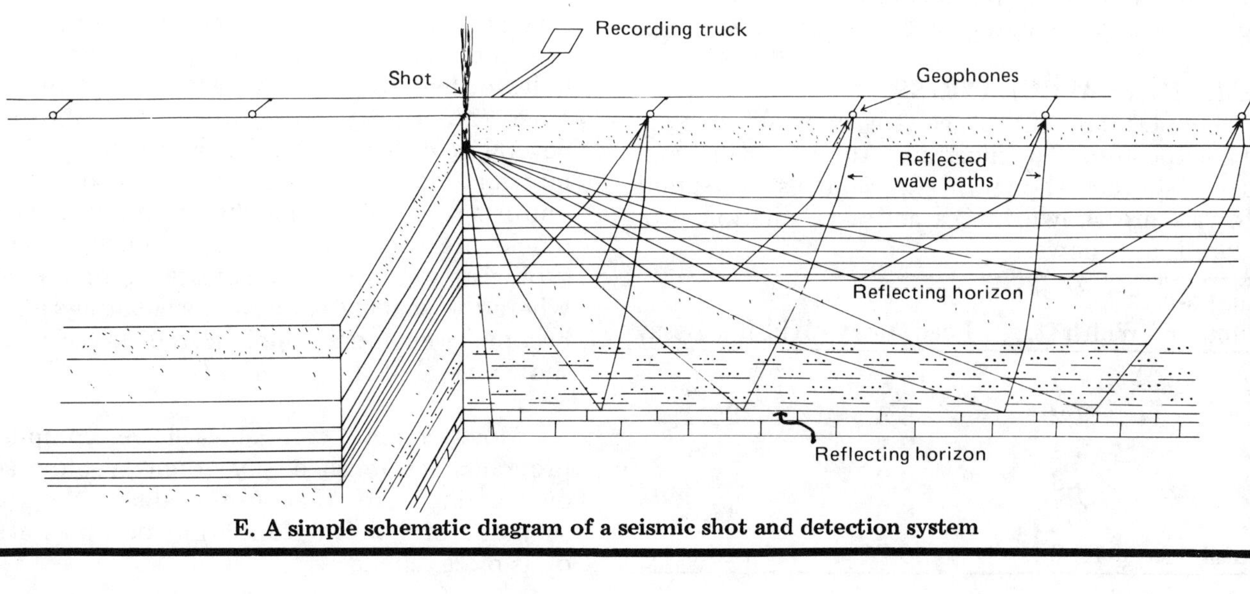

E. A simple schematic diagram of a seismic shot and detection system

DRILLING, SAMPLING AND ESTIMATING TONNAGES

Although the previously discussed methods of searching for ore may indicate an ore body, the deposit must be opened up, sampled and evaluated by some means. Past common practice consisted of driving drifts, crosscuts, and sinking shafts on the ore indicators so the deposit could be sampled. This practice is still followed in some cases, but it is expensive and time consuming. With improvements in drilling programs, some type of drilling is often done on a suspected ore deposit.

DRILLING PROGRAMS

An outcropping vein should be tested at depth before much money and time is committed to the project. A drilling program such as shown in A is often used. One hole is often angled so that it will intersect the projected vein perpendicularly. The other hole is vertical to intersect the projected vein at a greater depth. Two or more holes can be drilled from one surface location. In order that the core can be studied for geological features, diamond drills are often used for this type of work in initial exploration.

REPRESENTATIVE FACE SAMPLING

One of the very important steps in evaluating a mineral deposit is sampling. If a face of ore is exposed, a channel of equal width and depth is cut across the entire face or width of ore. Frequently the ore is banded or layered with waste, and it is necessary to cut a representative sample across the entire face including both the waste and ore.

SAMPLING CALCULATIONS

Samples are cut at certain intervals along a vein exposure. To illustrate, some widths and assay data are shown in C. Samples at regular intervals are shown in C,a and a calculation procedure is as follows:

Sample Number	Width ft.	Lead Assay	Width x Assay
1	1.2	16.8%	20.2
2	3.2	6.3	20.2
3	6.4	7.1	45.4
4	5.3	9.2	48.8
5	2.1	6.8	14.3
Totals	18.2	46.2	148.9

Average width = 18.2 ÷ 5 = 3.6'
Av. Assay = 148.9 ÷ 18.2 = 8.2% lead
A common error in evaluating assays is to add up the assays of the samples and divide by the number of samples. In this case it would be 46.2 ÷ 5 = 9.2% lead which is considerably higher than the width x assay method.

Uneven sample spacing is shown in C,b and the calculation which is slightly more involved is as follows:

Sample	Width ft.	Influence Interval	Width x Interval	Zinc Assay	Width x interval x assay
1	3.2	12	38.4	7.1	272.6
2	4.3	10	43.0	5.8	249.4
3	4.2	11	46.2	8.2	378.8
4	3.1	14	43.4	6.3	273.4
		47	171.0		1,174.2

Average width = 171.0 ÷ 47 = 3.64 ft
Average Assay = 1174.2 ÷ 171.0 = 6.8% Zn

The influence of a sample is generally considered to be half way to the next sample.

ESTIMATING RESERVES

There are various classifications of ore reserves and a common method of indicating reserves in narrow vein underground mines is shown in D. A block of ore, opened and assayed on four sides, is called proven ore. A block opened on three sides and sampled is called probable ore. If it is opened on one or two sides, it is called possible ore.

EVALUATING DRILL HOLE DATA

A large or massive ore body is more commonly evaluated from drill hole information. A buried ore body is shown in E with a series of sample holes drilled from the surface. The pattern of the holes can be regular or irregular as shown in F. Several schemes are used for evaluating drill hole information. For uniform rectangular spacing, uniform rectangular blocks or uniform triangular blocks are commonly used, but many exploration geologists use more sophisticated geometric forms. Where the spacing is uneven, evaluating techniques become even more sophisticated and E,c and E,d show several of these methods.

Most mining companies have computerized programs established by their geological and engineering staff that can evaluate the deposits as more drill hole information becomes available or as prices change.

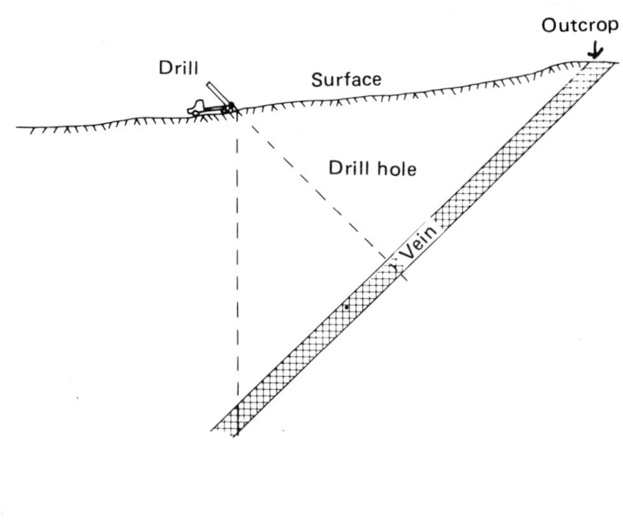

A. Cross-section of a drilling program on a vein

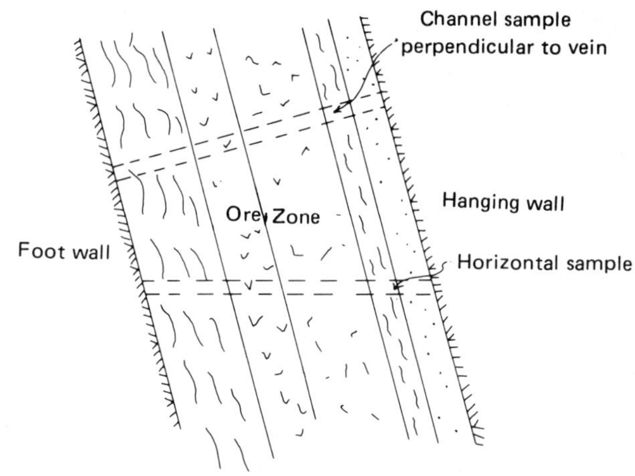

B. Sampling an ore face

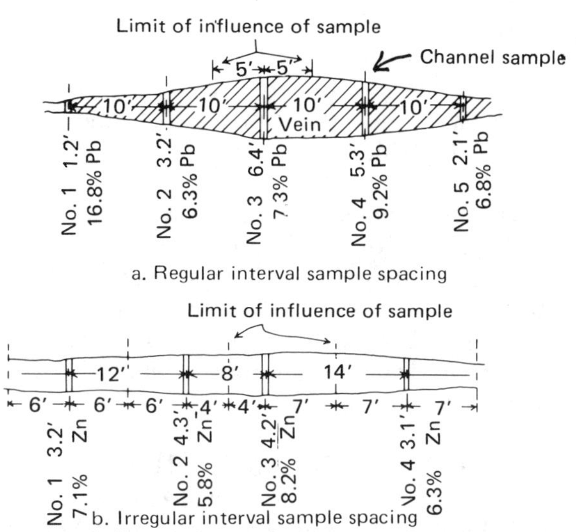

C. Sample intervals on exposed veins

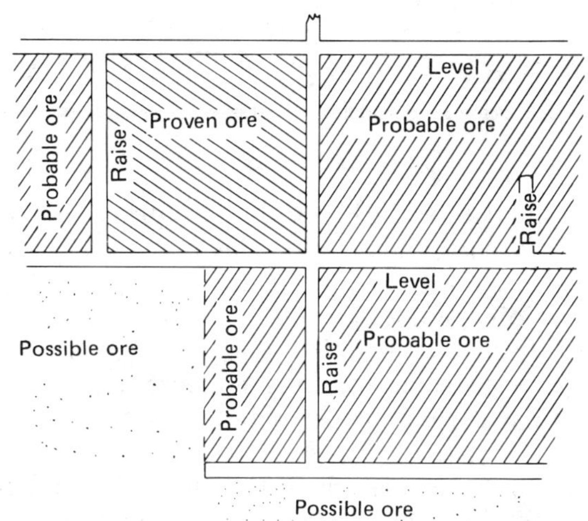

D. Long section illustrating class of ore reserves

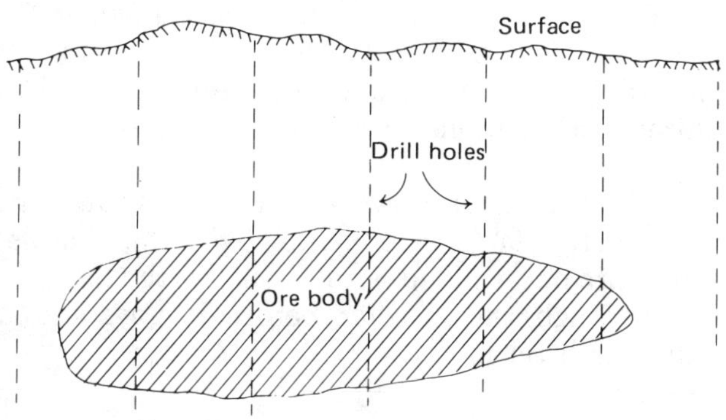

E. Long section of a drilling program for an orebody

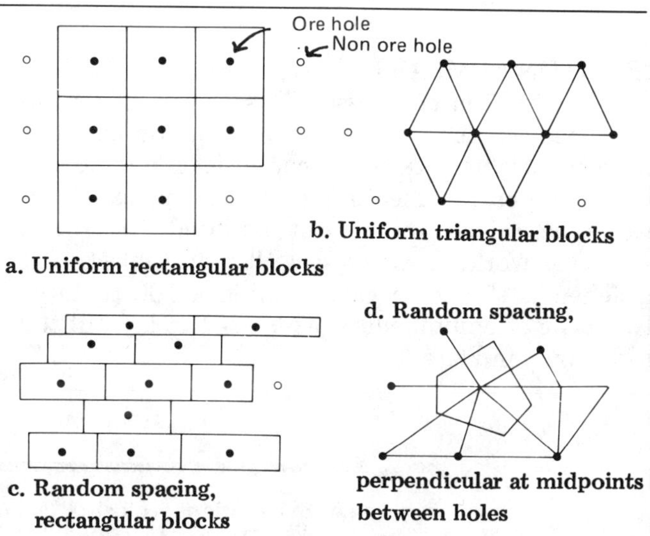

F. Geometric patterns for ore reserve calculations

EQUIPMENT USED IN MINERAL EXPLORATION

As mentioned in Plates 3-1 through 3-5, various types of equipment and machines are used in mineral exploration. The usual equipment that an exploration geologist uses is shown in A. The pocket transit or Brunton compass A,a is used by most geologists and mining engineers when engaged in geological field work. A tape is also frequently needed, as shown in A,b to measure distances, the widths of outcrops, etc. The small hand pick, often called a geologist's pick, is shown in A,c and a belt with attached pockets to carry the above mentioned pick and compass is shown in A,d.

BRUNTON COMPASS

The Brunton compass in B is a versatile instrument because it not only serves as a compass but has a line sighting device, a center-level bubble and a clinometer for measuring vertical angles. With care, these instruments can be used for rough surveying work. A Brunton with a portable tripod attached is shown in C, as well as a 100 ft. tape. This type of equipment is often used to lay out the grid shown in Plate 3-3.

BULLDOZERS, BACKHOES

When trenching, road building or drilling site preparation, the bulldozer is frequently used. D shows a large bulldozer with a ripper attached which will dislodge fairly large and tight rock without blasting.

If the trenches are not too deep, the combination tractor loader, or backhoe shown in E is widely used. After it has been dug, mapped, and sampled, the end loader can be used to refill the trench. For deeper trenches or large production digging, hydraulic backhoes similar to the one in F are commonly used.

SAMPLE SPLITTERS AND CRUSHERS

After the sample has been collected, it usually must be crushed and reduced in size before it can be analyzed. A sample splitter shown in G is frequently used. The crushed sample is poured into the funnel shaped lid at the top. Half of it flows into one pan, and the other half into the other pan. This can be done several times until the required sized sample is attained.

A small laboratory jaw crusher, shown in H, reduces the large sized chunks in the sample to some pre-determined size. Usually the sample is run through the crusher before it goes into the sample splitter in G.

A. Exploration geologists minimum field equipment. **B.** Close up of Brunton compass or pocket transit. **C.** Tripod for Brunton compass and tape. **D.** Hydraulic actuated bulldozer and ripper. *(Courtesy Anaconda Co.)* **E.** Tractor mounted backhoe and loader. **F.** Hydraulic actuated large capacity backhoe. **G.** Sample splitter. **H.** Small laboratory jaw crusher.

PLATE 3-6 EQUIPMENT USED IN MINERAL EXPLORATION

EXPLORATION DRILLING

The drill shown in A is versatile and can be used with a number of different combinations of drills and bits. The mast is shown in the vertical position, but it can be lowered so that it rests just above the truck cab. The whole drill is portable in this lowered position and can be driven on highways. In most drills of this type, the mast can be positioned at an angle so angle holes can be drilled.

SPECIALIZED DRILL BITS

Various bits can be attached to the drill stem, and often the tapered roller type bit or rotary bit, as shown in Plate 3-7, is used. Cuttings from the hole are sampled for mineral content. Through special rigging, water can flow down the outside of the hole and the cuttings can be brought up the center of the drill stem. This is called reverse circulation. Many geologists prefer this system for sampling cuttings because they claim there is less chance of dilution from sluff in the hole. In very hard ground a down-the-hole drill can be attached, and the cuttings recovered and analyzed for mineral content. A hollow-core diamond drill bit can also be attached for the recovery of core.

Because of their mobility and versatility exploration drill units, of the type shown in A, are in widespread use for exploration drilling. The drill unit is usually driven by hydraulic motors and pistons. The hydraulic pump is either driven from the truck engine or from an independent diesel engine.

HOLLOW-CORE DRILL BIT

Core recovered from a hollow-core drill bit is shown in B. This core is frequently split lengthwise in the mineral zone with half of it crushed and analyzed for mineral content and the other half saved for further geological study.

DIAMOND DRILL BITS

The drill in C is an older type of diamond drill which is air-powered. Drills of this type are still used when the holes are not too deep. Another type of surface diamond drill, either gasoline or engine powered, is shown in D. A core is being removed from the core barrel. When the length of the core barrel has been drilled, as shown in C, and D, all of the drill rods must be removed from the hole so that the core can be recovered.

HYDRAULIC DRILLS

The drill in E is a hydraulic powered drill used in underground drilling. The hydraulic pump, which supplies oil under pressure to drive the hydraulic motor and pistons, is driven by an electric motor. This is a wire-line drill where the core barrel can be recovered from the drill steel without removing the drill rods. The core barrel can be detached and lifted through the drill rods by a special mechanism attached to a wire line. The wire line hoist is the lower one with the smaller diameter cable. The time spent in removal, emptying and placing of the core barrel is greatly decreased by this system which in turn speeds up the drilling cycle.

The dream of every exploration person is to find a huge ore body. In F, a seam of coal up to 50 ft thick is exposed. This is the happy conclusion of a successful coal exploration program.

A. Truck mounted exploration drill. *(Courtesy Western Energy Co.)* **B.** Core from a drilling program. *(Courtesy Western Energy Co.)* **C.** Underground, air-operated diamond drill. *(Courtesy Mining Dept. Montana College of Mineral Science and Technology.)* **D.** Removing core from a core barrel. *(Courtesy Mining Dept. Montana College of Mineral Science and Technology.)* **E.** Underground diamond drill. *(Courtesy Anaconda Co.)* **F.** Exposed coal face in a Western U.S. Mine. *(Courtesy Montana Bureau of Mines and Geology.)*

PLATE 3-7 EXPLORATION DRILLING

PART 4

OPENING A MINERAL DEPOSIT FOR MINING

When a promising mineral deposit has been found production is the next thing to consider. Therefore, it is necessary to determine the extent and value of the deposit. In opening a mine, a prospector or small mining company usually digs any ore that is showing. The revenue from the sale of ore keeps the venture going and finances further development or prospecting work. A larger company, however, may be able to spend considerable capital before money is made from the mine. Frequently they will drill the deposit much more thoroughly to determine the extent and grade of the ore. After the ore body has been delineated and defined, a company may make several feasibility studies to determine the best way to mine the ore. A sum of money is then allocated to develop the mine in the most appropriate manner.

OPEN-PIT MINING

Being anxious for ore, the operator will often try an open-pit as shown in A. Stripping the overburden, shown in A, is always a problem because the depth to which the pit will go is unknown, and one does not want to strip more than is necessary. Here stripping is only done when necessary. The wider and richer the vein the deeper the pit. The availability and production capability of surface equipment makes open-pit mining a very attractive choice. Conversely, a well financed organization will determine how deep the pit will be and plan the stripping to insure that steady flow of ore will be maintained.

DRIFT MINING

Open-pitting is generally not practical for a narrow vein, therefore, if the terrain is favorable, a drift or adit on the vein is often used as shown in B. In this way, the vein can be assayed as progress is made and ore can be saved and sold to defray expenses. After the vein has been exposed in this manner, the ore above the level can be mined readily and some other method of developing the lower portion of the vein can be chosen.

INCLINED SHAFT MINING

In flat terrain, a shaft is almost always required in a narrow vein. The prospector or under-financed company usually follows the ore, and an inclined shaft is often dug similar to that shown in C. Again, this system allows the vein to be sampled, and any ore sold will help defray the costs of the opening. However, as the vein rolls frequently or changes dip and strike, the shaft may become very irregular in alignment and slope. Operational efficiency in such shafts is very low. If money is available an inclined shaft, properly aligned and sunk in the footwall, is often used as shown in C.

CROSSCUTS TO VEINS

In steep mountainous country, crosscuts to veins can be used. Crosscuts have the advantage of draining water and hauling ore on a near level horizon, and also of prospecting the country rock for parallel veins. However, because the crosscut is driven in barren rock, no ore is recovered to defray the cost of driving. In the western United States there have been several crosscuts driven which either didn't intersect a vein because there was none present, or because the operators did not have sufficient capitalization for completion of the project. Thus all the work was wasted.

VERTICAL SHAFT MINING

Many operators favor vertical shafts over inclined shafts because there now are more sophis-

PLATE 4-1 OPENING A MINERAL DEPOSIT FOR MINING

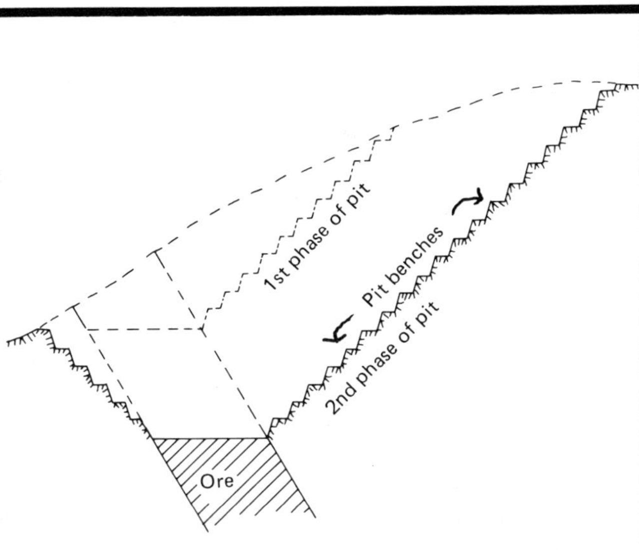

A. Opening a deposit with an open pit

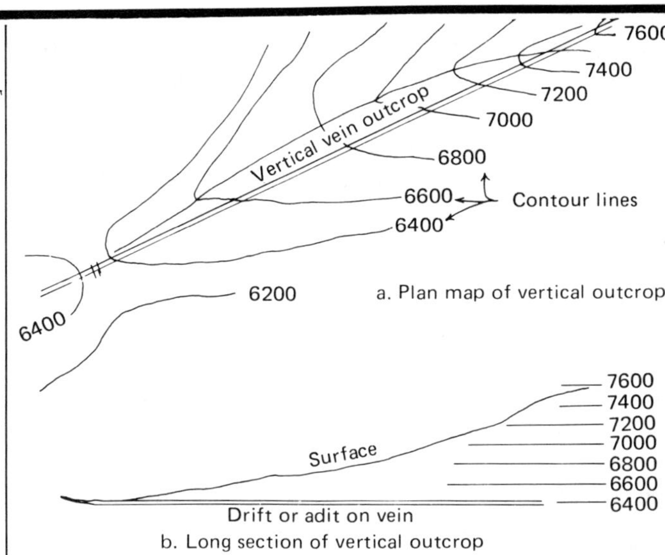

B. Drifting on a vein where topography permits

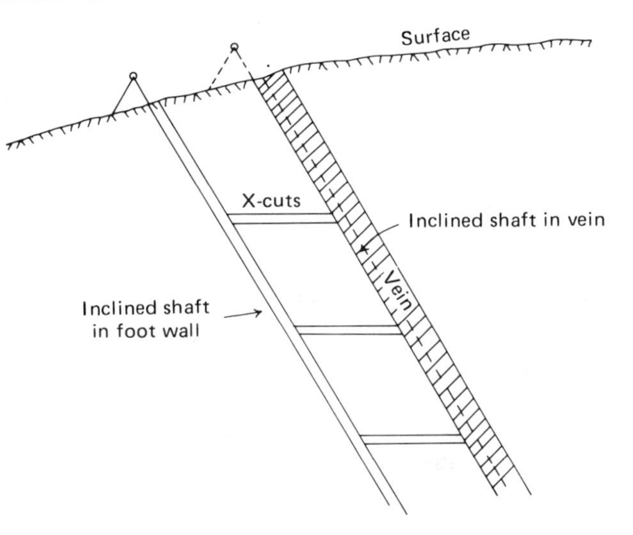

C. Position of inclined shafts for dipping veins

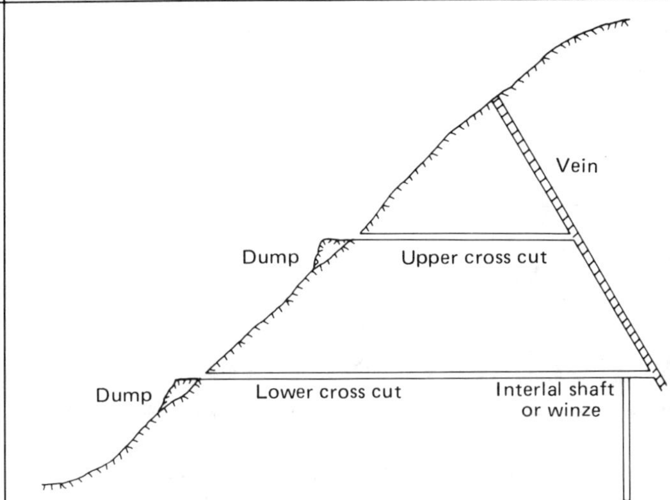

D. Cross-cutting to a vein in steep mountainous terrain

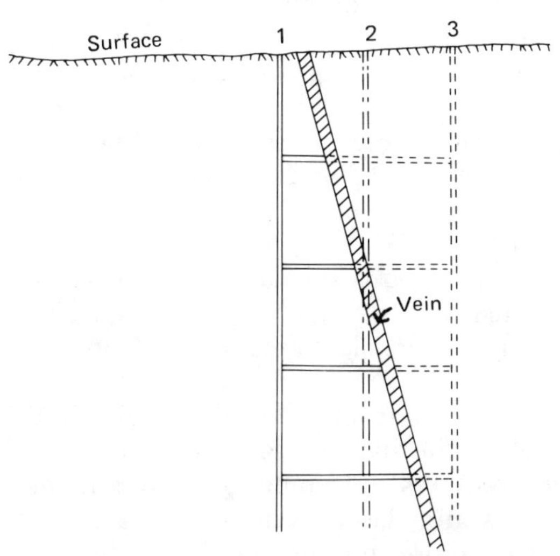

E. Position of vertical shaft with dipping vein

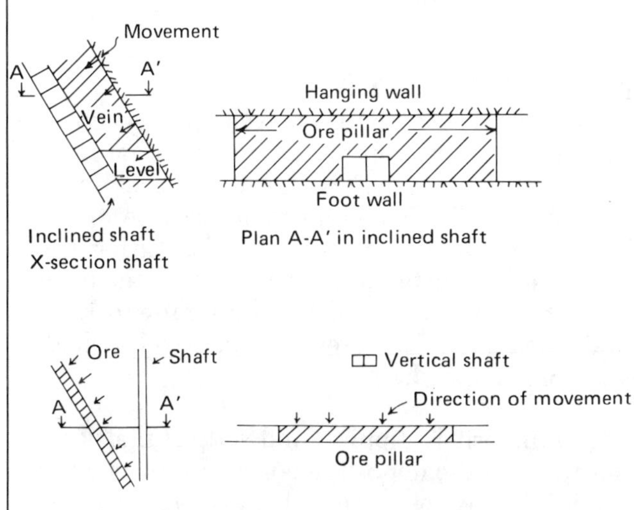

F. Leaving pillars to protect shafts

ticated systems to dig them, and as a rule the hoisting speeds in vertical shafts are greater. The choice of position of a vertical shaft in a narrow steep vein is shown in E. Position 1 is usually favored because in a dipping vein, the hanging wall is more apt to move into any mining opening than the foot wall. Hence maintenance is usually less in the shaft when it is sunk in the foot wall. The disadvantage to this method is the increase in length of the crosscuts as the shaft gets deeper. A shaft started in the hanging wall and intersecting the vein at depth is shown in E 2. This has the advantage of minimum crosscuts to the vein, but often the vein intersection is a high source of maintenance to the shaft because the ore zone is less competent than the wall rock.

If the shaft is dug in the vein or in the hanging wall, it is often necessary to leave a pillar of ore in the vein to protect the shaft, as shown in F.

OPENING A MINERAL DEPOSIT FOR MINING WITH VARYING CONDITIONS

If the surface topography is relatively level as the mine is developed, a shaft must be sunk if the ore can't be mined by open-pits. The optimum position for the shaft must be chosen so subsequent mining will not disturb or actually destroy its structural integrity.

DETERMINING SHAFT POSITION

In a vertical deposit, both walls may become weakened and move into the mined-out area as shown in A. If the shaft is located too close to the deposit it will become unstable because of the walls moving into the mined out area. Either the shaft should be located beyond the area influenced by the subsidence angle, or in an attempt to keep the wall rock stable the area will have to be filled-in as the ore is mined.

If the deposit is dipping, and vertical shafts are chosen to open the deposit, a configuration such as shown in B may be practical. Most hoists reach some vertical limit for operational efficiency; usually not over a mile deep, so it is necessary to sink a winze to mine the deeper levels. Often the ore hoisted from the winze is transferred along a haulage level to the bottom of the surface shaft. The haulage level is ordinarily designed and installed for efficient level ore transportation, and with the first winze located as shown in B the distance from the winze to the ore vein is not large until greater depth is reached.

DEVELOPING A DIPPING DEPOSIT

With the increased uses of diesel powered load-haul-dump equipment, declines are popular ways of developing a dipping deposit as shown in C, and D. The decline is normally driven in the footwall at a grade of from 10 to 15%. A grade of 10% means that the decline loses 10 ft. of elevation in 100 ft. of horizontal length. Therefore, to reach a depth of 100 ft. in a - 10% decline, it will have to be 1,000 ft. long. It is common to spiral the decline down so the long axis of the spiral is generally paralleled to the strike of the vein as shown in D. Crosscuts are driven from the decline to intersect the vein at various locations. The position of the crosscuts depends largely on the mining method employed.

Even though the mine may be developed with haulage levels and hoisting shafts, declines are often used where the mine is mechanized with diesel-powered mining machinery. The declines greatly facilitate movement of the machine from one place to the other, and operators justify the added expense by the increased efficiency in mining.

MOUNTAIN MINING

In steep mountainous country, a long adit may be driven to intersect the ore body at depth, as shown in E. In high mountainous country the weather is often very severe in the winter. A long adit, several thousand feet lower than the mountain summit, will often facilitate winter operations.

RAIL HAULAGE

Rail haulage over long distances is often economically advantageous. If a mill site can be found in a drainage area away from high population densities a long adit may be feasible. Usually the mine must be developed both above and below the haulage level as shown in E. This system requires a tremendous investment before any ore is mined due to the levels, raises and shafts required before mining begins. Only a well financed organization could consider this type of development.

PLATE 4-2 OPENING A MINERAL DEPOSIT FOR MINING WITH VARYING CONDITIONS

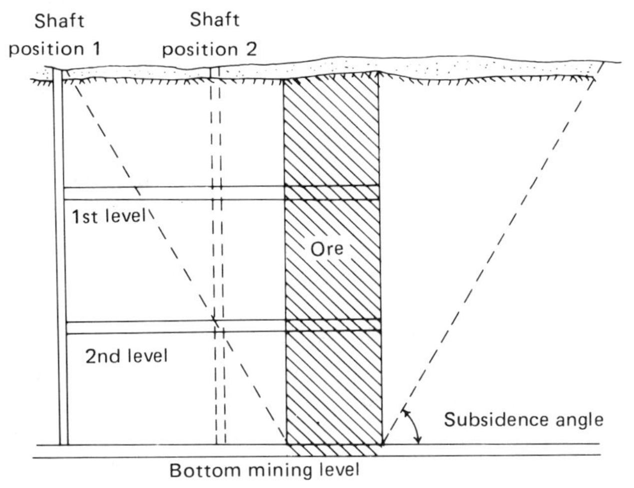

A. Layout of vertical shafts in vertical ore body

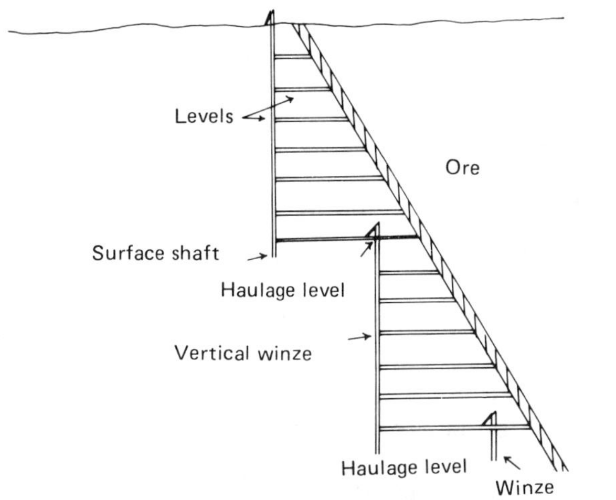

B. Vertical shafts and winzes for dipping ore body

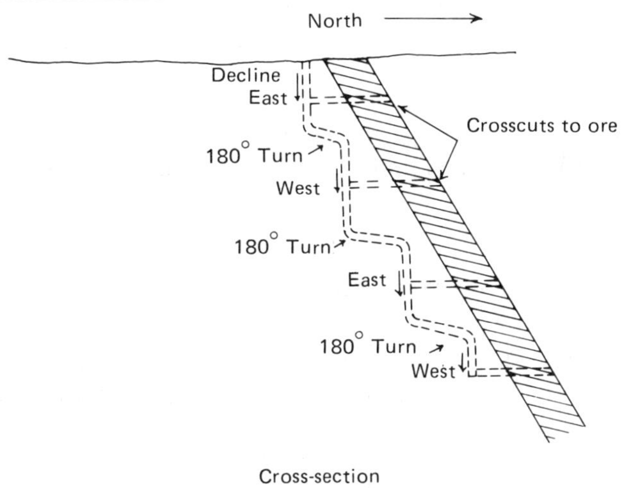

C. Ramping down a decline in the foot wall

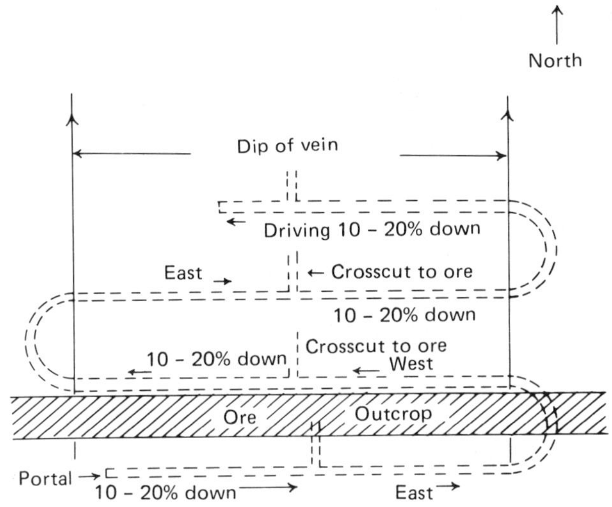

D. Plan of vein view of ramping down a decline

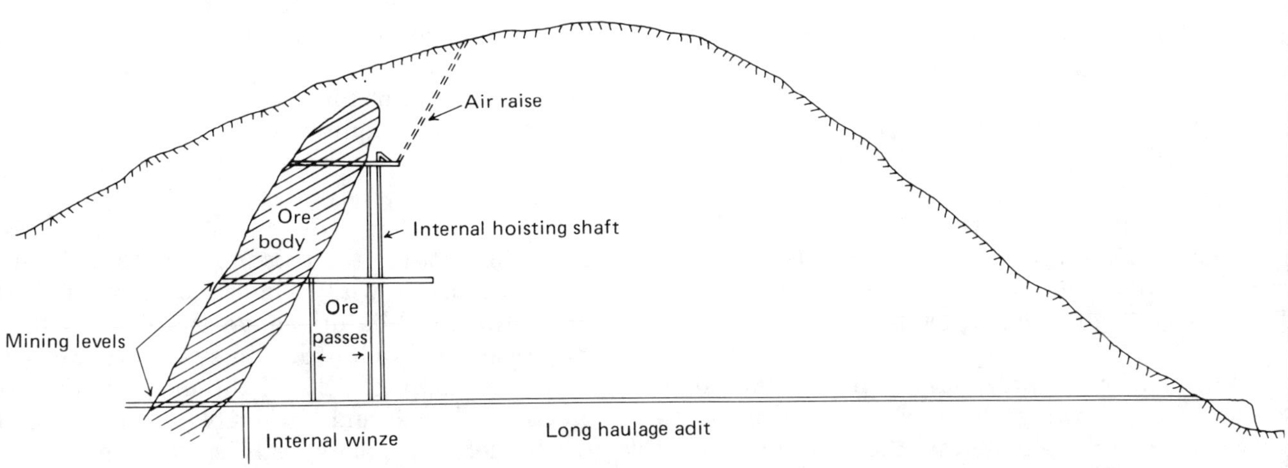

E. Developing an orebody with a long haulage adit and internal shafts and winzes

POSITION OF HORIZONTAL DEVELOPMENT OPENINGS

When following the vein with a drift, it is important to have the vein and the drift in the proper position. Three positions are shown in A. The ore in A,a is labeled the best position for at least two reasons. In most mining situations, it is better not to break the hanging wall if it can be avoided. A broken hanging wall is often difficult to support and can cause trouble as the entire stope is mined to the next level. Secondly, with the drift shown in position A,a, structures such as chutes, etc. that depend on gravity flow are easier to place in respect to the track in the drift.

BREAKING A HANGING WALL

Although it is best to not break the hanging wall if possible, sometimes the vein rolls, undulates and even faults off so that it is very difficult not to have the situation shown in A,b or A,c. When this does happen, the drift should be turned toward the foot wall as soon as possible to get back into the foot wall of the vein.

In some mining operations the entire face is drilled. Either the ore or waste can be blasted first. If the waste is blasted first, it is cleaned up and the ore is then blasted. This keeps the ore from being diluted with waste, but this practice takes more time because of the double blasting.

KEEP DRIFT IN FOOT WALL SIDE

In wide veins, unless it is for some special mining method, it is best to keep the drift in the foot wall side of the vein as shown in B. The main advantage of this location is that if gravity is to be an aid in the mining system, the drift is in the more advantageous position.

As mentioned previously, one of the main reasons for following the ore with a drift is to recover ore and help defray the cost of driving the opening. However, the vein material is often less competent than the surrounding rock, and also, the vein is commonly mined out. This action often produces high loads or forces on the drift, and heavy supports may be needed to keep the level open for the passage of trains.

PILLARS, SUPPORTS, SANDFILL

One common method of protecting the drift is to leave a pillar of ore both above and below the level as shown in C,a. If possible, as the area is mined out and abandoned, these pillars may be recovered. Another way is to provide very heavy supports to the drift such as shown in C,b and filling above and below the timber with sandfill. These methods either result in some loss of ore, or considerable time and money is spent in timbering the levels.

As mines get deeper and if rock burst conditions exist, the pillar shown in C,a may become very heavily loaded and may burst causing much damage to the level.

LATERAL SYSTEM

Some mine operators use what is called a lateral system as shown in D. A drift parallel to the vein is driven, usually in the foot wall side of the vein, and crosscuts are driven every so often (for example 200 ft apart) to intersect the vein. The crosscuts are driven through the vein and into the hanging wall so there is tail room for a train of cars to be loaded. A cross section of the system is shown in E.

When the lateral crosscuts are driven in the walls, the ground is usually better, and support and maintenance is not such a problem as is a drift in the vein. All of the vein can be mined and pillars are usually not required.

The main disadvantage of this system is the cost. As most of the laterals and crosscuts are driven in waste, there is no ore to help defray expenses. Even with a lateral system of drifts, a drift still must be driven on the vein so that the stope can be developed. This system requires much more development. But if the maintenance cost is greatly reduced, and if valuable ore does not have to be left in pillars, this system may be economically better than the haulage drift driven in the vein.

PLATE 4-3 POSITION OF HORIZONTAL DEVELOPMENT OPENINGS

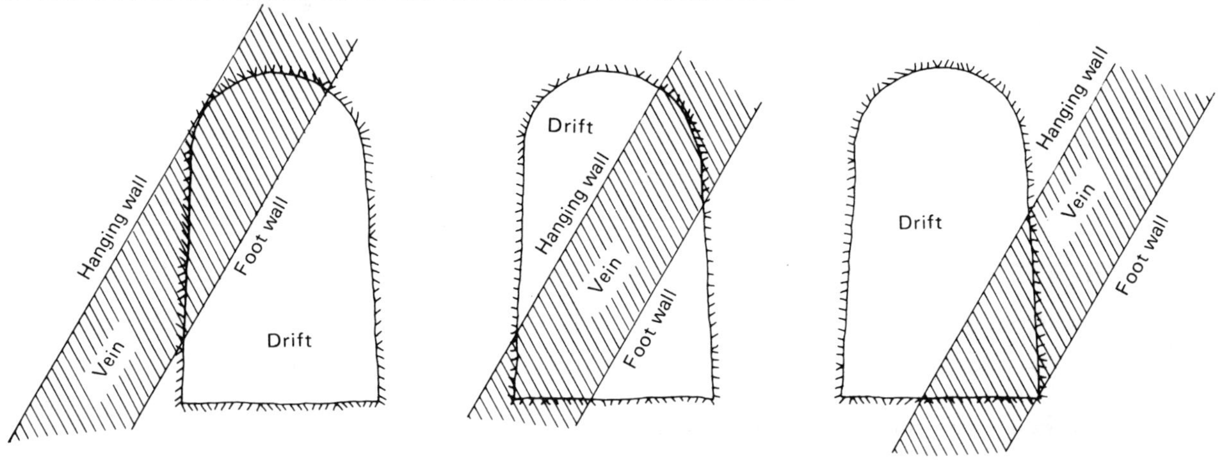

a. Best position

b. Poor position

c. Poor position

A. Sketch showing drift in relation to position of vein

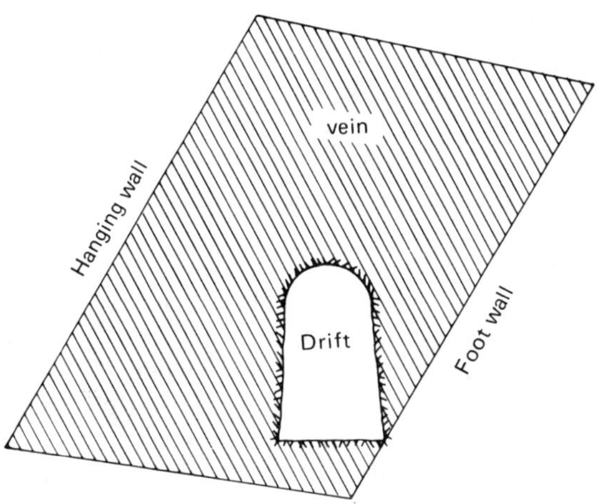

B. Position of drift in wide ore body

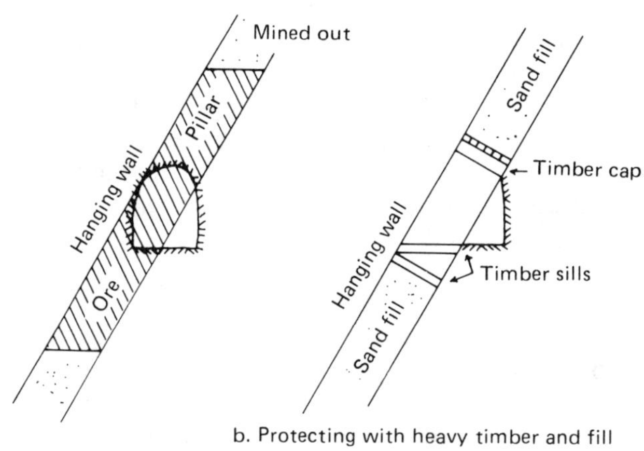

b. Protecting with heavy timber and fill

a. Leaving ore pillar

C. Protecting drift from heavy ground

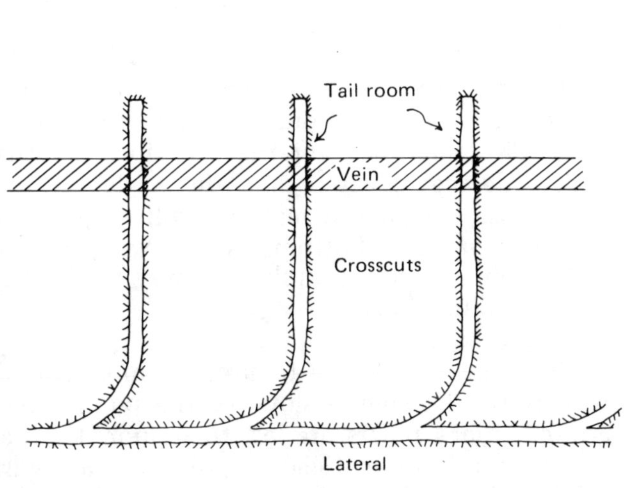

D. Plan view of lateral and crosscut system

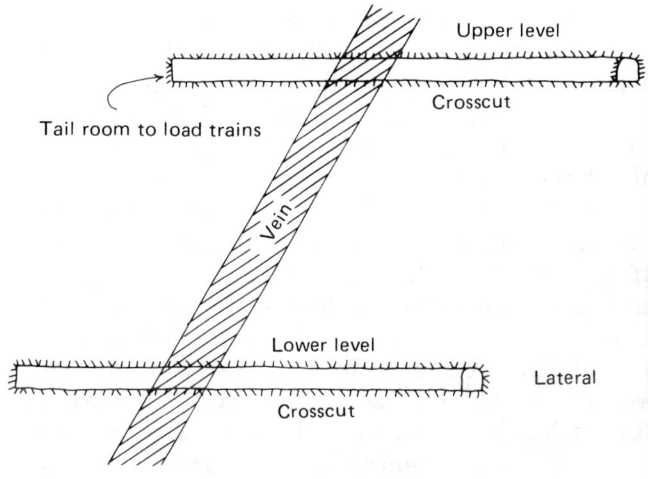

E. Cross-section of lateral and crosscut system

SUBSIDENCE AND PREVENTION OF SUBSIDENCE

Before any underground opening is excavated, the earth stress or force lines are established. In most cases they are fairly uniform in magnitude and are evenly spaced. But when an opening is made, these forces or stresses are disrupted because they can only occur in the earth and not in the space of the opening. There are many factors involved in the stability of openings in the earth's crust, but eventually if the opening gets big enough the earth's structure will cave into the opening, and the surface or other nearby openings will frequently be affected.

SUBSIDENCE ZONE RATIOS

The shape of a subsidence zone is shown over a horizontal deposit in A. The depth, area, and thickness of the mined zone are all important criteria. If the area is inhabited, subsidence will cause considerable damage to property on the surface. Assuming the length of the opening is great, somewhere above 1.4 times the depth, the width of the opening becomes an increasingly important parameter in determining subsidence. The British Coal Council has done considerable work on subsidence, and has found if the width divided by the depth is less than 0.25, there is negligible subsidence. If this ratio is 1.3, subsidence occurs which is up to 90% of the thickness of the mined seam. The maximum subsidence occurs in the center of the trough, as shown in A, and some subsidence occurs to the dashed lines. The area affected is larger than the mined out zone, and is defined by the width of the opening and the angle of the draw as shown in A. If the width is increased, the subsidence profile tends to flatten out over the mined out zone. The angle of draw in Europe has been found to be as great as 35°, but 25° is the value generally used for calculations.

If the mineral seam is pitching as shown in B, the angle of draw varies at the two ends with the greater angle at the deeper or lower end of the seam.

BLOCK CAVING SUBSIDENCE

Subsidence usually occurs over block caving zones as shown in C. When the ore caves, the overlying surface also caves. The maximum subsidence commonly occurs vertically over the mining zone. The subsidence angle is measured from the horizontal rather than on the vertical, and it is equal to 90° - the angle of draw. The angle of break is defined as the angle extending from the mining level to where the surface shows some evidence of tension cracks which can be harmful to surface structure. Although subsidence may not be measurable, the surface tends to move toward the opening. This angle has been measured at about 45° in several caving operations.

SOIL MECHANICS

The final slope angle design of large open-pits has been influenced in the past by the angle of break in caving operations. However, in several cases it has been found in large open-pits where the slopes are unstable at this large an angle. In large open-pits, the walls, although they look like solid rock, are often broken by faults and other fracture phenomena. Hence, the principles of soil mechanics are often used to predict slope failure. Some soil parameters are shown in D such as angle of internal friction and angle of repose. The angle of repose is the angle that broken rock will make with the horizontal when it is dumped from the top of a pile. The angle of internal friction is usually a few degrees greater than the angle of repose and is the angle the material will stay at as it is dug from the bottom. A pit wall, however, will often fail in a complicated way as shown by the dotted line in D.

To control pit wall failure, the most common way is to flatten the slope of the pit. In underground mines pillars are left to support the walls, as shown in E, or artificial supports such as hydraulically placed back fills as shown in F. These systems shown in E, and F will be discussed more in the sections on mining methods.

PLATE 4-4 SUBSIDENCE AND PREVENTION OF SUBSIDENCE

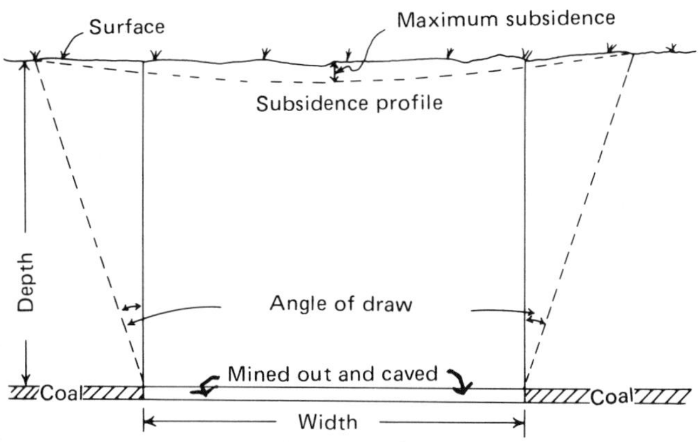

A. Shape of subsidence zone over a horizontal seam

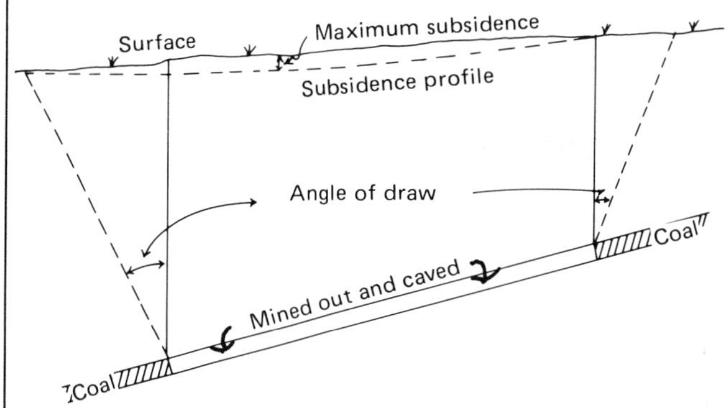

B. Shape of subsidence zone over a pitching seam

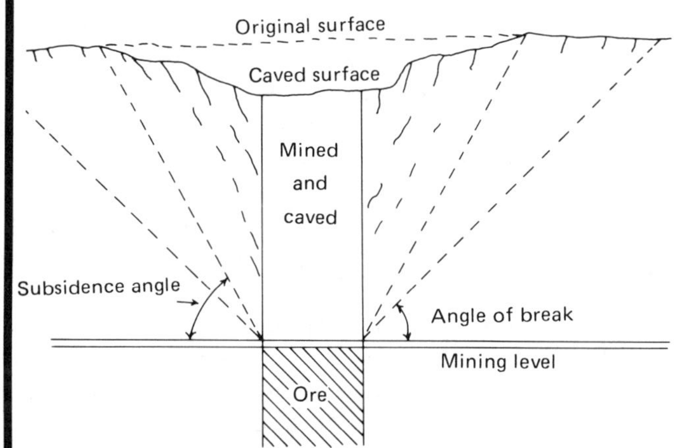

C. Angle of subsidence and break around a caved zone

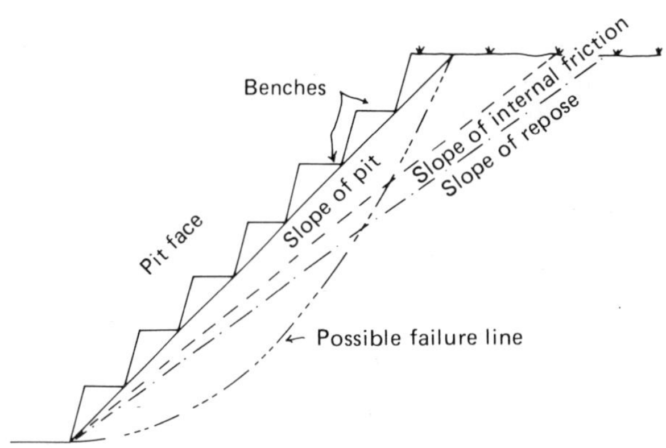

D. Slope stability factors in an open-pit mine

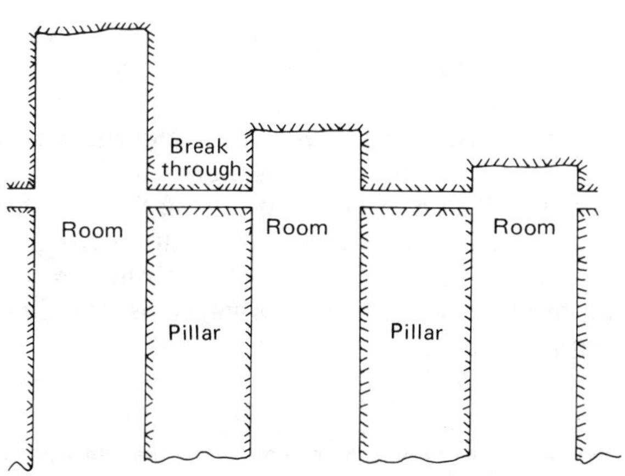

E. Leaving pillars for support (plan view)

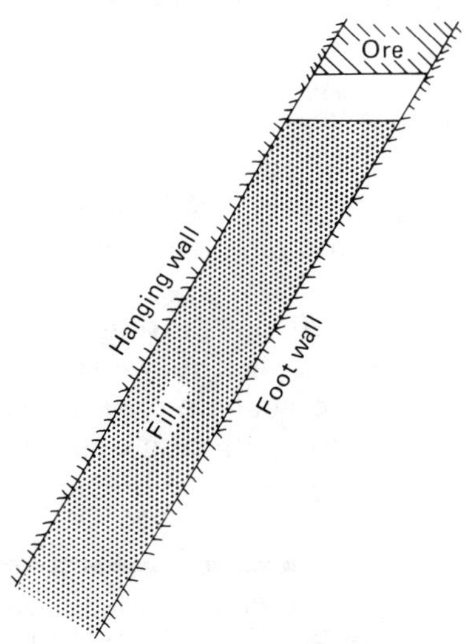

F. Cross-section of filled stope

SURFACE SUBSIDENCE

As discussed in Plate 4-4, surface subsidence often occurs when minerals are removed by underground methods. Means of preventing subsidence were shown to be leaving pillars and filling the mined out portion with waste or supportive timbers. These methods may be only temporary because time has an influence on the supportive power of pillars and timber. Waste fill is permanent, but it may compress before it acts as a good support.

CREEP DEFORMATION

Creep is defined as the permanent deformation of a material subjected to a load for a long time, even though the material is not loaded to the elastic limit. Rocks and soils under load exhibit creep. In some room-and-pillar mines, the heavy loaded pillars may hold the roof for years, but eventually they will fail.

Examples of failure and subsidence appearing on the surface are shown in A, B and C. These are old abandoned coal mines that were mined fairly close to the surface. As the timbers rotted and failed, the pillars exhibited creep. The roof over the rooms eventually failed and caved to the surface. This produces a very pockmarked surface which inhibits agricultural development.

When the energy crunch started to develop, and the large coal seams in the western part of the United Sates became economic to mine, many environmental and conservation groups did not want them strip mined. They clamored for underground mining only.

NEAR SURFACE UNDERGROUND MINING

However, if one looks at the possible result of near surface underground mining shown in A, B, and C and the land reclamation results shown in Part 9 of this publication, it appears that the land is more useful after strip mining than it would be by near surface underground mining. Pit wall failures sometimes occur as discussed in Plate 4-4D. Tension cracks appearing on a stope failure are shown in D. Also, some parts of the wall have pulled away and started to subside.

Caving methods as discussed in Part 8 of this publication destroy the surface, and a surface cave over a block caving system is shown in E. Supporters of underground caving systems say that less surface is destroyed by caving than by open-pitting, see Plate 4-4, because of the steeper angle of subsidence and the resulting pit slope angle to recover the same amount of ore. Overburden is not stripped in caving; it falls into the caved area. For this reason it does not have to be piled in dumps which also require surface area.

Even when stopes are filled with waste rock, sand or hydraulic backfill, subsidence is not completely stopped. The void space or volume in this material when it is deposited underground is often equal to the volume of the solid material; therefore considerable compression of the fill can take place before it becomes effective in supporting all of the loads imposed on it by the mining process. The amount of subsidence, however, is greatly reduced with waste fill.

A. Coal mine subsidence appearing on the surface. *(Courtesy Montana Bureau of Mines and Geology)* **B.** Another view of surface subsidence. *(Courtesy Montana Bureau of Mines and Geology)* **C.** Close up of surface subsidence. *(Courtesy Montana Bureau of Mines and Geology)* **D.** Slope wall failure in an open pit. *(Courtesy Anaconda Co.)* **E.** Subsidence over a block caving operation. *(Courtesy Anaconda Co.)*

PLATE 4-5 SURFACE SUBSIDENCE

PART 3 & 4 CONCLUDING REMARKS

Part 3 shows some of the work and processes involved in mineral exploration. As can be seen, modern day mineral exploration is costly and very time consuming. Part 4 shows some of the planning that must be considered after an ore body is found. With all of this information, a feasibility study must be made to insure that the venture will be profitable.

In the United States, most state agencies, and in some instances, agencies of the federal government must approve contemplated mining plans before the venture can start. The agencies are exacting in their demands for information, and the mine operator is required to supply the required information before a mining permit is granted.

FEASIBILITY STUDIES

The feasibility study not only includes all of the factors considered in parts 3, and 4, but it must include an economic forecast of the rate of return that can be expected from the mine at some rate of production. Common factors to be considered are:

Cost of exploration to find deposit
Cost of permitting and meeting government requirements
Tons and grade of ore in the deposit
Expected extraction efficiency in mine and mill
Cost of mining per ton of ore
Cost of milling per ton of ore
Capital cost of mine and mill
Mining and milling rate per day
Depreciation allowance
Depletion allowance
Taxes to be paid
Royalties to be paid
Amount of capital required
Transportation costs
Miscellaneous costs
Contingency costs

Before a mine is developed, several feasibility studies may be made as developments arise, such as changes in prices for the minerals, in cost of equipment or in the tax structure.

For most investors in a mining venture the rate of return on that investment is of prime importance. Because a mine is speculative at best, most companies and investors will not consider a rate of return of less than 15 per cent. Usually 20 to 25 per cent is the required return.

Thus, planning is very important in the orderly development of a mine. The person responsible for the feasibility study needs all relevant information available, especially when it comes to estimating the tonnage and grade of the deposit. However, obtaining this information through drilling or other exploration methods is costly, and decisions are often made on the size and grade of a mineral deposit based upon a minimum of information. Estimating costs is even more difficult, especially when planning several years ahead.

Feasibility factors are often programmed into a computer, and as more reliable information becomes available, revised calculations are made.

PART 5

MAJOR OPERATIONS OF MINING

The three major operations that create problems in almost any mining operation are:

1. Breaking the ore and waste rock.
2. Supporting the excavation.
3. Loading and transporting the ore and waste.

These three major operations may not have equal importance in all mining operations, especially in open-pit, because the ore may be soft or the sides may not need support.

For example, in a shallow open-pit mine, supporting the sides of the pit is not a problem if the sides stand well, but in deep pits the sides may not support themselves at steep angles. The mine operator must choose the proper slope of the pit sides so wall rock will not slide into the pit. In a placer mine, support and breaking are minor problems compared to transporting, washing and disposing of the gravel. In most underground mines, each of the three operations requires much attention, though at any one time one may require more attention than the others.

Most underground mines have three additional operations which require considerable attention along with the above mentioned ones. These are:

4. Hoisting the material from the mine.
5. Controlling and pumping water.
6 Providing a suitable working ventilation environment.

Additionally in most mines, other operations are not only important but necessary to carry out the main operations. In all mines, both surface and underground, management must supply necessary materials and tools to the working areas, supply compressed air, water, fuel, electric power, and provide adequate maintenance and machine service.

In this part, these six operations will first be explained by examining the fundamentals of the operations, then pictures and sketches of the operation will follow. The builders and suppliers of mining equipment are ingenious people and provide versatile and workable equipment. No matter how complicated looking the machine is, it still is designed to carry out one or more of the basic operations listed above.

BREAKING ORE AND ROCK

Drilling and blasting are still one of the major ways to break hard rock or ore. Although now, giant heavy duty machines are capable of breaking and loading softer rocks and ore rocks replacing drilling and blasting. These machines, called continuous miners or borers, also load the broken material. Loading and transporting are now combined in machines called load-haul-dump (LHD) units which are adaptable to many situations.

SUPPORTING EXCAVATION

Support, especially in underground mines, still receives considerable attention since falls of ground are one of the major cause of death and injury. New mining methods are instituted primarily because they reduce personnel exposure to rock falls, helping to improve performance safety.

HOISTING, CONTROLLING WATER, AND VENTILATION

The operations of hoisting, controlling water and providing a suitable work environment are all necessary. Hoisting capacity limits the size of the mine. Water must be handled or no mining can occur. Normally the hoisting and pumping requirements are chosen when the mine is opened. This may also be true of the main ventilation requirements for the mine. Air supply must be continually changed as work progresses in different areas. Large quantities of ventilation air may be wasted through leakage or air short circuiting in a mine. For that reason ventilation requires constant attention by the mine operators.

FUNDAMENTALS OF DRILLING

Drilling is a very important part of most mine exploration, development and production operations. Small diameter holes, usually under one foot in diameter, are used primarily for sampling a mineral deposit or for placing explosives in a rock mass. Drill holes several feet in diameter are used directly as shafts, raises and other types of development work. There are several different drilling systems used in the industry.

HAND PERCUSSION DRILLS

The hand percussion drill is illustrated in A. This was a common way to drill holes in relatively hard rock in the early mining days. The steel had a sharpened end or chisel bit, and the steel was struck with a hammer. This action drove the steel into the rock a short distance, chipping off some of the rock as shown in A,a. The steel is rotated about 1/8 of a turn, see A,b, as the hammer is lifted so that when the steel is struck again, it can chip off another portion of rock. This is a very laborious procedure by hand; however, the modern rock drills, as illustrated in Plate 5-1 still use this basic principle, except that the hammer or piston and the turning mechanism are actuated by compressed air or high-pressure hydraulic fluid. Water is forced through a hole in the steel to the bit where it keeps dust from forming.

SOFT ROCK APPLICATION

Another type of drill that is used extensively in coal mines and mines where the rock is relatively soft is shown in B. The steel frequently has a spiral configuration similar to the ordinary steel or wood type drill. The drill steel is continually rotated with a force applied to the steel which enables it to bear hard on the bit. The bit peels off or breaks off small layers of the rock or coal thereby deepening the hole. The cuttings are carried up the hole by the spiral action of the rotating drill steel. Water is also commonly used in the steel and bit to keep dust to a minimum.

DIAMOND-CORE DRILL

The core-type diamond drill is primarily used for exploration work because it produces a cylindrical core of rock which can be recovered from the hole and examined and assayed for mineral content. A bit of this type is shown in C,a. Occasionally a solid diamond bit is used to drill holes in a very hard rock or ore formation for blast holes. The drill and steel are continually rotated, often with a thrust or force applied, and the very hard diamonds wear away the rock particles. Water is usually forced down the inside of the drill steel and the cuttings are carried up the hole by the water as it flows around the drill steel and up the side of the hole.

CHURN DRILL

Larger sized holes, from 4 to 12 inches in diameter, are sometimes drilled with a churn drill. A churn drill bit is sketched in D. The bit weighs several hundred pounds, and a rope or cable attached from the bit to the drill alternately lifts and drops the heavy bit. This action cuts and spalls rock similar to the percussion drill shown in A. It is necessary to rotate the bit back and forth which is often done by twisting or rotating the cable in shallow holes. In deep holes, the natural twisting of the rope usually causes enough rotation of the bit as it is alternately raised and dropped. Water is poured into the hole, and the cuttings form a mud which are removed from the hole with a bailer shown in D,c. The bit must be hoisted from the hole before the bailer can be put in.

ROTATING TRICONE BIT

The rotating tricone bit shown in E is used for oil and gas, blast hole, and exploration drilling. The drill steel is rotated, and a heavy force or thrust is applied which causes the sharp teeth to penetrate into and chip the rock. The cuttings are removed by compressed air, water or mud. In some exploration drilling, water is forced down the hole on the outside of the steel, and the cuttings are returned through the hollow drill steel. This is called reverse circulation.

Other miscellaneous drilling means are shown in F. The wash pipe shown in F,c is often used to penetrate water bearing unconsolidated material to bed rock so that some other type of drill can be used in the bed rock.

PLATE 5-1 FUNDAMENTALS OF DRILLING

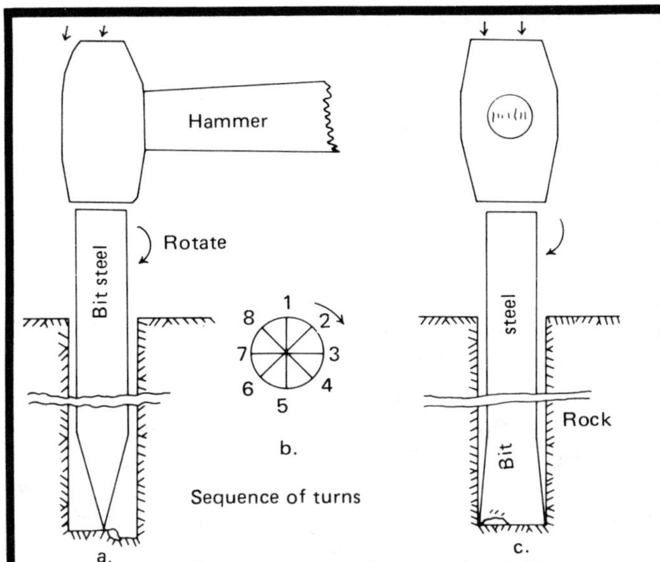

A. Cut-away view of percussion drill

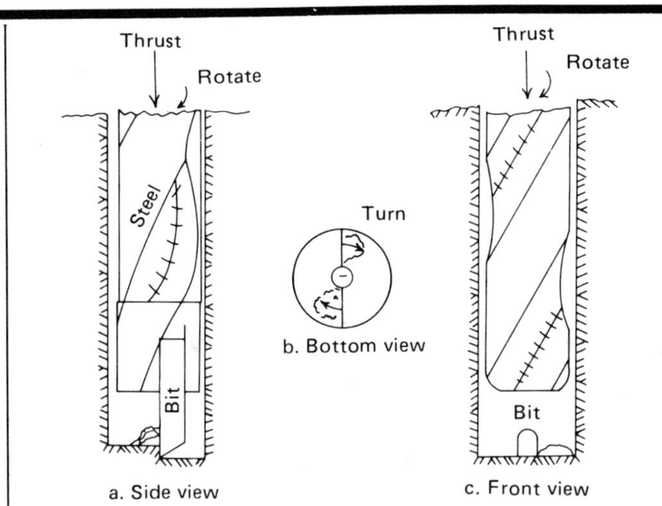

B. Cut-away view of auger drill for soft rock

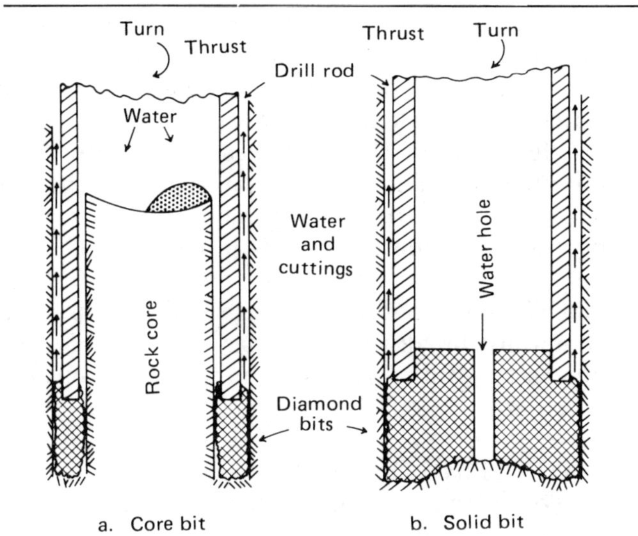

C. Diamond drill bits

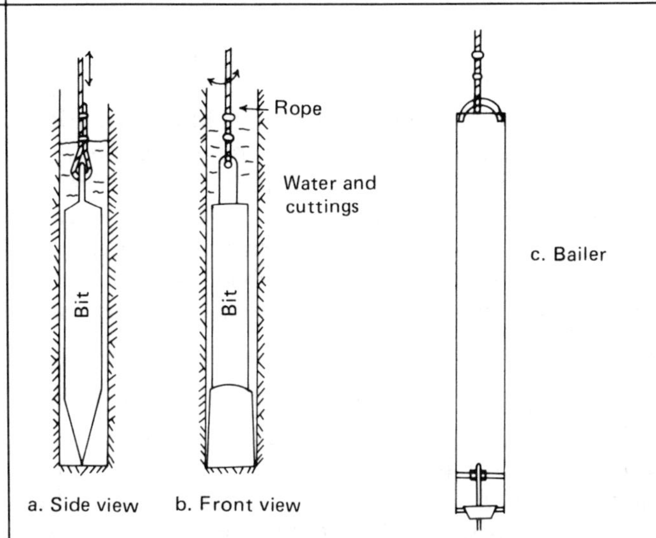

D. Churn drill bits and hole bailer

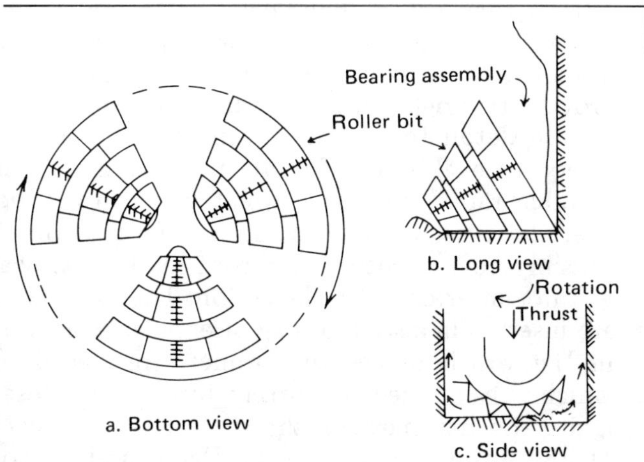

E. Tapered roller bit (Rotary bit)

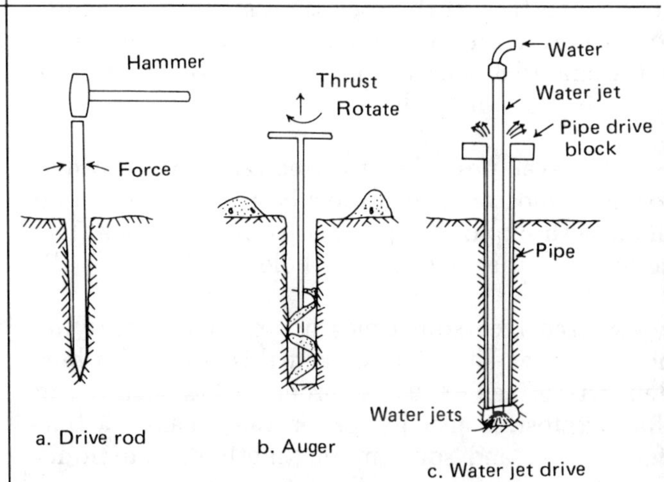

F. Miscellaneous drilling methods

ACTION OF EXPLOSIVES

Explosives are used extensively in mining to break and fragment solid rock and ore so loading and haulage equipment can move the material.

In recent years, progress has been made in drill type, hammer type, and boring or auger types of machines called continuous miners that also break the rock without using explosives. They have not, however, replaced explosives in hard competent ground except in a few experimental cases. These developments are working well in weak rock but haven't been as successful in hard rock. However, progress is being made on these machines, and undoubtedly the future will see more use of continuous mines in hard ground.

BLAST HOLES

In most mining operations, explosives are placed in blast holes. The action of explosive action is shown in A. If the explosive is placed in a hole and detonated, the result is usually a crater of varying size. The bottom of the cone is formed at the free face. If the rock is resistant to breaking, sometimes hardly any cone is formed. Conversely, if the material is easily broken, the cone may start at the bottom of the hole.

Wherever possible, two free faces are provided for an explosive. It is always best to have the hole filled with explosives parallel to one free face. The action of the detonating hole is shown in B.

A free face sometimes has broken rock lying next to it as shown in C, and this is called a burdened free face or just a burden. It has been found that explosives usually break well against a burdened face, and some mining methods commonly use this practice. Usually broken rock has considerable void space. The force of the explosive must compress it slightly so the solid rock can expand and break. In fact, it has been found that even though a free face does not exist, a detonating hole will still break and fracture the rock to a certain extent around the hole. However, it does not fragment it very well. This practice, called destressing, is often used in relieving the high stresses in pillars of ore or other highly stressed zones so that they will not burst. See Plate 5-2.

TYPES OF ROUNDS

Two types of round, or hole sequence and spacing, are used in development headings. One is the angle-cut type shown in D,a. It is usually difficult to break ground with this type of round any deeper than the narrowest dimension (W) of the opening. The other type is called the burn cut where two or more parallel holes are drilled together and one of them is left unloaded so that the loaded hole can break toward the unloaded hole. These types of rounds, if drilled properly, commonly break much deeper than the narrowest dimension. Both types of rounds are discussed in more detail.

DETONATING DEVICES

To detonate a hole filled with explosives it is necessary to use a primer or some type of detonating device. Explosives generally are made safe so they can be handled and transported with safety. The main bulk of the explosive is quite insensitive to most shock. A detonator is used to provide the right shock so the bulk will explode. A fuse detonator or cap is sensitive to fire, and a safety fuse allows a fire to burn at a slow predictable rate. The fuse detonator or cap is slipped over the fuse, crimped so it won't come off and possibly coated with waterproof material so water will not interfere with the action. The fuse and cap are inserted in a stick or bag of explosives as shown in E,e which makes the primer. An electric cap which uses an electric current instead of a fuse to ignite the detonator is shown in E,d. The placing of the primer is shown in F,a,b, and c. Some like to put the primer in the bottom of the hole while others place it in some other part of the hole. If the hole is very long detonating cord can run the entire length to insure complete explosion.

PLATE 5-2 ACTION OF EXPLOSIVES

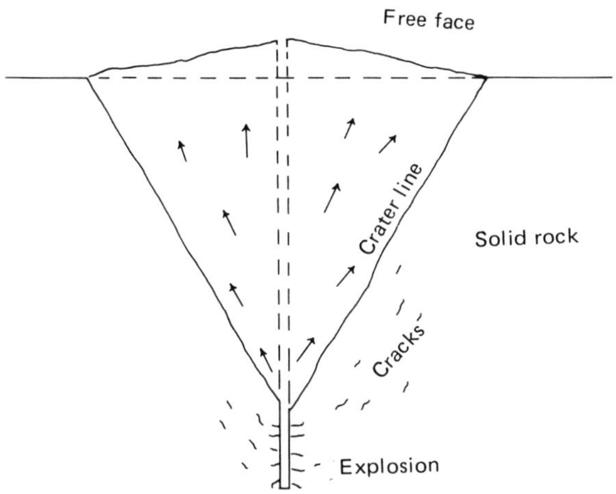

A. Cratering theory of explosives

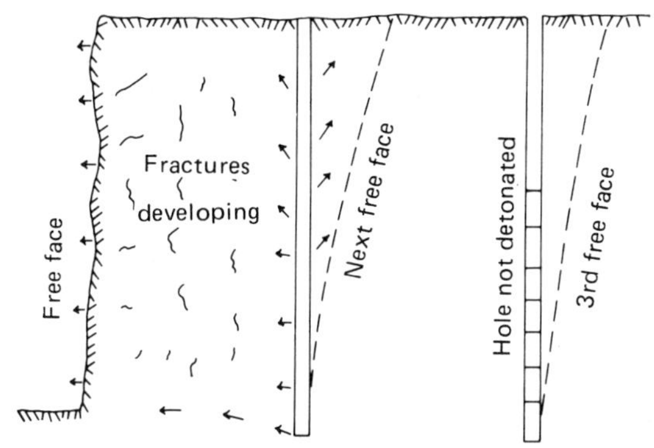

B. Blasting toward a second free face

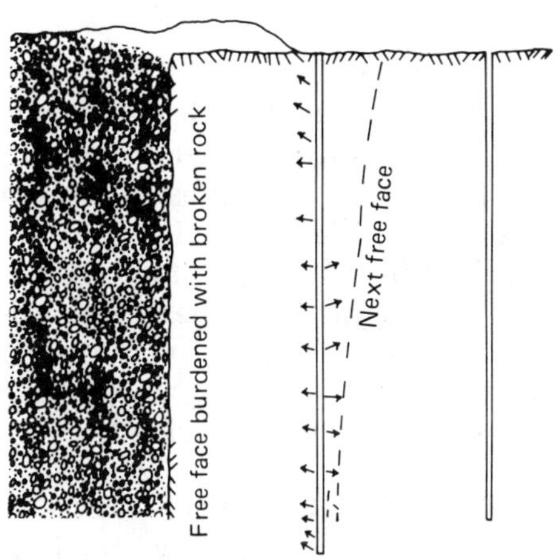

C. Blasting toward a second burdened free face

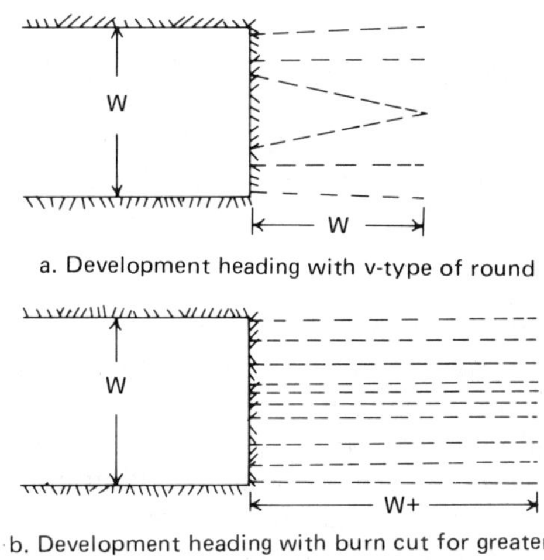

D. Development headings with two types of rounds

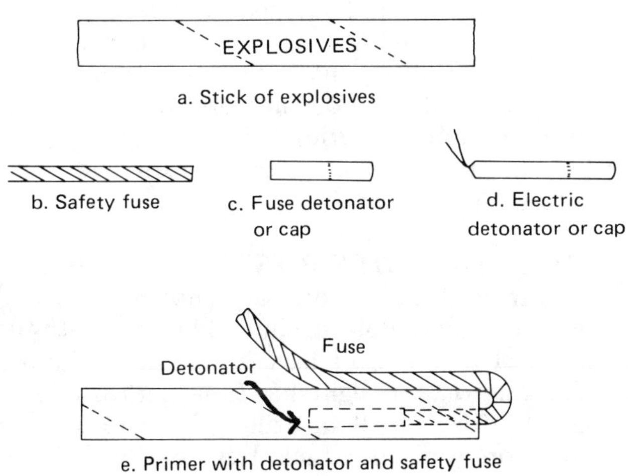

E. Components of a primer

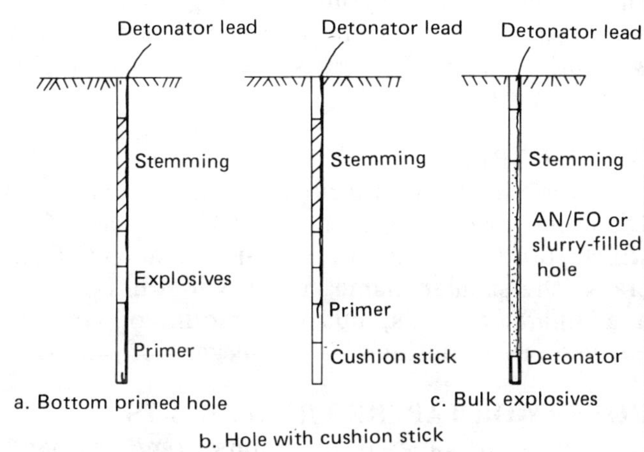

F. Various ways of charging drill holes with explosives

DRILL BITS

There are many shapes, sizes and styles of drill bits. The ones shown in Plate 5-3 are representative of the types of bits commonly used in the mining industry. The cutting action of each fundamental type of bit drilling action was explained in Plate 5-1.

The diamond bit in A has industrial diamonds set in a matrix so they will be held firmly in place as the bit is rotated. The diamonds, being very hard, will abrade the rock as they are turned against it. It is necessary to bathe the diamonds in water while drilling so they won't overheat. Being composed of carbon the diamonds can actually burn if not cooled by the water. Diamond drill bits are designed for different types of ground, but the one shown in A is a common design. The bits do wear out and have to be replaced, but compared with other drilling materials, the life of a diamond bit is very good.

DRILL CORES

Short sections of drill cores are shown in B. Diamond drills come in several standard sizes which produce cores of various diameters. Generally, the smaller diameter drill will drill faster on a comparative basis, but the amount of core recovered is sometimes too meager for analysis.

ROTARY OR TAPERED ROLLER BITS

Rotary or tapered roller bits come in many styles and designs. The buttons shown in the left bit in C are made from very hard tungsten carbide and give good performance in harder rocks. The chisel teeth type shown at the right in C are used a great deal and work well in drilling softer shales, limestones, sandstones and igneous rocks.

BEARING ASSEMBLIES

These bits rotate as the drill rod is turned. A disassembled roller bit is shown in D. The bearing assemblies for these bits present special design problems because they turn in mud, water and abrasive rock chips. The bit must be designed to spray air, water or mud on the teeth and to carry away the cuttings. The bearings are subject to heavy loads because of the weight of the drill column and the thrust provided by the drill.

DOWN-THE-HOLE DRILL BITS

A bit assembly for a down-the-hole drill is shown in E. The buttons are made of tungsten carbide, and as can be seen, the bit is worn out. Down-the-hole drills are used in hard ground. They can be attached to the drill column on a rotary drill. The drill is right at the bottom of the hole so the drilling shock does not pass down through the drill column. Rotation of the drill is provided by the mechanism used to turn the roller or rotary bits.

SMALL DIAMETER BITS

Small diameter bits are shown in F, and are used for blast hole drilling. The bit on the left is all steel and doesn't have a long life in hard rock. The one on the right has tungsten carbide inserts and performs well in drilling hard rock. Another common way to attach bits to the drill steel is by a tapered friction fit. To remove the bits a special tool is used. Usually these bits can be changed faster than the threaded ones.

PLATE 5-3 DRILL BITS

A. Diamond drill bit. Diamonds are in the dark Matrix at the top

B. Cylindrical rock core from a diamond drill

C. Tapered roller bits showing different styles of teeth

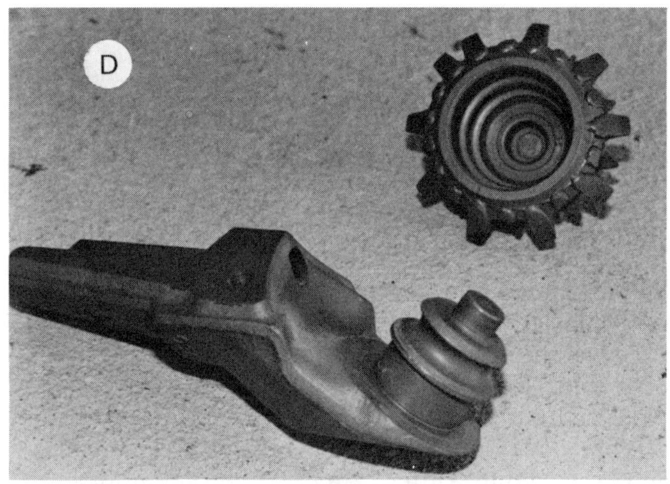

D. Disassembled roller bit showing construction

E. Bit assembly for a down-the-hole drill (greatly worn)

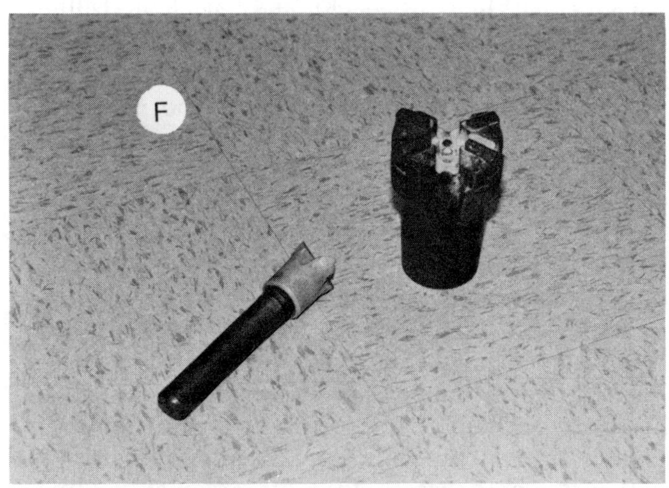

F. Threaded steel bit (left) - tungsten carbide insert bit on the right

DRILL ROUNDS

The fundamentals of drilling and blasting have been examined in Plates 5-1, and 5-2. Plate 5-2 explained the cratering and free face concept of explosives. A group or pattern of drill holes is called a round, and Plate 5-4 shows various types of rounds used in mining. Not only must the explosives be placed in drill holes for proper rock or ore breakage, but the distance the holes are spaced and the firing order are also very important.

Air operated drills are used extensively in underground drilling because they are comparatively small, light in weight, and have high drilling rates. Because of lower noise levels and higher penetration rates, hydraulic drills are being used more frequently. Those drills are jumbo mounted. For drilling in open pits some churn drills and wagon drills are used, but electric or diesel powered rotary drills are used in larger numbers.

HOLE PATTERNS

As mentioned previously, the pattern of holes is important, and depends on the type of drill being used to make the holes. For example, with a long feed drill being used in a narrow drift, angle holes are difficult or impossible to drill. When solid rock is broken, its bulk may become as much as 200 per cent greater than its original volume. Unless this broken rock can move out of the way of the next hole to be blasted, the effectiveness of the explosives is greatly reduced. Therefore, wherever possible, holes are placed so that the rock broken by the explosives will be thrown clear of the solid unbroken rock. Two free faces are desirable, and this situation occurs mostly in stopes and open-pits B, C, and D.

In most development headings, the explosives can break rock toward only one free face. The holes must be so placed that the explosives can break toward that free face. Two general systems are in use. For long advances per blast, the most common one is the burn-cut round. In the burn-cut round, the holes are all drilled parallel or nearly parallel to the direction of the heading, and the central or burn holes are drilled close to one another. Not all of the holes are filled with explosives, the theory being that the neighboring hole when blasted will break into the unloaded hole. A full burn-cut round is shown in E, and various arrangements of burn-cuts are shown in M, N, O, and P.

The burn-cut round is easily drilled with long feed jumbo type drills shown in plate 5-5 D, E, F, and plate 5-6, A, B, D, E, and G.

The other system is the angle-cut round. The cut holes are drilled at an angle with the free face. Various patterns of angle cuts are in use. The draw cut is shown as used in a drift F and in a shaft H. V-cuts are used in shafts and raises G. Various patterns of angle cuts are shown in I, J, K, L.

PLATE 5-4 DRILL ROUNDS

A. Air-operated rock drill (Courtesy Archibald Co.)

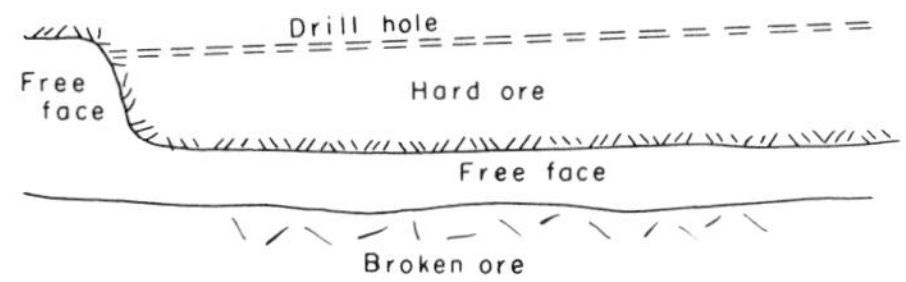

B. Long-section of drill holes

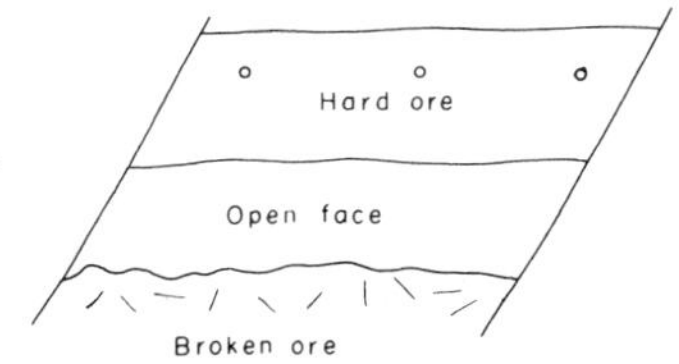

C. Cross-section of drill holes

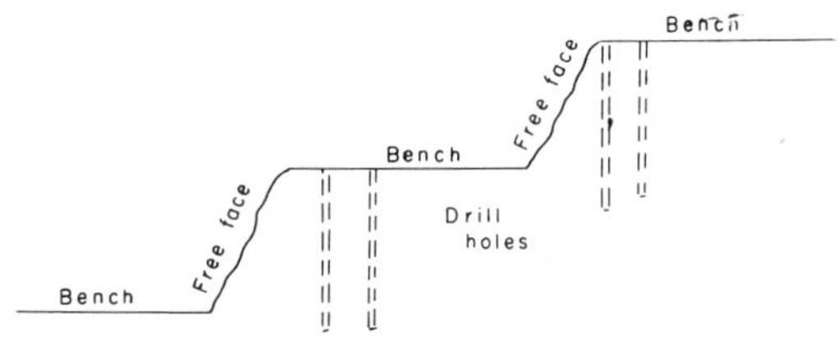

D. Cross-section through open-pit showing drill holes

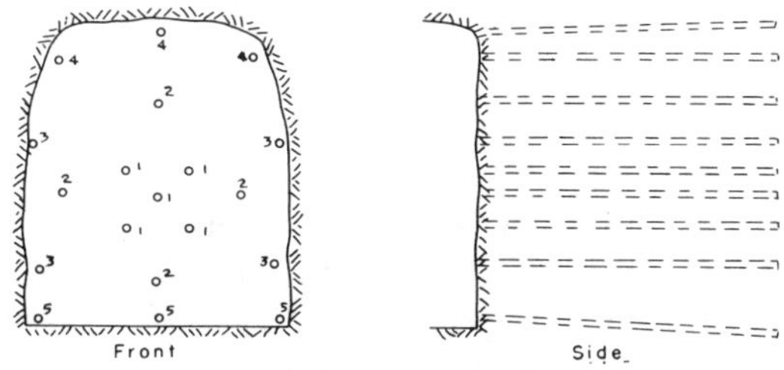

E. Burn-cut drift round (1. Cut holes 2. Reliever holes 3. Trim holes 4. Back holes 5. Lifters)

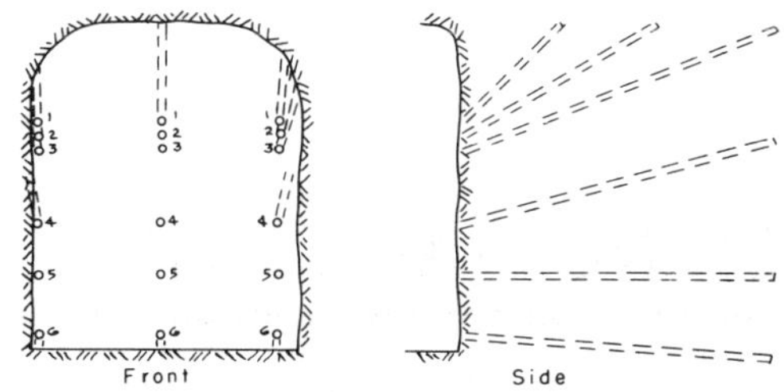

F. Drift round with a top-draw cut (Commonly drilled with stoper type drill)

Note: In figures E, F, G, and H the order of hole blasting follows the number sequence

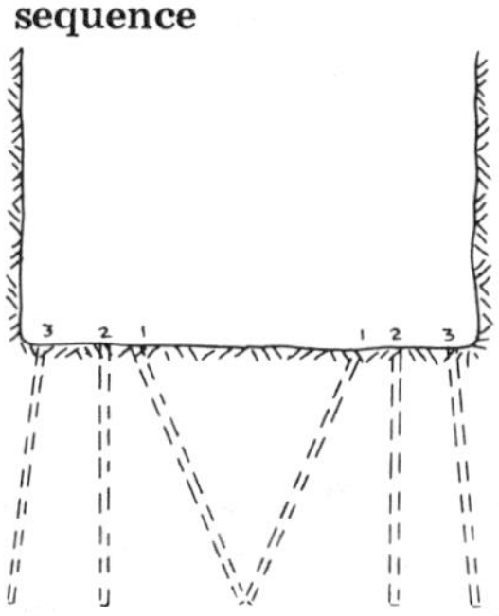

G. V-cut shaft round (side view)

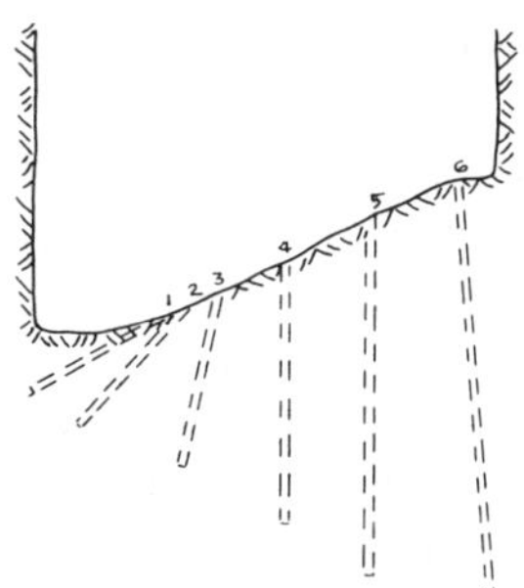

H. Sump or bench type round in shaft (side view)

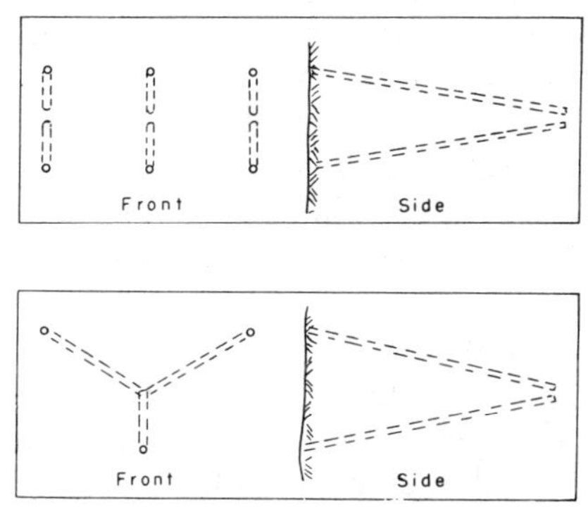

I-J. Pyramid cut (3-hole)

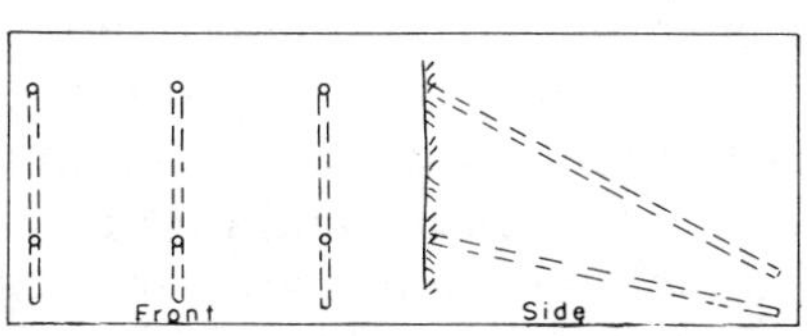

K. Bottom draw cut

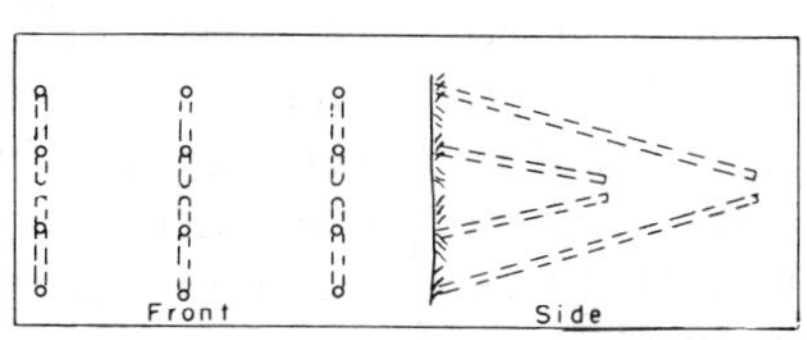

L. Baby V-cut

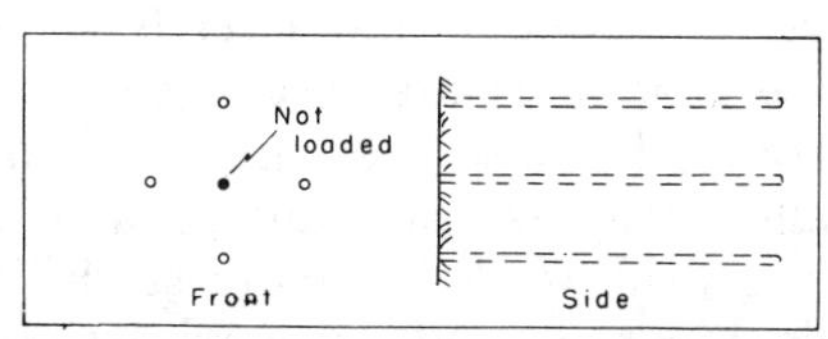

M. 5-hole burn cut (moderately soft ground)

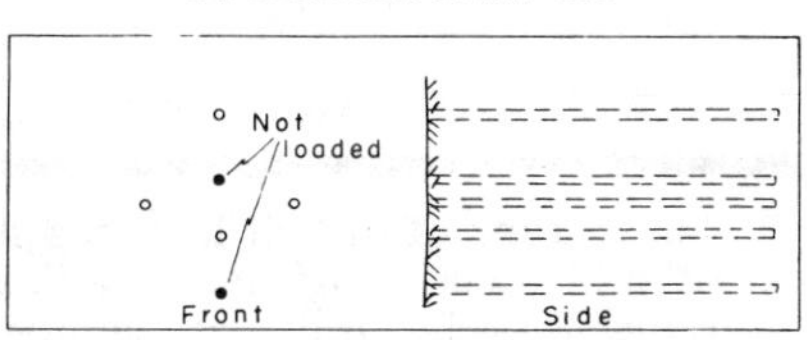

N. 6-hole burn cut (medium ground)

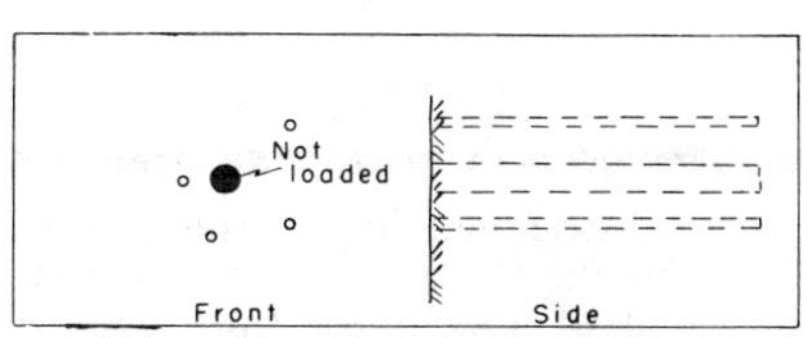

O. Large-center-hole burn cut

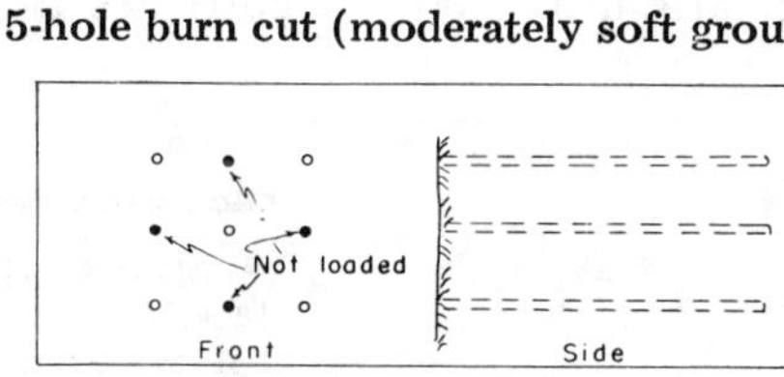

P. 9-hole burn cut (hard ground)

DRILLING MACHINES

Air-operated drill machines are classified as stopers, sinkers, air legs, drifters, and special machines. Each is especially adapted for a certain type of drilling. The stoper drill A is specially adapted for drilling overhead holes. It is used most in stopes and raises. The air leg drill B is designed for drilling horizontal holes, but is flexible and works well for other types of drilling. These drills are most used in drifts, crosscuts, and stopes, but they find some use in raises.

A sinker drill is designed for drilling down holes. The sinker is similar to an air leg drill without the leg. The sinker drill is hand held and is most used in shaft sinking.

Drifter drills are in general use for driving entries, crosscuts, and laterals to develop a level. These machines are also used for drilling stopes or rooms because they are more powerful than the smaller stopers and airlegs and usually have a longer drill feed. To increase drilling rates, these drills are mounted on carriages called jumbos C, D, E, and F. The drill is attached to a moveable arm mounted on the jumbo so the drill can be rapidly put in position. The drills shown in E are called underslungs and are pushed into the rock by an air piston. The drill may be advanced by a screw feed F or by a sprocket chain E. The jumbo shown in E is used in highspeed tunnel driving and it runs on tracks.

Many jumbos are now mounted on rubber tires or crawler tracks which make them very flexible and portable.

The drill shown in C is an air drill used for high-speed drilling of holes greater than 20 feet in length. It is called a rotary percussion drill. In an ordinary drill, air causes a piston to strike the drill steel inserted in the drill. The force of the blow is transmitted down the steel to the attached bit, which cuts and chips the rock. After the blow is struck, the steel is turned by the machine to a new position for another blow. The rotary percussion drill uses the same hammering action, but the steel is rotated much faster and rotates while hammering. This action, plus much greater pressure on the bit, greatly increases the drilling rate.

A. Stoper type of drill designed for overhead drilling. **B.** Air-leg type of drill drilling in a stope face. *(Courtesy Archibald Co. Cleveland.)* **C.** New style of drill used in high-speed drilling—a rotary-percussion type of machine. *(Courtesy Ingersoll-Rand Co.)* **D.** Drill and jumbo drill carriage in a drift heading. *(Courtesy Anaconda Co.)* **E.** Another style of jumbo using an underslung type of drill *(Courtesy Archibald Co., Cleveland).* **F.** Long-feed drill jumbo used in tunnel drilling *(Courtesy Joy Mfg. Co.)*

PLATE 5-5 DRILLING MACHINES

JUMBO DRILLS

As mentioned in Plate 5-5, the trend in jumbo type drills is to mount long feed drills on self-propelled carriages, either track or rubber-tired mounted. Plate 5-6 shows several of these types of jumbos. If the holes are to be drilled to some depth such as 12 ft, the drill carriages are long enough so that a steel change does not have to be made to drill the required depth of hole. See A, B, D, E, and G. These pictures show that the drill carriages are held in position with hydraulic actuated booms attached to moveable carriages.

The booms can be positioned almost any place in the face. All of the controls for positioning the boom and operating the drill and the carriage are grouped in front of the operator as shown in A, D, and E. Because the booms, carriages and drills are heavy, most jumbo carriers have hydraulic feet that can be actuated to hold the whole assembly in a more stable position as drilling occurs.

DRILL STEEL SUPPORTS

In A, B, and G there is a drill steel support at the end of the carriage and one along the carriage. The one at the end of the carriage holds the steel in position so the hole can be started more easily. Starting or collaring a hole on a slanted face is difficult because the bit tends to ride down along the slant. The support on the carriage holds the bit in place so the hole can be started. The one in the center keeps the drill steel from whipping around during drilling. As the drill machine moves up to these supports, they open up so the machine can pass.

VALVELESS PNEUMATIC DRIFTER

Drill manufacturers are always trying to improve their equipment, and the drill shown in C is a valveless, pneumatic drifter which is a relatively new development. The end of the machine has a threaded drill steel which is used to drill long holes where sections of steel must be attached to it by a threaded collar.

The jumbos shown in B, D, E, and G are all hydraulic which means the drill is actuated by hydraulic fluid rather than by compressed air. The advantages of these machines are given as being less noisy, producing less oil and water mist at the face, and giving good penetration rates. The jumbos are usually self-contained for these types of machines, that is, the jumbo carriage has the hydraulic pump and power source attached. Often electric motors turn the pump, but some jumbos have diesel engines instead.

STEEL TURNING RATES

Most pneumatic rock drills originally had constant steel turning rates as the hole was drilled. In some types of ground it is advantageous to vary the steel turning rates. The machine shown in F is of this type, and called an independent rotation machine.

The large jumbo shown in G is in a mine that has high and wide headings. Note the ventilation tube at the top of the drift, and the rock bolts in place on the wall on the right side. The modern jumbo is a sophisticated piece of machinery, and it greatly facilitates the process of drilling holes. Their mobility, power, and ease of operation make them cost competitive with other types of drills.

A. Jumbo showing boom and drill mechanism. *(Courtesy EIMCO Corp.)* B. Long feed mechanism for jumbo drill. *(Courtesy Gardner-Denver Co.)* C. Valveless pneumatic drifter. *(Courtesy Gardner-Denver Co.)* D. Hydraulic rock drills drilling large cavern. *(Courtesy Atlas Copco Co.)* E. Underground hydraulic drill jumbo. *(Courtesy Gardner-Denver Co.)* F. Independent rotation pneumatic drill. *(Courtesy Garder-Denver Co.)* G. Collaring a hole with a jumbo. *(Courtesy Ingersoll Rand Co.)*

PLATE 5-6 JUMBO DRILLS

F

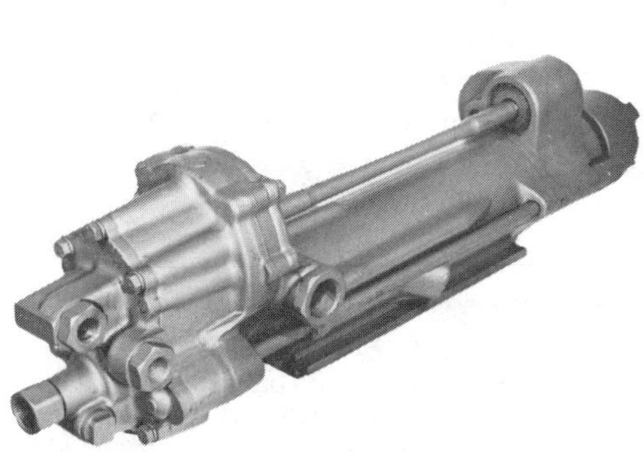

LOADING AND BLASTING

Ammonium Nitrate, mixed with fuel oil in correct proportions, makes an explosive mixture which can be detonated with the proper intensity. This type of explosive, designated AN/FO, has become popular in mining because neither the ammonium nitrate nor the fuel oil is explosive in ordinary quantities, and can be mixed together at or near the mine. The large diameter holes used in pit drilling require considerable amounts of explosive; which is brought in by truck load, and loaded into the hole.

LOADING AND PREPARING FOR DETONATION

Plate 5-7 shows techniques for loading and preparing for detonation holes filled with explosives. The system shown in A is a pneumatic delivery type of truck. Since the hose can be moved easily from hole to hole this unit is quite flexible.

GRAVITY DISCHARGE TRUCK

A gravity discharge truck is shown in B. Here the truck discharge must be made directly over the hole which, in some cases, may require much maneuvering of the truck. AN/FO may also be loaded in plastic bags and lowered into the hole by hand as shown in C. Sometimes water will interfere with ordinary AN/FO so waterproof bags are used to insure proper detonation of the explosive. In some cases a plastic stocking is inserted into the hole and the AN/FO is placed inside it to keep the explosive dry.

SIDE AUGER DISCHARGE TRUCK

A side auger discharge truck, which also has considerable flexibility is shown in D. The loading boom can be swung around to fill the hole.

An underground coal face, loaded, wired and ready to be fired is shown in E. The preparation of the primer for large diameter holes used in strip mines or open-pits is shown in F. The detonating cord or wire is shown being prepared in G.

PNEUMATIC PLACER

AN/FO is also used underground in small diameter holes. Usually an ordinary stick of dynamite with a detonator acts as a primer. The hole is usually filled with AN/FO by a pneumatic placer. Aluminum, an additive to AN/FO, has produced good blasting results in some types of ground.

A system to shoot coal is shown in H. This is called a non-explosive airdox system of shooting. The airdox tubes are inserted in drilled holes, and pressured to about 10,000 psi with air. Breaking the coal is controlled by blowdown valves and sequence valves. Since there are no hot explosive gases, only air, the environment is not harmed. This type of breaking is used frequently in gassy coal mines.

A. Pneumatic delivery (discharge) AN/FO truck. *(Courtesy of Amerind—MacKissic Inc.)* **B.** Gravity feed discharge AN/FO truck. *(Courtesy Anaconda Co.)* **C.** Hard loading drill holes. *(Courtesy anaconda Co.)* **D.** Side Auger Discharge AN/FO truck. *(Courtesy Amerind—MacKissic Inc.)* **E.** Coal face loaded with explosives, wired and ready to be fired. *(Courtesy Montana Bureau of Mines & Geology)* **F.** Inserting primer or cap in detonator. *(Courtesy Anaconda Co.)* **G.** Preparing detonator lead system. *(Courtesy Anaconda Co.)* **H.** Loading a coal face with Airdox cartridges. *(Courtesy Long-Airdox Co.)*

PLATE 5-7 LOADING AND BLASTING

BREAKING ROCK BY NATURAL FORCES

If we dig an opening underground, the weight of the overlying rock will produce stress or pressure effects around the opening. Plate 5-8,a shows a small cube of rock deep within the earth. The arrow labeled Vfd represents the force caused by the weight of the overlying rock. A force does not exist by itself in such a situation but must be balanced by an equal and opposite force, Vfu. These forces tend to compress the cube, and this action in turn causes it to exert a force on the neighboring cubes of rock.

This lateral force varies considerably depending on the type of rock. For a loosely filled sand or a very porous rock the lateral force may be near zero. For most medium to hard rocks this lateral force varies from 10 to 30 per cent of the vertical force. However, if the rock is plastic in character, such as may occur in a dense water saturated clay, the lateral force may approach the vertical force value. In this situation the rock acts as a fluid. Great pressure and high temperature may cause a rock to act like a fluid.

STRESS AND PRESSURE

There is some confusion between the words stress and pressure. Both are defined as force per unit area. A pressure or stress of 20 pounds per square inch means that there is a total force of 20 pounds distributed and acting over the 1 square inch area. The total force on a mine opening is the product of pressure or stress times the area. Generally, the term pressure is used for fluids, and stress is used for solid materials. However, in mining terminology the term pressure for heavy rock forces is frequently used as it is in B, and C.

In some areas of the earth the lateral or side stress is greater than the vertical stress. This is probably due to mountain building forces in the crust of the earth. This side stress has caused unusually heavy ground effects in some relatively shallow mines.

When an opening is dug in the stressed earth the balancing forces or stresses around the opening are disturbed, and they must compensate in some way or thc rock will flow or push into the opening. Sometimes the rock will do this.

However, the most usual pressure effects around a square opening are shown in B. Rock and ore are weak in tension, so the rock tends to fracture in the top and bottom of the drift. Also, rocks will fail in shear at the corners and sometimes on the sides of the opening. The rock stress effects in stopes are shown in C, a, and b. The rock tends to break in tension at the top of the stope. The lateral pressure effects in the solid rock may also cause the rock to break in the walls of the opening.

This breaking effect caused by high stresses aids in the system of cave mining. To prepare a block of ore for caving the block is undercut D, a. Pressure effects cause the ore in the top or back to break and fall into the opening D, and E,a. Because solid rock increases in volume when broken, the opening is soon filled with broken ore E,b. This broken ore supports the back and stops the caving action. When the broken rock is pulled away from the back F, caving starts again.

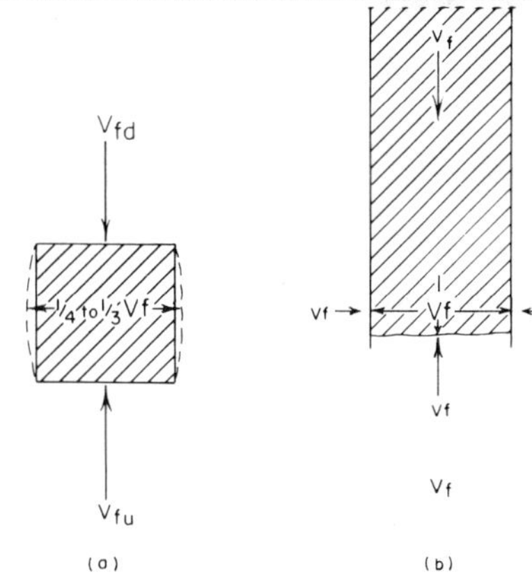

A. (a) Force effects on a solid block of material
(b) Force effects of a tube filled with fluid

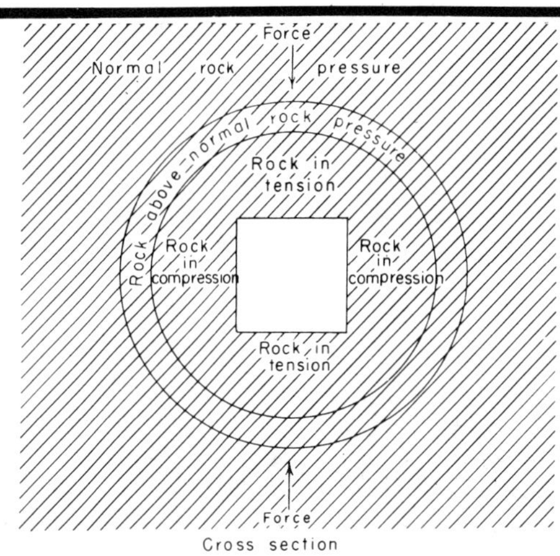

B. Effects of one-direction force on a square opening in solid rock

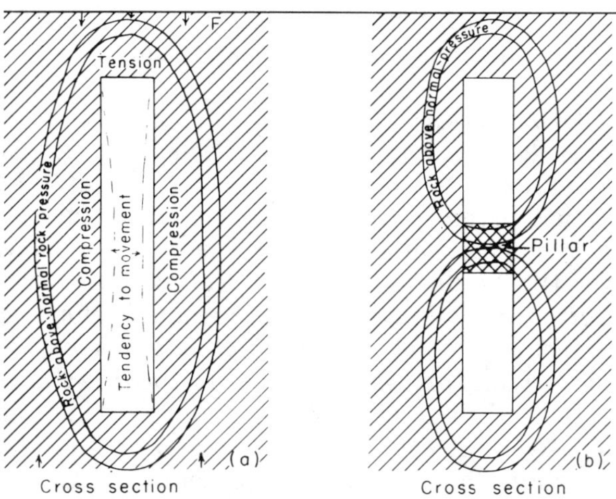

C. (a) Effects on one-direction force on a tabular opening in rock (b) Change of pressure effects caused by pillar or support in the stope

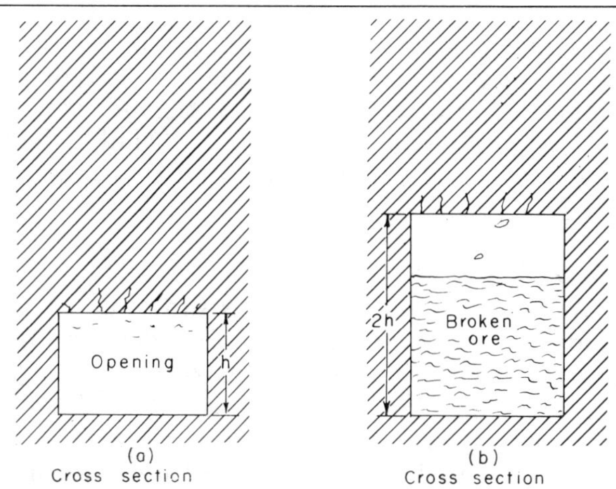

D. Opening in solid rock filled by caving material. Broken material increases in volume

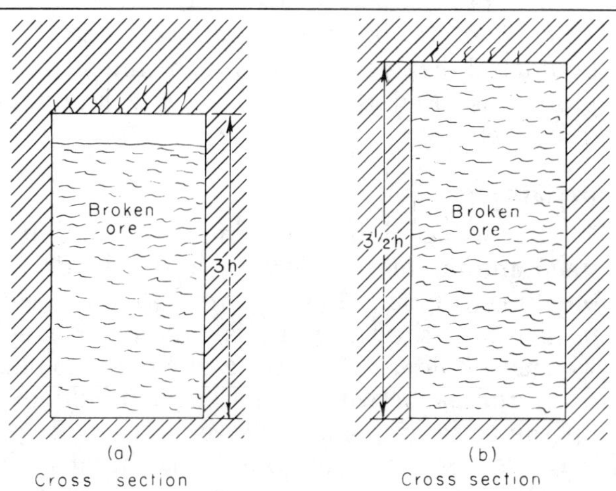

E. (a) Further development of caving (b) Broken rock completely fills the void and gives support to the back

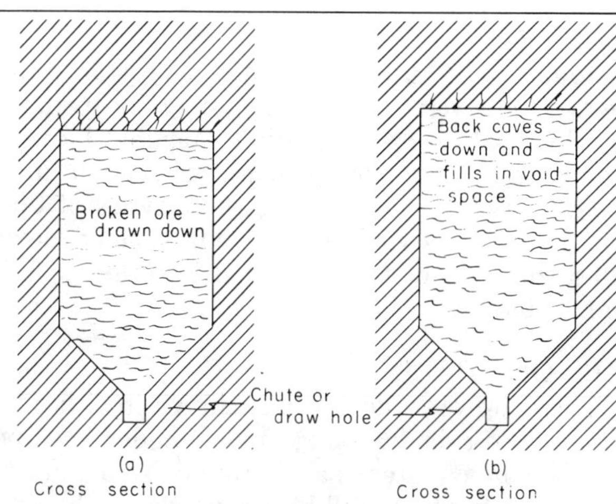

F. (a) Drawing down or removing broken ore causes the back to cave again (b) The void fills again

BREAKING LARGE BOULDERS

In most mining operations large boulders often result from the rock or ore producing process. Frequently they must be reduced further in size so loading and moving equipment can handle them. This is called secondary breakage, and it is often a production problem. The trend in the industry is to use larger equipment which can handle larger boulders. Yet large boulders must occasionally be broken.

GRIZZLY RAILS

An underground situation is shown in A where boulders rest on grizzly rails. Grizzlies are rails, beams, or bars that stop oversize boulders from going into loading chutes or crushers. A rather common way is to use a hand actuated sledge hammer. In breaking some types of boulders firmly supported on grizzly rails, hitting them on a flat face will often cause them to fragment. This procedure works well on hard, brittle boulders.

HAMMER AND GAD

Some rocks have more cohesive resistance than others, and even though they may have cracks, they do not break well by direct hammer blows. A hammer and gad arrangement, as shown in B, may be effective in breaking these types of boulders. The gad is a chisel or wedge-shaped device with a sharp point that is forced into the crack by hammer blows.

WRECKING BALL

The drop or wrecking ball shown in C has had mixed success in breaking large boulders in open pits. The heavy steel ball is raised by a portable crane and allowed to fall by gravity on the boulder. Often the boulder is lying on a dirt surface which absorbs much of the impact energy of the ball, thereby reducing the effectiveness of the method.

EXPLOSIVES

One of the most effective ways of breaking boulders is to place a small charge of explosives inside the boulder. To do this it is necessary to drill a hole in the boulder several inches deep, as shown in D. It is surprising to see how large a boulder a one-quarter to one-half stick of explosive will fragment. The main objection to this system is the length of time involved with setting up the drill, drilling and tearing it down again.

MUDCAP TECHNIQUE

Although not nearly as efficient an explosive, the mudcap technique, shown in E, is often used. A stick or more of explosives is unwrapped, and the explosive material is spread like a pancake over the boulder. If part of the explosive can be inserted in minute cracks, so much the better, but usually a flat surface is chosen when placing the charge. This is very similar to the one shown in A when hitting with a sledge hammer. A detonator is inserted into the formed explosive pancake, and the whole thing is covered, preferably with mud a few inches deep. Hence the name mudcap. However, if mud is not available, sand or other fine detrital material can be shoveled over the explosives. The theory is that this covering material reflects some of the explosive force into the boulder breaking it. Although much more explosive is used than in D, the technique works quite well.

BOMBING

In caving or bulk mining operations it is not uncommom for boulders to wedge in draw raises blocking off the flow of ore. Bombing with explosives, as shown in F, is one way to dislodge wedged boulders. If possible, the bomb is placed in open spaces between the wedged boulders as far up the raise as possible. Several sticks of powder may be used in one bomb. If bombs are not effective in breaking up the wedged rock, a drill is sometimes used under the boulders, and holes are drilled in the boulder. This is only done when the safety of the miners is assured.

PLATE 5-9 BREAKING LARGE BOULDERS

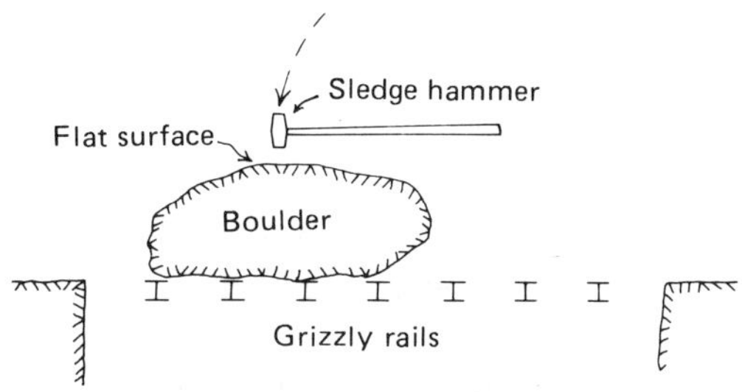

A. Breaking boulders on grizzly with sledge hammer

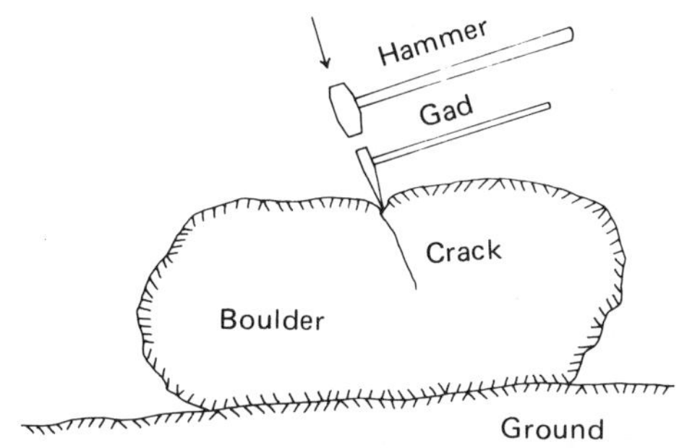

B. Breaking boulders with gad and sledge hammer

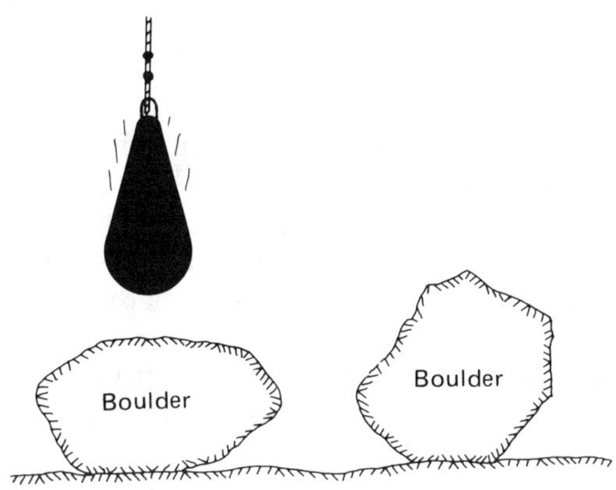

C. Breaking boulders with drop ball

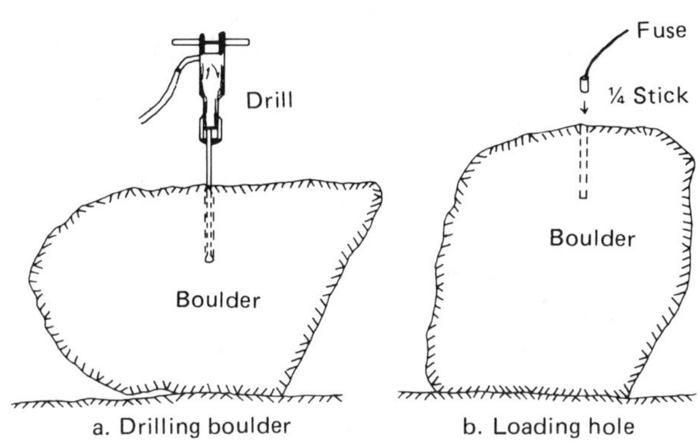

D. Drilling and blasting boulders

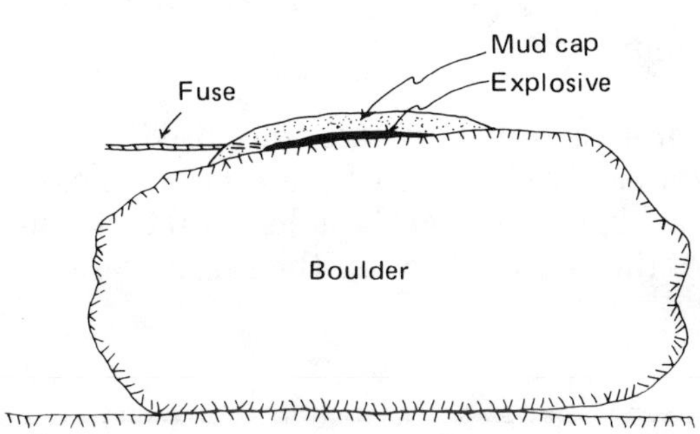

E. Breaking boulder with explosive mud pack

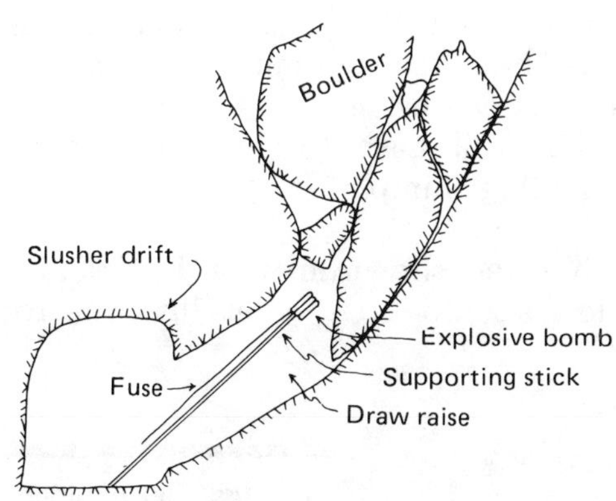

F. Bombing plugged draw raises

BREAKING LARGE BOULDERS

The problem of breaking large boulders, lumps or chunks was discussed in Plate 5-9. As can be seen by studying that plate, most methods of breaking boulders required considerable effort and time. Mining equipment manufacturers have developed machines, a few of which are shown in Plate 5-10, that are effective in breaking large boulders or lumps.

Conveyor belts are much more efficient transporters of material when oversized material is kept off the belt. A feeder-breaker is shown in A, and C. This reduces large chunks of coal, or material that is friable and not too hard, to an optimum size for the conveyor belt. This breaker unit is mounted on crawler tracks so that it can move about the mine.

ROTARY PICK BREAKER

The actual breaking mechanism is a rotary pick breaker. The rotating breaker may be stationary or it can be made so that it floats on hydraulic cylinders. The breaker will adjust to the height of the lump entering the breaker, and the unit will break the lump to optimum size for the belt conveyor. As can be seen in A, it looks as though the breaker will have to do some adjusting to handle that large lump.

The breakers shown in B, D, E, F, and G use a hydraulic force to drive a bit into the rock for breakage. Hydraulic pressure can deliver large forces and high power when properly applied.

The boom mounted machine, shown in B, is attached to a narrow crawler carriage so it can be used in grizzly drifts in a block cave mine. See Plate 8-19. Not only does the machine break boulders on the grizzly rails, but it is flexible enough to enter the draw raise, some distance and dislodge boulder hang ups. See Plate 5-9F.

Large boulders often occur in pit operation, and often a bulldozer will push them out of the way to be broken at a later date. The machine in D is moving along one of the boulder piles breaking the large pieces. An end-loader or shovel will pick up these broken chunks and load them into haulage trucks. The breaker shown in E is mounted on a commercially built wheeled tractor.

The trend in mining is to use larger equipment which can handle larger boulders. At some place in the mining or milling process these large pieces must be broken to some maximum size. This usually occurs at the head end of a conveyor belt, at the skip loading pocket, or at the entrance to the mill. A grizzly which consists of rails spaced at equal intervals to keep oversize material from flowing into the opening is usually put over these feed points.

Hydraulic breakers are mounted over these grizzly areas as shown in F, and G. This expedites the breaking of large boulders that do not go through the spaces between the rails.

A. Large coal boulders entering a rotary pick breaker. *(Courtesy Long Airdox Co.)* **B.** Underground hydraulic breaker. *(Courtesy Joy Mfg. Co. Hard Rock Division)* **C.** Another view of a rotary pick breaker. *(Courtesy Long Airdox Co.)* **D.** Hydraulic breaker on a self propelled vehicle. *(Courtesy Joy Mfg. Co. Hard Rock Division)* **E.** Hydraulic breaker mounted on a wheeled tractor. *(Courtesy Joy Mfg. Co. Hard Rock Division)* **F.** Hydraulic breaker mounted over a grizzly. *(Courtesy Joy Mfg. Co. Hard Rock Division)* **G.** Close up of breaker in action. *(Courtesy Joy Mfg. Co. Hard Rock Division)*

PLATE 5-10 BREAKING LARGE BOULDERS

SUPPORT

Plate 5-11 shows the distortion of a circular opening in solid rock under a vertical force. This distortion causes the rock to break by tension overhead or in the back and by shear at other points, as shown in B. This is a theoretical concept. Actual distortion is difficult to observe because walls are not smooth, and some distortion probably takes place as soon as the solid rock is broken.

ROCK OR ROOF BOLTS, SHOTCRETE

To support the broken rock several methods are in use. The old standby is timber. C, and D show the cross section and long section of a timbered level set. A development in the last 35 years is the use of rock bolts. E shows a haulageway supported by rock bolts. The bolts are anchored into a solid rock and the nut is pulled tight against the rock. Tightening the nut on the bolt tends to pull the rock back in place, thereby providing support to the opening. Another method of support used where permanence is required is concrete, as shown in F. Dry concrete mix containing certain additives can be moved with compressed air to a nozzle. It can then be mixed with the proper amounts of water at the nozzle and blown against a rock face. The wet concrete mixture will usually adhere very well, and when dry forms a strong supporting shell. This shell may be several inches thick and it is called shotcrete. This shotcrete system has been successful in many applications where a permanent opening is required underground. The area may be bolted first, and the shotcrete covers and adheres to the bolts very well.

YIELDABLE SETS

Steel is also used, especially in very heavy ground. Termed yieldable sets, when the rock stresses become too great they yield but do not collapse. This relieves the high stress level in the rock for a period of time until it builds back again. These types of sets have given good performances in certain types of application.

TIMBER STULLS

Stopes need not be kept open as long as development headings. After a stope is mined out, it is allowed to cave. In large stopes the stress effects may cause the stope to collapse. Supports in stopes tend to keep the stress effects at a safe level. See Plate 5-9. If the ore occurs in a narrow vein, timber stulls can be used for support, as shown in G. If the ore is not high-grade, parts of it may be left as pillars to help support the walls H. If the ore is too good to leave unmined, the mined-out section may be supported by filling it with broken mine waste or sand fill, as shown in I and J. Square-set timber is used in high-grade stopes where the ore is weak K. L is an isometric view of a stope timber set labeled to show the various parts.

Rock bolts are used in stopes to support the walls and back. Often a combination system is used where rock bolts, timber and backfill are all used. In some cases the stopes may be backfilled and the pillars mined out at a later time.

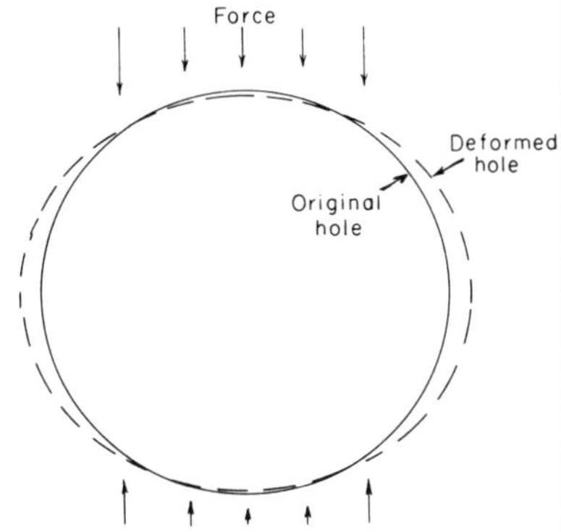

A. Deformation of a round hole by vertical pressure of force

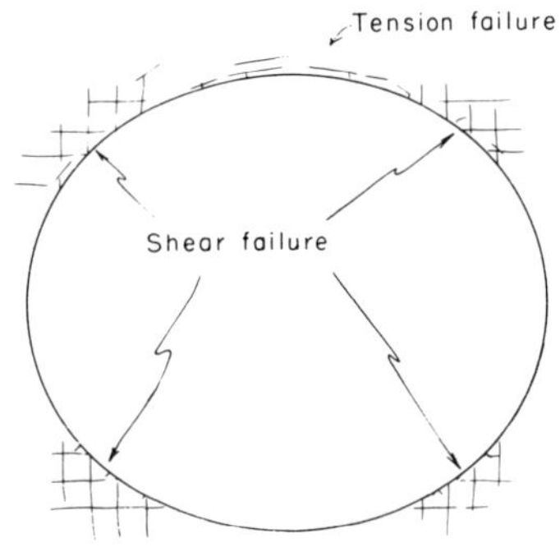

B. Typical rock failure about an opening as a result of vertical force

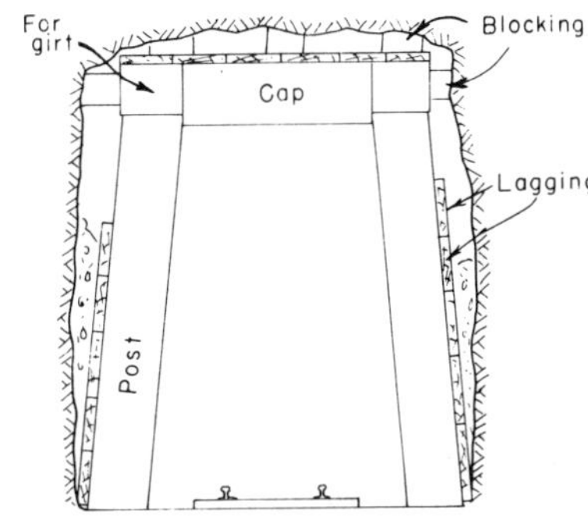

C. Cross-section of a timbered drift set

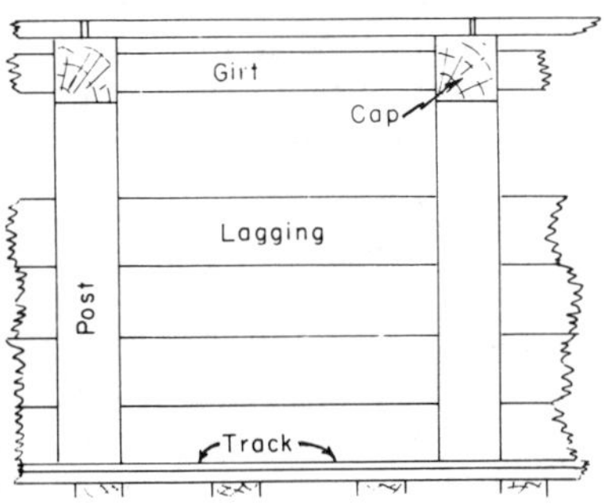

D. Long section of a timbered drift

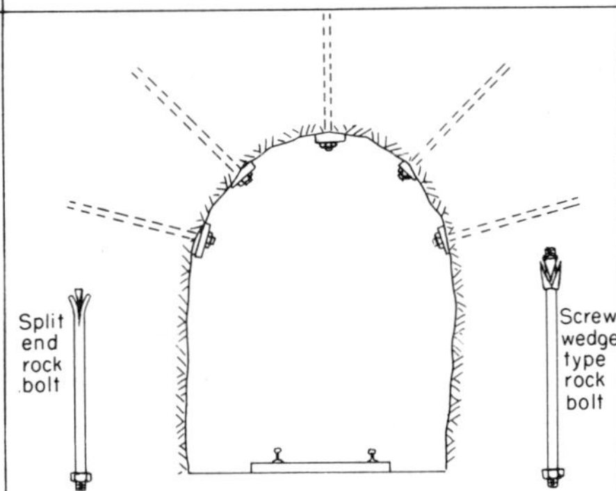

E. Cross-section of haulageway supported by rock bolts

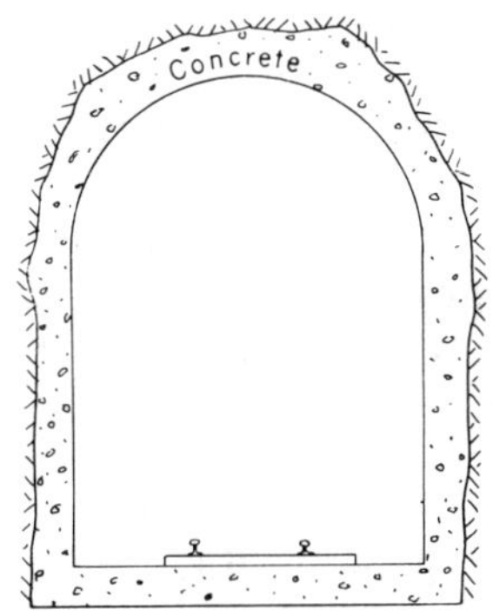

F. Cross-section of haulageway supported by concrete

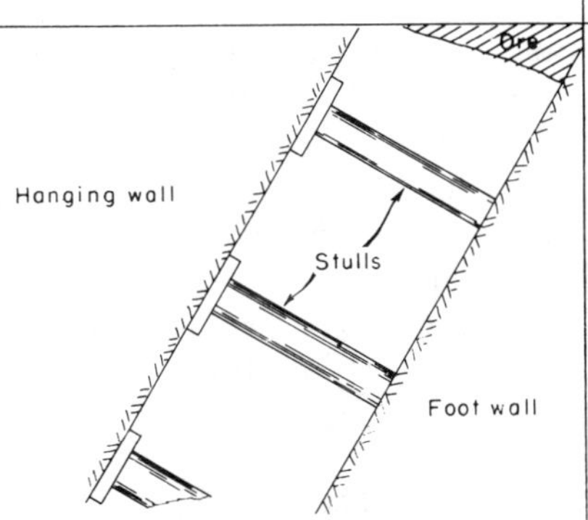

G. Cross-section of a stope with stull support

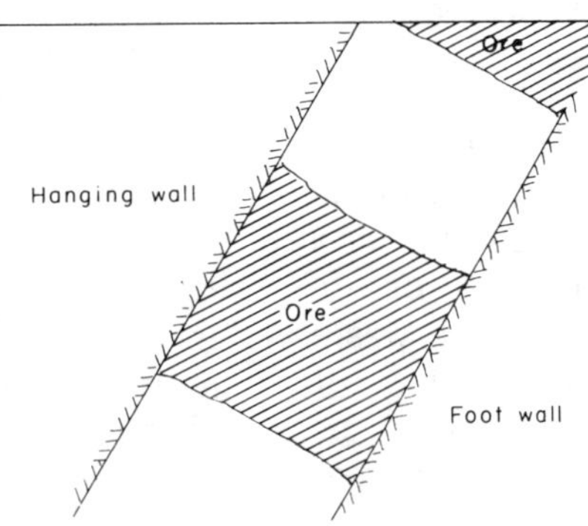

H. Cross-section of a stope with ore pillar for support

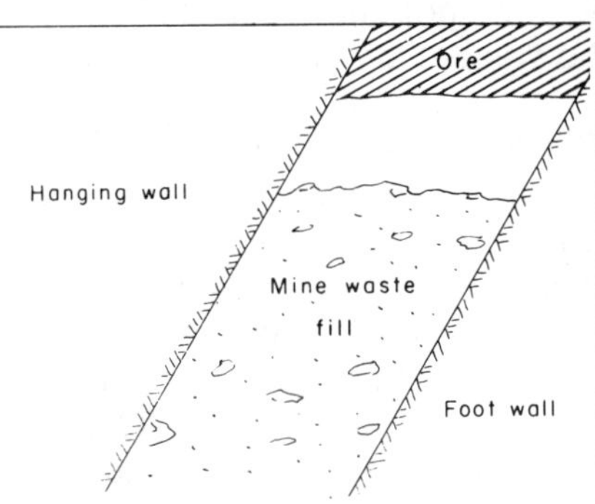

I. Cross-section of a stope with mine waste fill for support

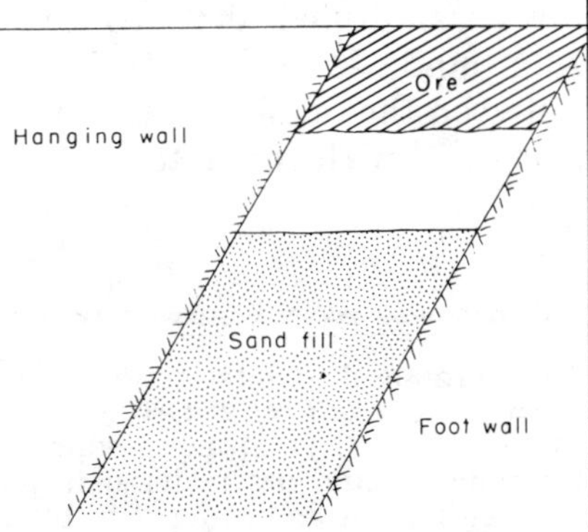

J. Cross-section of a stope with sand fill for support

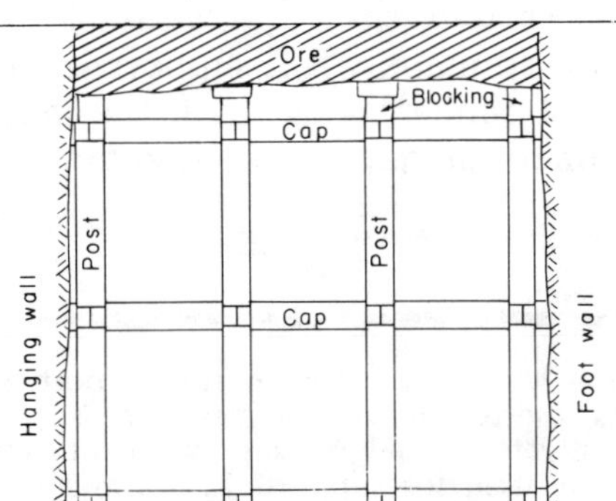

K. Cross-section of a stope with square-set timber for support

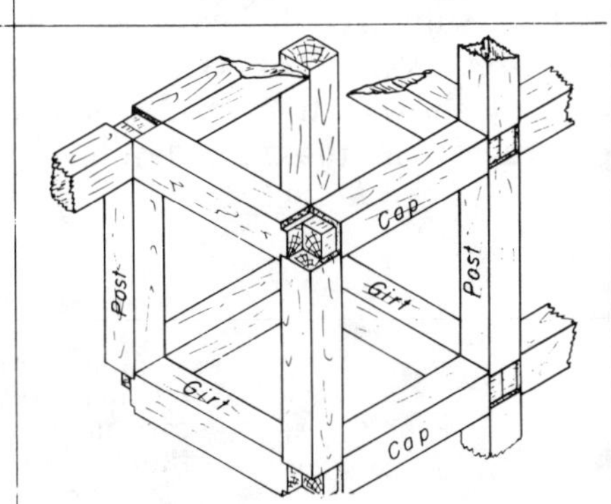

L. Isometric view of a timber set

SUPPORT

Plate 5-12 shows various types of supports used underground. A is an underground station at the shaft in a small mine in Montana. The support is timber of square cross sections. B shows the start of a curve in an underground haulageway supported by rock bolts. Note the wood headboards used on these bolts to hold more ground in place. C is a drift section concreted where the ground is extremely heavy and ground support is a difficult problem.

Materials commonly used to support stopes are natural pillars, mill tailings, sand or mine waste fill, and timber. D shows a stull-supported stope. Each individual stull is cut to length to fit into a particular spot and is wedged tightly in place. Note the wedges at the top of the post.

ROOM-AND-PILLAR, SAND-FILLED STOPES

Natural pillars are left for support in most coal mines and other room-and-pillar mines. E shows a haulageway in a room-and-pillar mine. Coal or ore has been left in place along each side. Mine waste and sand may be used to support mined-out portions of a vein. F shows a sand-filled stope ready to be filled with sand to close to the back of ore. Notice the back is supported by rock bolts.

SQUARE-SET TIMBER

For support of heavy or weak ground the square-set system of timber is most commonly used. G shows, in a model, square-set timber as constructed in the mined-out portion of a vein. The spaces within the timber set may be filled with mine waste, rock or sand fill at some later time.

ROCK OR ROOF BOLTS

Rock or roof bolts are used to a great extent in mining. The early bolts were the type shown in Plate 5-11E i.e. the split end bolt and the screw wedge bolt. The hole depth was critical in the split-end type to properly place the bolt. The screw-wedge bolt gained in popularity because the hole depth was not critical. A screw wedge end is shown in H. It is still a popular type of bolt.

These types of bolts have two flaws. One is if the rock breaks away from the headboard, see B, and F, which frequently occurs, the bolt doesn't hold anything. Also, since the bolt does not completely fill the hole it doesn't provide resistance to shear or side forces.

To overcome these shortcomings, resign type bolts have become popular. A steel rod is inserted in the hole with a quick setting resign or glue which completely fills the hole. When dried, the steel supplies tensile strength and the resign gives lateral support. The tube steel bolt shown in I is a split set bolt. This is a new bolt which gives support such as a resign bolt does, i.e. shear resistance and pull out resistance.

TIMBER, CONCRETE, AND STEEL

Timber, concrete, and steel are used for shaft supports, and a reinforced steel tube is being lowered into a vertical drilled shaft to give wall support in heavy ground. After the tube is in place in J, the space between the outside of the tube and the ground may be filled with concrete.

A. Mine station with square timber for support. **B.** Underground drift intersection supported by rock bolts inserted in wooden headboards. *(Courtesy Anaconda Co.)* **C.** Grizzly drift in a cave mine; concrete is used for support. *(Courtesy Anaconda Co.)* **D.** Stope in which round timber posts are used for support. **E.** Room-and-pillar mine, natural ore material in place is used for support. **F.** Sand-filled stope, the sand is mixed with water and brought in by a pipe line. *(Courtesy Anaconda Co.)* **G.** Mine model showing labeled timber sets. **H.** Screw-wedge bolts end. **I.** Split-set steel rock bolt. **J.** Reinforced steel-tube shaft support. *(Courtesy Teton Exploration Drilling Co.)*

PLATE 5-12 SUPPORT

SUPPORT

Roof and rock bolts are used for support in most mines. The various types of bolts were discussed in Plates 5-11, and 5-12. There isn't any difference between a rock and roof bolt. In coal mines and other mines where they mine horizontal or near horizontal beds, they are called roof bolts. In metal mines they are called rock bolts.

ROOF BOLTERS

Equipment manufacurers have developed machines that are self-propelled, can drill bolt holes, insert the bolts, and tighten them. The machines shown in A, B, and D are roof bolters. One type of machine is shown in A. The one shown in B has two roof drilling and placing machines. Note the roof bolts and the steel headboard plates used to keep the roof in position.

Rock bolts are often used with wire mesh to keep fractured rock in place as shown in C. Rock, as it is exposed to ventilation air, frequently spalls off. The purpose of the mesh is to keep the rock from falling to the bottom of the working area.

The roof bolter in D has a protective cover over the operator's head. These machines are designed so the operator has either a protective cover, or the controls are located so he stands or works under a bolted roof section.

HYDRAULIC SUPPORTS

Longwall and shortwall mining will be discussed more in Part 8, but the more recent developments in these methods utilize hydraulic supports. These supports protect the people and equipment working in the face. They move up as the face advances, and the roof behind them is allowed to cave. The support, shown in E, is called a shield, and two large hydraulic cylinders hold the canopy against the roof. The end has a hinge arrangement so the support will resist lateral movement of the roof. The support in F is called a 6-post chock. This design will allow some lateral movement of the roof. Longwall systems are being used more frequently because of the recent development of hydraulic support and longwall mining machines.

A. Roof bolter. *(Courtesy FMC Trona Mine)* **B.** Roof bolter in action. *(Courtesy Long Airdox)* **C.** Rock bolts and wire mesh holding rock. *(Courtesy Anaconda Co.)* **D.** Starting to drill roof bolt holes. *(Courtesy Long Airdox)* **E.** Hydraulic 2-leg shield longwall support. *(Courtesy Thyssen Corp.)* **F.** Hydraulic 6-leg chock. *(Courtesy Gullick Dobson)*

PLATE 5-13 SUPPORT

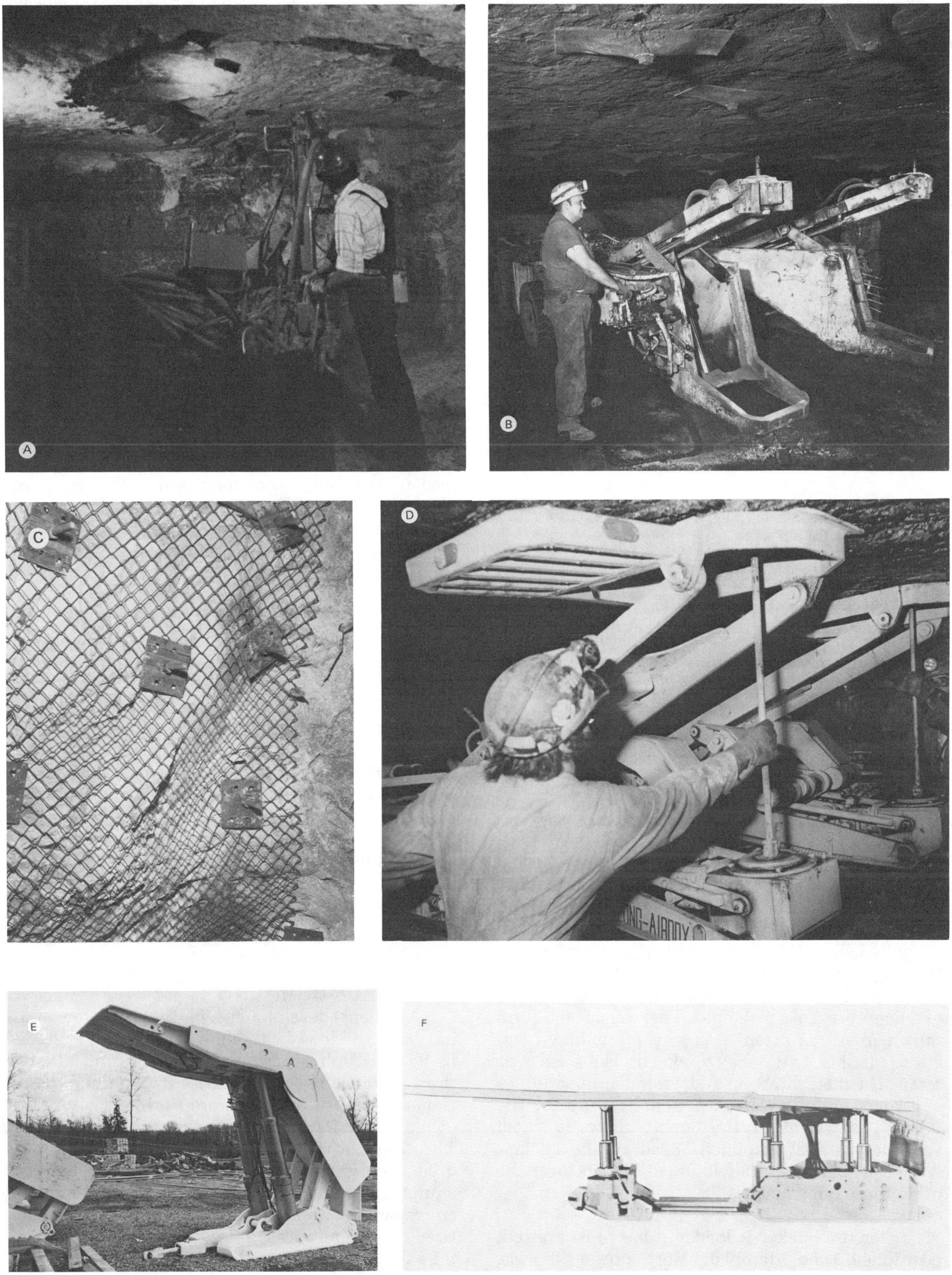

PRINCIPLES OF LOADING

After the material is broken from place, it must be loaded on some type of conveying device. Combined equipment, which not only loads but hauls and dumps, is becoming quite popular in mining. These are called load-haul-dump (LHD) units. Various types of loading actions are shown in A.

SHOVELS, BACKHOES, AND DRAGLINES

Shown in A,a is the typical person-operated shovel. The shovel is pushed into the pile, called crowd, and then it is lifted. Most mechanical power and hydraulic shovels use this action. Instead of pushing the shovel into the pile it can be pulled. See A,b. The weight of the bucket and the angle of the teeth will provide the crowd necessary to fill the bucket. Most backhoes and draglines use this principle. The clamshell action shown in A,c is often used in shaft sinking operations. It digs from the top down. The slusher or scraper bucket, shown in A,d, is similar in action to a garden hoe. The bucket is pulled toward the hoist. The weight of the bucket allows it to fill and the material is pulled ahead of the bucket. After the bucket is full it usually floats over the other loose lying material.

After the bucket is loaded it has to be hoisted, positioned, and dumped. Most power shovels, backhoes and draglines hoist the bucket and swing to a dumping position, as shown in B. Often the two actions, hoisting and swinging, are combined once the bucket is loaded.

A popular loading action in underground metal mines is the overhead cast type of loader, as shown in C. It is necessary to have the conveyance either attached to the loader, or in a relatively close position, so the material cast from the bucket will land in the conveyance.

CHAIN-CONVEYOR

Another system often used in coal loading and other non-metallic minerals is shown in D. A gathering pan or lip is mounted on the end of a chain type conveyor belt. The pan or lip has gathering arms or some type of rotating wheel or chain attached to it which gathers, scrapes, or pushes the broken material onto the conveyor belt. Two common types of gathering systems are shown in D a, and b. The chain-conveyor elevates the broken material and deposits it into the hauling conveyance.

SLUSHER OR SCRAPER

The slusher or scraper system shown in E is also quite often used in underground loading operations. Ore is fed to the slusher lane through draw points by gravity. Large chunks can be broken on the lane floor by various means, the material scraped to the loading point, and loaded into a haul conveyance by gravity. Slushers are often used in cut-and-fill types of stopes to scrape the ore into gathering chutes and raises. Gravity is often combined with a slusher system. The slusher bucket gathers the broken material and deposits it in a vertical or upward inclined raise which acts as a storage area for the chute. The broken ore can then be loaded in a conveyance, as shown in Plate 5-19 G.

LOAD-HAUL-DUMP UNIT

A recent development in the past few years has been the load-haul-dump (LHD) unit sketched in F. This unit loads a large bucket and lifts it a short distance and then carries it to a dump point. Generally these machines are powered with diesel engines and have great flexibility and versatility. The trend now for short hauls up to several hundred feet is to use the flexible load-haul dump equipment instead of laying track for track bound conveyances. Sometimes these units load rubber-tired, diesel powered, large capacity trucks which take the material from the mine.

PLATE 5-14 PRINCIPLES OF LOADING

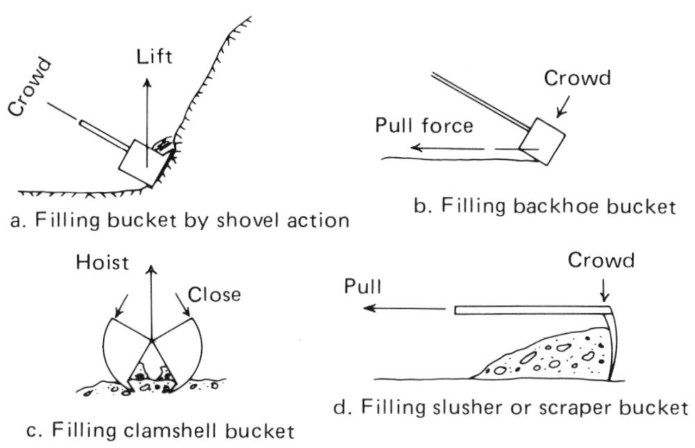

A. Various types of loading action

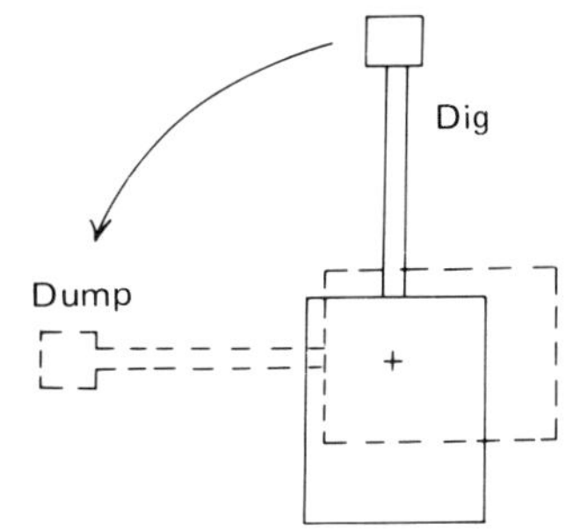

B. Swing type of elevating and dump action

C. Overhead cast type of loader

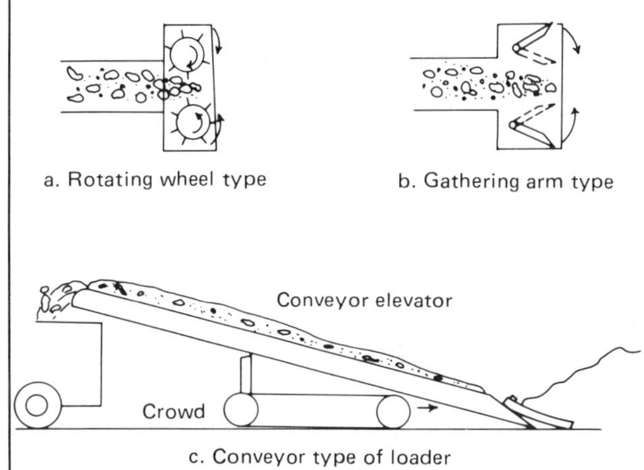

D. Gathering and conveyor loader

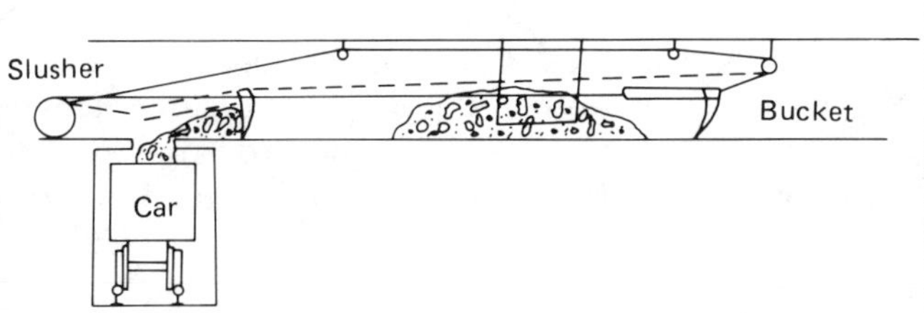

E. Combined scraper and gravity loading

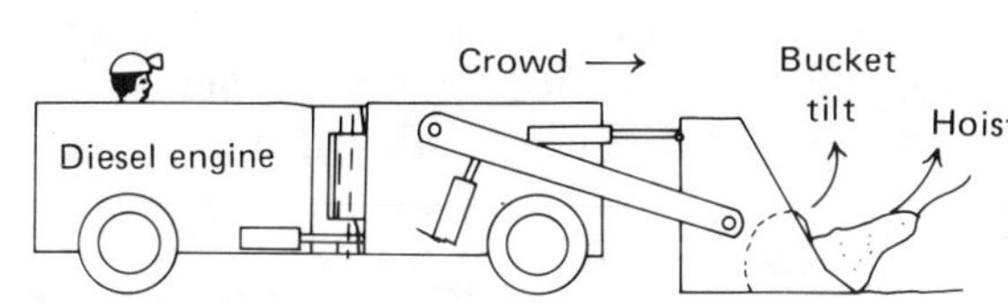

F. Load-Haul-Dump unit

LOADING AND HAULING

The principles of loading were discussed in Plate 5-14. Plate 5-15 shows some of the equipment that loads and hauls material. Hydraulic shovels, shown in A, are becoming popular. These machines not only give good performance, but they have high digging and loading rates.

The power shovel is still used where large production is needed. This machine, shown in B, is loading coal in a strip mine. The buckets may be up to 25 cubic yards in capacity for coal loading shovels. Shovels are also used for stripping overburden from the coal, and in this case, the bucket has up to 100 cubic yards in capacity. As can be seen in A, and B, the machines are huge. Compare the size of the people with the machines.

BIG-BUCKET, LONG-BOOM DRAGLINE

The big-bucket, long-boom dragline shown in C is also used for stripping overburden from coal. Power shovels are not efficient in digging below their track or feet position and must have a bank to dig. The dragline can only dig efficiently below its feet position and is often used to dig where water is present below the surface of the ground. As can be seen in C, the bucket fills well. These buckets are up to and over 100 cubic yards in capacity. Draglines can be equipped with long booms and can deposit the material well away from the machine.

DIESEL POWERED TRUCKS

In underground mines, diesel powered trucks are replacing track haulage systems. Two different types of trucks for underground use are shown in D, and E. The diesel engine must be modified slightly for underground use and equipped with exhaust conditioning devices. The diesel trucks are versatile and can be driven where needed. They are designed to work in narrow widths and low heights, turn very sharp corners, and climb steep roadways. They can haul large loads, and move at a good speed.

HYDRAULIC BACKHOE

A hydraulic powered backhoe is shown in F. Like draglines, these machines dig below the ground level and are good for digging trenches. Mine operators also use them for digging ore in situations where shovels or bucket loaders are not efficient.

A. Hydraulic powered shovel (scoop action). *(Courtesy O & K Orenstein & Koppel AG)* **B.** Electric powered shovel loading coal. *(Courtesy Western Energy Co.)* **C.** Dragline filling bucket. *(Courtesy Western Energy Co.)* **D.** Diesel powered truck emerging from a mine portal. *(Courtesy Wagner Co.)* **E.** Diesel powered underground truck. *(Courtesy EIMCO Corp.)* **F.** Hydraulic backhoe. *(Courtesy International Harvestor Co.)*

PLATE 5-15 LOADING AND HAULING

LOAD-HAUL-DUMP UNITS (LHD)

The adaption of the diesel engine for underground use started the trend toward using load-haul-dump (LHD) units in underground mining situations. They are maneuverable, and can load and carry large loads. They can carry ore efficiently up to a thousand feet or more in some cases. The units shown in Plate 5-15, A, D, E, and F are all diesel powered. They are all articulated in the center so they can turn sharp corners. The units are powerful enough to climb inclines, see D, within the mine. The diesel powered, rubber-tired mounted (LHD) units are flexible, can go where they are needed, and have given excellent service. Most of them are four-wheel-drive and therefore have good power for loading.

DIESEL AND BATTERY POWERED

Not all (LHD) units are diesel powered. The one in B is a battery powered electric scoop used in low coal mines. Note that because of the length of the scoop, a telescoping scraper board discharge arrangement is used to empty the scoop. The scoop does not tilt down for emptying like the ones shown in A, D, and F. Diesel powered equipment is not allowed in all coal mines, therefore, electric power is widely used.

Small loaders of the type shown in C have been used successfully for cleaning out, and driving prospect adits. The machine turns like a crawler tractor, i.e. the wheels stop on one side while the wheels drive on the other for turning. The turning radius is very short which makes the machine highly maneuverable so it can negotiate narrow low headings. These machines make good loading and hauling machines for driving adits up to several hundred feet in length.

The machines shown in E, and G have separate loading attachments to fill the boxes. The whole machine carries the loaded ore to the disposal point where the box is dumped.

The unit shown in G is air operated and is remote controlled. The operator can stand in an optimum position for observing and directing the action of the machine. These machines are replacing slushers in stopes because they are highly maneuverable and can muck out hard-to-get-at places.

A. Loaded LHD unit taking out ore. *(Courtesy EIMCO Corp.)* **B.** Scoop-type LHD used in low coal mines, battery operated. *(Courtesy Fairchild Inc. by Charles Ryan Assoc.)* **C.** Small LHD unit used in exploration work. **D.** LHD unit negotiating a decline. *(Courtesy Wagner Co.)* **E.** LHD unit loading. *(Courtesy Atlas Copco Co.)* **F.** LHD unit turning a corner. *(Courtesy Rome Industries)* **G.** Remote controlled underground loader. *(Courtesy Atlas Copco Co.)*

PLATE 5-16 LOAD-HAUL-DUMP UNITS (LHD)

MUCKERS AND SLUSHERS

Loading or moving ore by means of a hand-held shovel must be kept to a minimum. Two kinds of machines used underground to load and move waste and ore are the mucking machine and the slusher-scaper combination.

MUCKING MACHINE

A is the front view of a mucking machine in a drift. The bucket is in the down position ready for loading. The whole machine moves forward and as the bucket is driven into the ore pile (muck) it is raised and pushed forward at the same time so that the bucket will be filled. The filled bucket is raised over the top of the machine and throws the muck into a car attached behind, see Plate 6-6B. E is a rear side view of a mucking machine used in a stope loading operation as shown in Plate 5-17. Mucking machines may be operated by compressed air or electricity, but air operation is more common.

SLUSHER-SCRAPER COMBINATIONS

A slusher is essentially a double-drum hoist. Slushers may have either electric or compressed-air motors of 5 to 125 horsepower. A 50-horsepower electric slusher is shown in B. One cable of the slusher is attached to the front of the scraper so that it can be pulled toward the slusher. The other cable goes around a block securely fastened in the far end of the slusher drift and is attached to the back of the scraper D. Pulling this cable brings the bucket toward the end of the drift for another load. As the scraper or bucket C is dragged across the muck pile, the ore is pulled into an ore pass (the dark hole in front of the scraper in C) or into a car, see Plate 5-14e.

A. Mechanical air-operated mucking machine. *(Courtesy Anaconda Co)* **B.** 50-horsepower electric slusher in use underground. **C.** Bucket 5 feet wide used with the slusher pictured in B. **D.** Slusher bucket in use in a flat-lying stull-supported stope. **E.** Mucking machine in a drawhole loading point.

PLATE 5-17 MUCKERS AND SLUSHERS

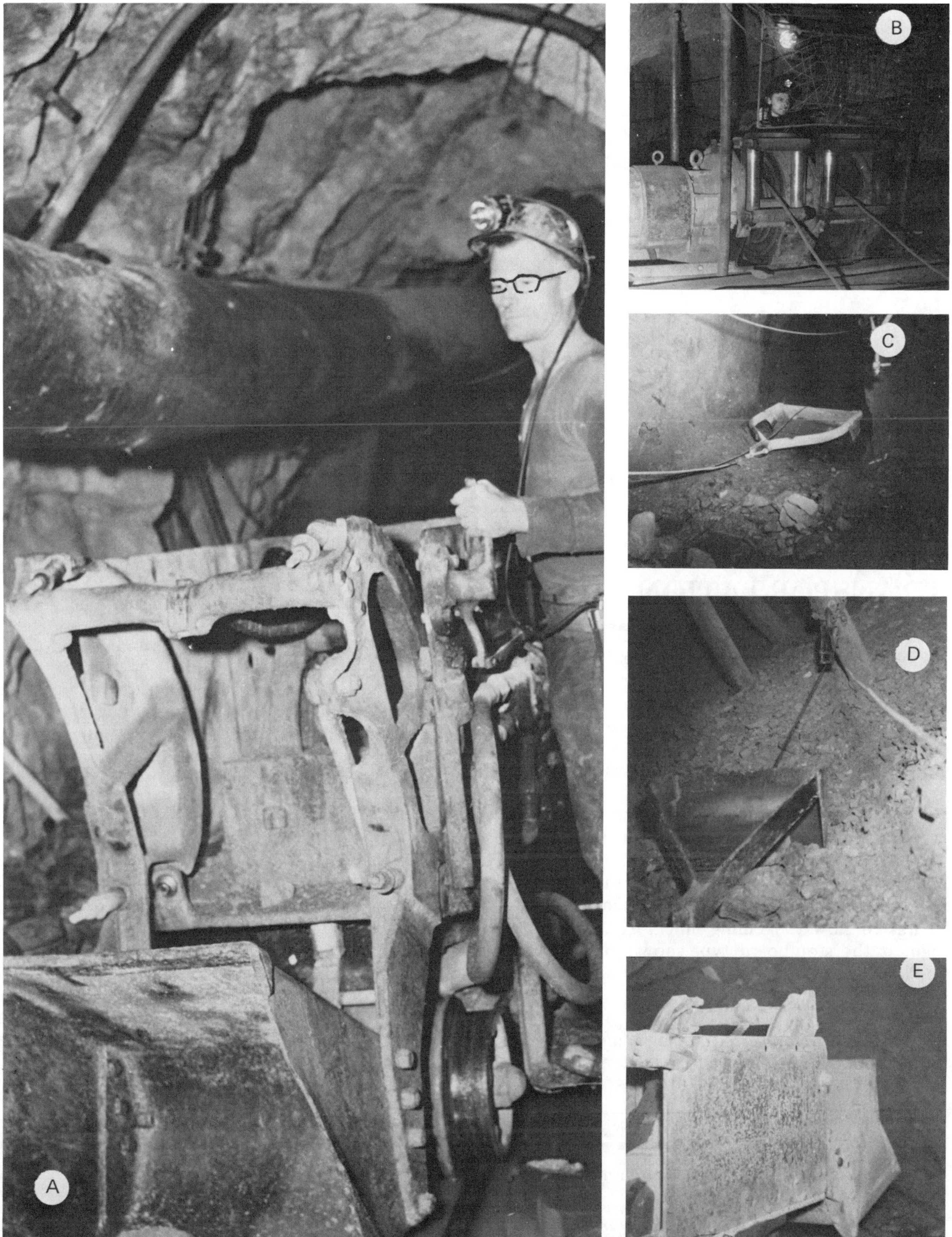

UNDERGROUND LOADING AND TRANSPORTATION

Numerous styles of chutes are in use. Plate 5-18A is a picture of a composite model showing four different styles of chute mouths and gate arrangements. Gates are used to control the flow of ore from the chutes into the ore cars. The gate on the far left is an arc type that swings up from underneath the mouth to shut off the ore flow. Second from left is an overhead type of arc gate, which swings down from above to cut off the ore flow. Third from left is a straight swing gate similar to the overhead gate. The gate fourth from left is not strictly a gate but has heavy hanging bars or chains to slow the falling muck. The gate on the right is the stop-board type shown in more detail in Plate 5-19E.

ORE CARS

Most ore is transferred underground in ore cars. The cars shown in B are 5-ton capacity, side-dump cars. A side-dump car of different make is shown in dumping position in C. Ore cars used underground range from ½-ton to 30-ton capacity, but 1-ton, 3-ton, and 5-ton are most common.

Cars are pulled by underground locomotives powered by electricity, diesel engine, or compressed air. Plate 5-18D shows a large underground trolley-type electric locomotive pulling a train of cars; electricity is supplied to the motor through the overhead trolley wire. Electric locomotives that carry their own batteries for electric power are termed battery locomotives.

SHUTTLE CARS

In many flat-lying mines, electric- or diesel-powered trucks called shuttle cars are used to transport material. The one shown in F hauls approximately 15 tons of material per trip. Loading machines provided with raking arms and a conveyor system are used to load the shuttle cars. These machines find widespread use in mining coal and bedded nonmetallic minerals such as gypsum

A. Model showing various types of chute mouths. **B.** Train of cars used for underground ore haulage. *(Courtesy Anaconda Co.)* **C.** Car in dumping position at an underground dump. *(Courtesy Anaconda Co.)* **D.** An underground trolley-type electric locomotive pulling a train of ore cars. *(Courtesy Anaconda Co.)* **E.** An underground loader using a raking and conveying action to load shuttle cars. **F.** Underground ore hauler known as a shuttle car.

A

B

C

D

E

F

REMOVING ORE AND LOADING CARS

Loading and transporting ore is the last major operation to be discussed in this section. Methods of removing ore other than the ones shown on this plate are included in the part on mining methods.

A common method of development by shrinkage or open types of stopes is shown in A. The zone above the chutes is funneled so the broken ore is directed toward one of the chutes at the bottom of the stope. E shows a car in position to be loaded from the chute.

In some types of stopes, the ore cannot pass through the mined-out portion, so it must be directed toward ore passes. B shows a slusher dragging the ore to an ore pass in one such system. The ore pass leads to a chute where the cars can be loaded.

Another system used in shrinkage or sublevel stopes is the slusher or scram drift. C shows this system under a type of open stope. Chutes may be a bottleneck if the ore contains large boulders; this system is designed to load directly into cars G.

If the dip of the vein is so flat the ore will not run down the stope by itself, it is necessary to scrape or slush the ore from the mining face. A system may use a scraper and slusher both to run the ore down the stope, and to load the cars D.

If chutes cannot be used because of boulders of ore, and if the physical layout is such that the installation of a slushing system is impractical, a mucking machine and draw-point system may be used. The bottom of the stope is belled out, but the ore is discharged into a small crosscut F. The mucking machine then loads the cars directly.

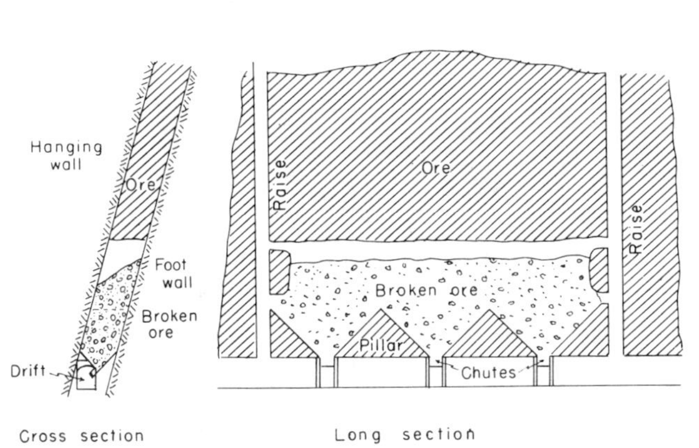

A. Removing ore from a shrink-type stope by a system of chutes

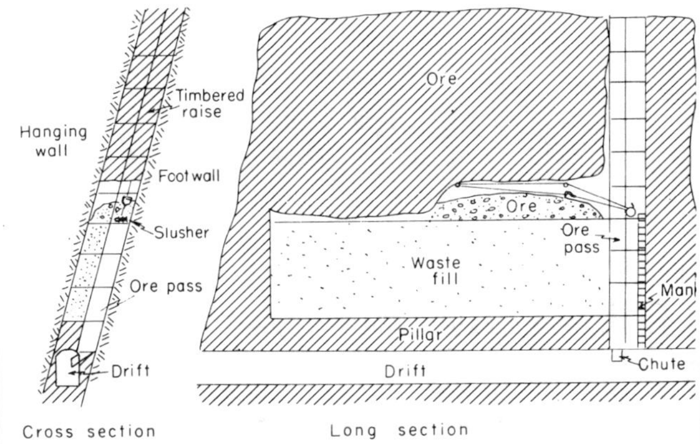

B. Removing ore from a cut-and-fill stope by a slusher pulling ore to an ore pass

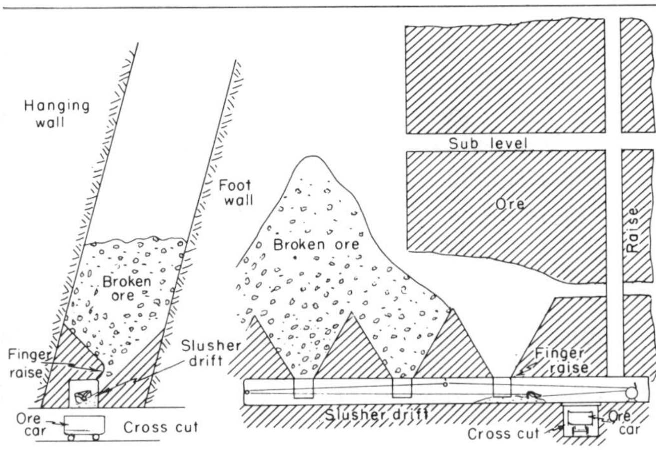

C. Removing ore from a large sublevel stope by a slusher drift and slusher loading directly into cars

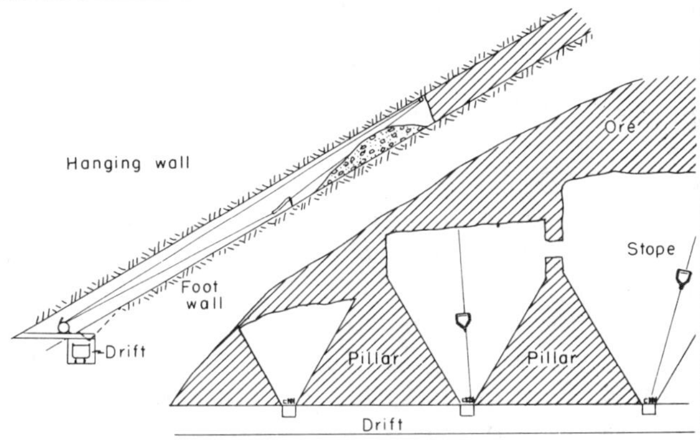

D. Removing ore from a flat-lying stope by slushers pulling ore to a chute

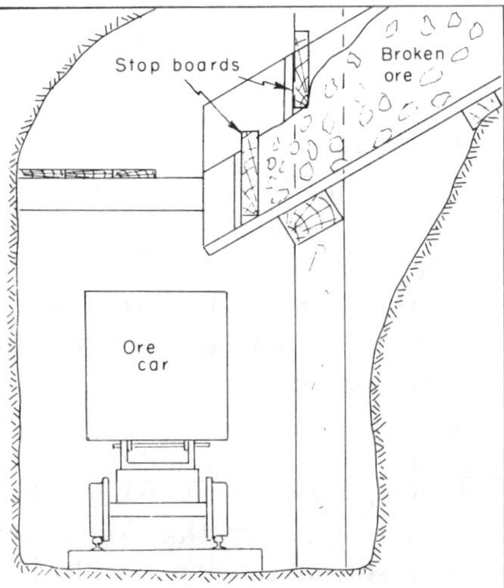

E. Chute loading into an ore car (cross-section of a stop-board chute)

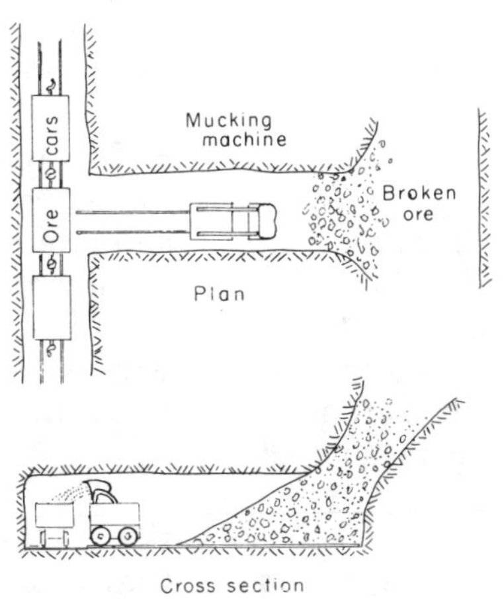

F. Loading ore from a stope by a mucking machine

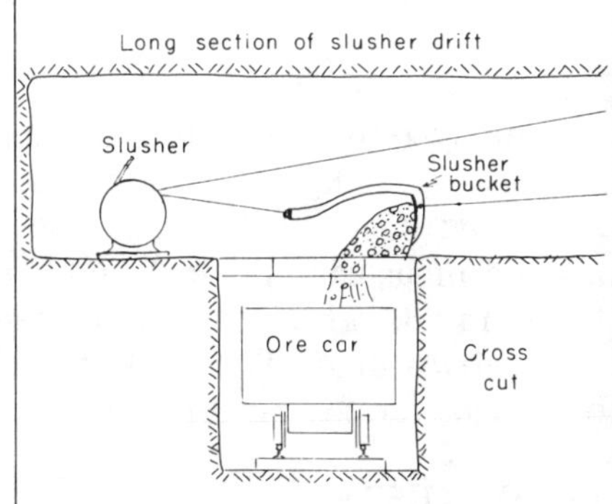

G. Direct loading of ore car by a slusher in a slusher drift

HOISTING ARRANGEMENTS AT SHAFTS

Ore-carrying devices called skips are employed in most mines except for those using trucks, conveyor belts, pneumatic, or hydraulic systems to remove ore. A few very small mines still hoist ore cars in cages. The cars are dumped on the surface and lowered into the mine for refilling.

SKIP LOADING

Skip loading, if it is to be efficient, has to have certain underground facilities. One style of skip layout for a vertical shaft is shown in A. The ore is brought by train and dumped over the grizzly where oversize chunks are broken before passing through. Some mines use large crushers instead of grizzlies so only the proper sized material will go to the skip. The pass and storage pocket act as a surge so several trains can be dumped in case the skips are not operating. A measuring pocket is usually installed at the lower end of the storage pass. This holds only one skip full. There are protective devices to prevent this pocket from over-filling and spilling ore down the shaft. When the skip gets into position, the measuring pocket dumps rapidly into the skip.

SLUSHER POCKET

The system shown in A works well for dry ore, but for wet sticky ore a slusher pocket and lane has been used successfully as shown in B. The slusher fills the measuring pocket, and it can move the wet sticky muck which tends to hang up in the ore pass, shown in A. The slusher pocket is big enough to hold serveral trains of muck.

DRUM HOISTS

Two types of drum hoists are shown in C. The single drum hoist is used on service cages to raise and lower supplies and personnel as shown in C,a. The double drum hoist in C,b is used for skips. As one drum is wound the other is unwound, so when one skip is hoisted the other one is lowered into the mine. This is called counter-balancing because the weight of the skip being lowered offsets the weight of the one being hoisted. The drums are clutched enabling their positions to be changed, facilitating the skips going to different levels in the mine.

FRICTION WHEEL HOIST

Long used in other parts of the world, a hoist that is gaining in popularity in the United States is the Friction Wheel. Two styles of these hoists are shown in D. The hoisting rope or ropes instead of being rigidly fastened to the hoist drum or wheel is held in place by friction between the rope or ropes and the drum. A skip or cage and counter-weight is shown in D,a. Notice that a tail rope hangs down the shaft from the cage to the counter-weight. This rope weighs the same per-foot as the hoisting rope so the weight is equal on each side of the friction wheel. The counterweight usually weighs as much as the cage or skip plus one half the weight of the maximum load. The friction force required between the wheel and rope is only one half of the maximum load. These hoists are much lighter in weight than the big drum hoists and can be installed in the headframe as shown in D,a. Ground mountings are also popular as shown in D,b. This hoist has two skips mounted on it with a tail rope.

SELF-DUMPING SKIPS

Skips are usually automatically self-dumping, and one common arrangement for an inclined skip is shown in E. The narrow front wheels follow the curved rails into the dump while the wide rear wheels ride up the headframe causing the skip to overturn and dump its contents.

There are many designs for dumping vertical skips. One method is shown in F where the skip uses a side door. Occasionally the doors stick in a closed or open position causing problems. One system turns the skip upside down to dump. Other styles use various systems to allow the skip to empty completely and efficiently.

PLATE 5-20 HOISTING ARRANGEMENTS AT SHAFTS

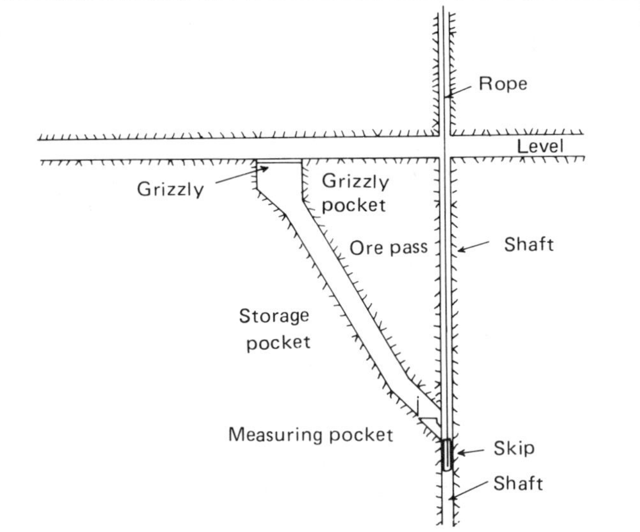

A. Skip loading ore pass and measuring pockets

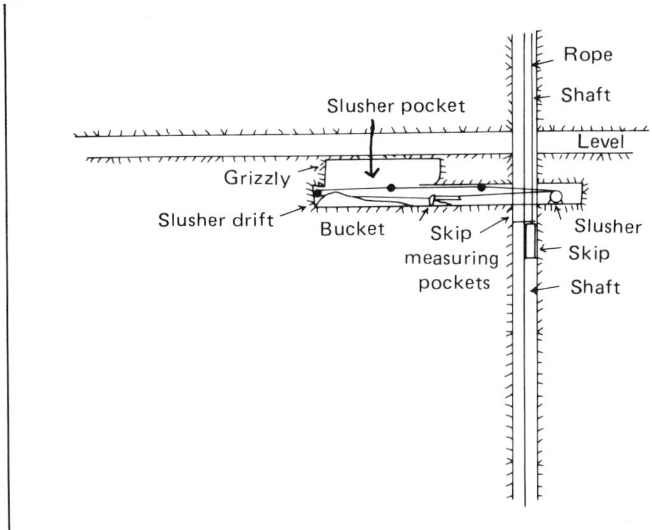

B. Slusher lane for loading skips

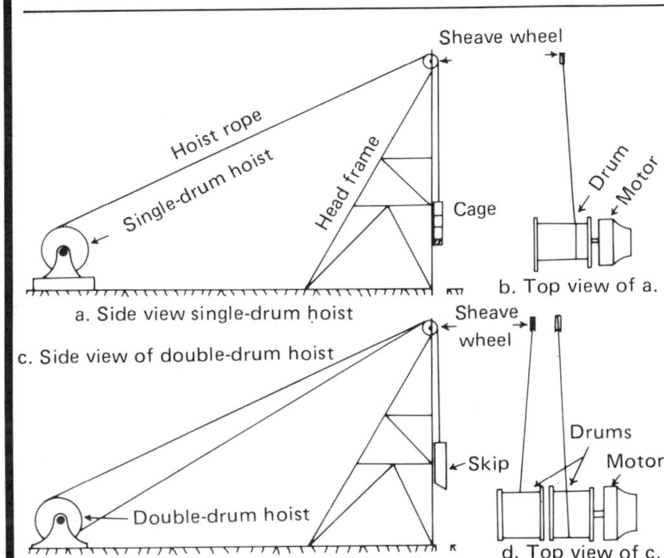

C. Views of single and double drum hoists

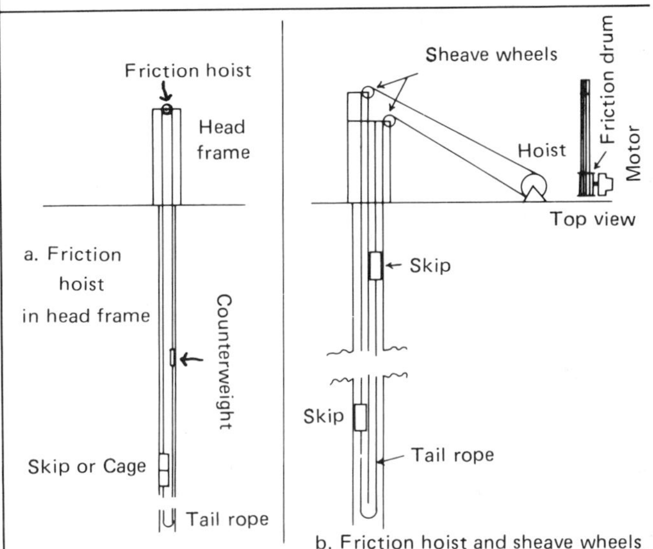

D. Friction type hoists

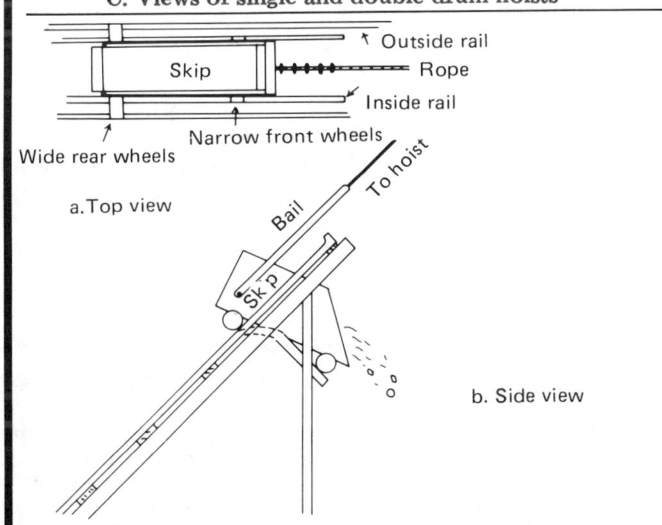

E. Dumping arrangement for inclined skip

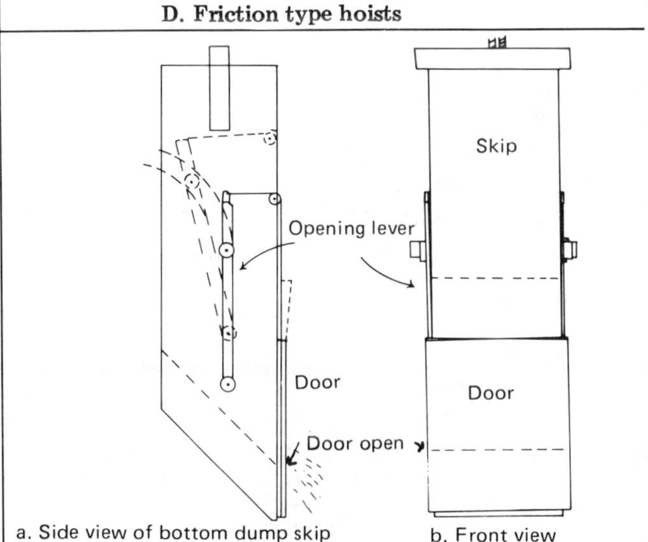

F. Dumping arrangement for vertical skip

HOISTING

Plate 5-21 shows various types of hoists along with surface arrangements. The principles of hoisting and their associated underground systems were discussed in Plate 5-20.

HORSE-WHIM

In the very early days of mining, people carried ore up ladders on their backs. The horse-whim was an early invention which harnessed animal power to the tedious chore of hoisting. A whim is on the right in A. A horse, or team of horses, was hitched to the long arm sticking out, and walked around a circle. The arm, attached to a shaft and gear, meshed with a gear on the drum causing the drum to rotate. The rope ran from the drum along the ground to sheave wheels in the headframe. To lower the bucket, the animal had to back up. Later whims had clutches and brakes on the drum to enable it to be released and controlled when lowering the bucket without the animals backing up.

STEAM HOISTING ENGINES; HOISTING ROPES

With the invention of the steam engine, it wasn't long before steam was utilized in hoisting engines. A steam engine is shown in B, and a close up view is shown in C. Most hoisting ropes are circular in cross sections, but the rope in C is flat. Part of it can be seen sticking out in front of the right drum or reel in C. Flat ropes and reels have two advantages. One is that the radius of the rope to the center of the reel is shorter when the skip is deep in the mine than it is when the skip is on top. By the principle of mechanics, it takes less force to raise the skip with the shorter radius or lever arm. The other advantage is that with round ropes and drums, the hoist must be positioned far enough from the head frame so that the rope will wind or spool properly on the drum. This can require a large surface area. With a flat rope and reel, the hoist can by positioned near or under the head frame.

COMPRESSED AIR AND ELECTRIC DRIVE HOISTS

Compressed air is often used to run hoists underground, and one type is shown in D. Compressed air hoists have good slow-speed control, and are used where precision hoisting is required.

Modern electric drive, double-drum hoists are shown in F. A direct connected electric motor attached to the drum shaft is shown in I. Direct current motors, as shown in I, are often used on hoists because of their precise speed control. However, improvements are being made in controlling alternating current motors, and they are used more frequently on new hoist installations.

HEADFRAMES

Headframes, sometimes called gallows frames, shown in A, E, H, and J, are necessary to position the rope in the shaft. In A, when the bucket is hoisted, it is wheeled away and dumped. H shows a big steel headframe with a skip dump arrangement and ore bins. A close up of the skip dump is shown in J. The towers shown to the left in H are called rope or idling towers. They are necessary to keep the rope from whipping between the large sheave wheels in the headframe and the hoist drum.

A special construction headframe, as shown in E, is used to lower heavy steel shaft supports down a shaft. A cage is shown in G which is used to transport people and equipment in a shaft with a small cross sectional area.

A. Old horse-whim hoist. *(Courtesy World Museum of Mining)* **B.** Reel type air-stream hoist with flat cable. *(Courtesy World Museum of Mining)* **C.** Close up of flat cable on reel. *(Courtesy World Museum of Mining.)* **D.** Old air driven hoist. *(Courtesy World Museum of Mining)* **E.** Temporary headframe for drilling a shaft. *(Courtesy Teton Drilling Co.)* **F.** Modern double-drum electric hoists. *(Courtesy Anaconda Co.)* **G.** Cage used to lower men and material into a mine. *(Courtesy World Museum of Mining)* **H.** Headframe and idling towers for an underground mine. *(Courtesy Anaconda Co.)* **I.** Large electric motor and drum on hoist shown in F. *(Courtesy Anaconda Co.)* **J.** Skip dump bin in headframe. *(Courtesy World Museum of Mining)*

PLATE 5-21 HOISTING

HYDROLOGY AND WATER PROBLEMS IN MINING

Water is usually present in the ground at some depth below the surface. Generally the sands and gravels lying over the bedrock contain most of the water. It is not uncommon, however, for the ores and surrounding rocks to be faulted, broken and fractured, with water filling these void spaces. Water then flows into the workings, and it must be pumped from the mine. Some mines don't have much water so handling and pumping is of minor consequence.

The movement of water in the underground environment depends on the permeability of the soil or rock. Permeability is the extent to which the voids in the rock or soil are connected to allow the flow of water. The drawdown of the water surface pertaining to a mine shaft is shown in A. The configuration shown is for rock with relatively low permeability because the drawdown curve is steep. If the rock was more permeable, the curve would be much shallower, and greater amounts of water would have to be pumped to keep the mine dry. Ground water tends to be recharged from rain and snow making pumping in a mine with a shallow water-table a continual problem.

DRAWDOWN ZONE

As the mine is extended, the drawdown zone increases in area, as shown in B. Some mines collect the water as near the surface as possible and pump it out of the mine. This reduces pumping costs because not all of the water will flow to the bottom levels of the mine. Due to the large drawdown zone in an operating mine more water will have to be pumped during wet seasons because the ground absorbs and tends to collect the excess moisture.

WATER SEALING

Sometimes the permeability is so great in underground layers (aquifers) that it is necessary to seal off the area around the mine to diminish the flow of water. Special types of cement, bentonite and other types of materials are used to seal the permeable zones. Holes are drilled into the permeable zone, and the water sealing material is pumped through the holes under pressure into the voids of the aquifer.

When water is encountered underground, it must be collected and directed toward the mine pumping plant. In wet mines it is common to carry a ditch in the level headings, as shown in C. The water is usually muddy and carries suspended solids. Often the suspended solids settle out in the ditches, filling them up and causing the water to flow over the ties and tracks. Cleaning the ditches is laborious, but if they aren't the water at the tie and rail level will wash out the road bed as trains run over the track, resulting in bad track and derailments.

SUMPS

Sumps are commonly installed in mines with the sediment problem just mentioned. A sump, as shown in D, is an underground pond where the suspended material can settle out in the quiet water. Eventually the sump must also be cleaned out, but it is much easier to keep the ditches cleaned and road bed maintained when they are used. Sumps are generally used in connection with a pumping plant so that the larger solid particles will not enter the pump.

MINE PUMPS

Two systems of installing mine pumps are shown in E, and F. Centrifugal pumps, the ones most commonly used to dewater mines, must be kept primed when starting or they will not pump. In the suction type shown in E it is necessary to have a foot valve on the intake side so the water does not drain from the pump. Sometimes this valve doesn't seat properly, and the water will drain from the pump making it difficult to start. Other mines use the head type shown in F. The pump is always supplied with water because it is lower than the water surface.

PLATE 5-22 HYDROLOGY AND WATER PROBLEMS IN MINING

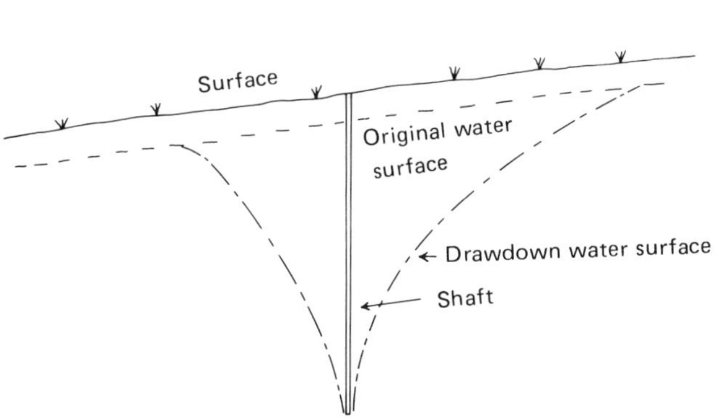

A. Ground water profile as shaft as sunk

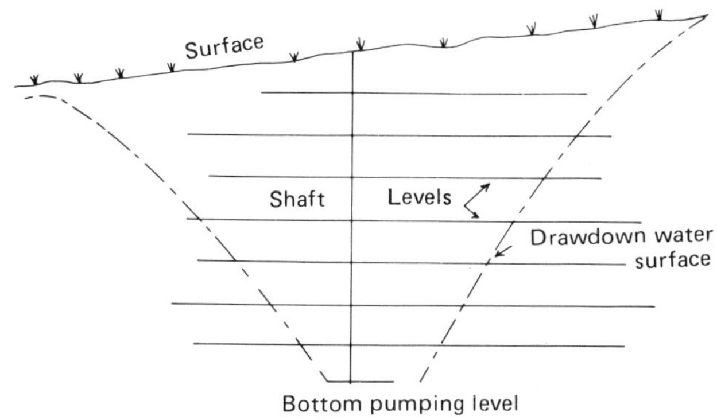

B. Ground water profile during mining and pumping

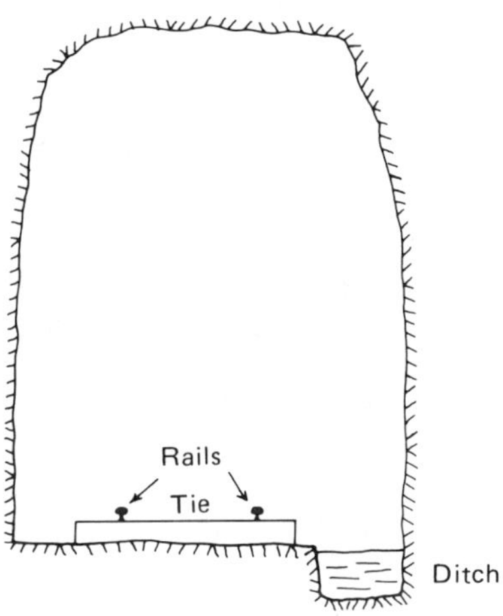

C. Cross-section of drift showing ditch

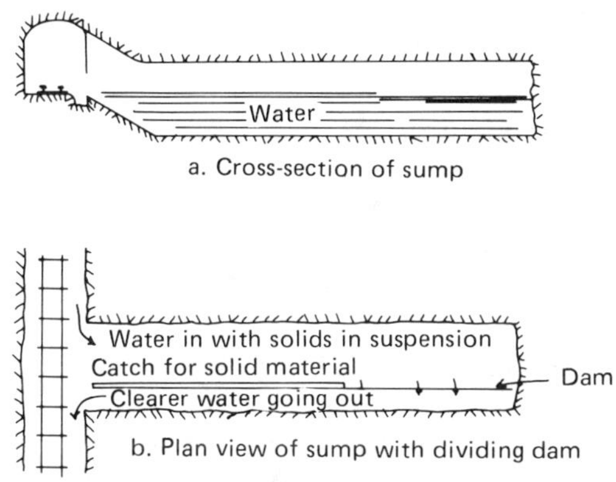

D. Sump used for collecting suspended solids

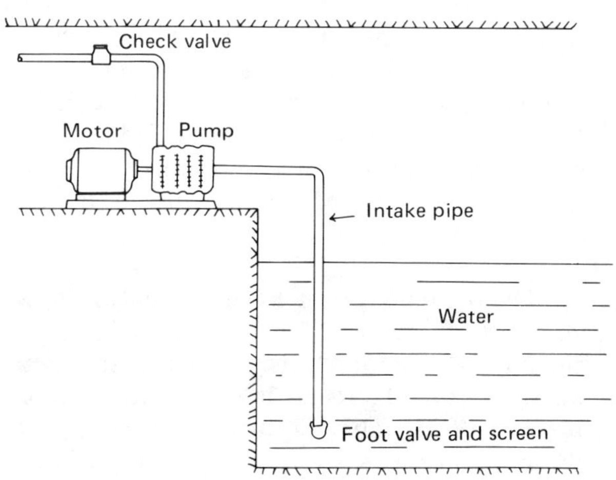

E. Suction type of pump intake

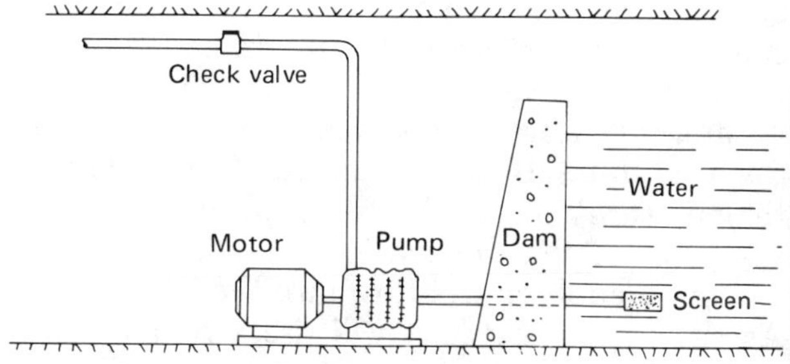

F. Head type of pump intake

NATURAL AND FORCED VENTILATION

In underground mines sufficient air is required for personnel to breathe and for equipment to operate. Not only must adequate air be supplied for breathing, but contaminants must be diluted and carried harmlessly from the mine. The following problems may be encountered in mining, but fortunately no one mine will have them all.

Explosive gases
Radio active emissions
Poisonous gases
Depletion of oxygen
Stratification of air
Uncomfortable work environment
Dust
Objectionable fumes

The method of controlling these problems is to prevent their formation, if possible, or to dilute the contaminants with fresh or conditioned air so they are no longer dangerous.

Natural ventilation is still used whenever possible. As air is warmed it becomes lighter per unit volume and tends to rise; conversely, if it is cooled it gets heavier per unit volume and tends to sink. Shallow mines in mountainous country can often take advantage of these phenomena for natural or near natural ventilation of the mine. Such a mine is shown in A. The underground temperature stays almost constant throughout the year. If the outside air is colder than that in the mine cool air will flow into the adits as shown. As the air is warmed it will rise vertically through the raises and stopes as shown, providing good movement of air. If the outside air is warmer than that in the mine the circulation will reverse and flow the other way because the cooler air sinks. However, when the mine and outside temperature are the same, very little, if any, air will flow in the mine.

VENTILATING HORIZONTAL LEVELS

Horizontal levels can also be naturally ventilated if proper conditions exist, as shown in B. The warmed air tends to rise and flow along the roof or back as shown by the arrows in B,a. However, the longer the opening becomes, the less effective the natural ventilation because the air will short circuit and not flow into the face.

Raises, especially long ones, are difficult to ventilate naturally. As the raise is blasted, hot gases form and tend to stay in the back of the raise. Eventually the gases may cool and settle out of the raise, but some form of auxilliary ventilation is usually required if the raise is to be worked continually. A shaft or winze will usually ventilate relatively well if cool air is brought to the collar of the winze. The cool air will flow down the bottom, and the warm air along the back, causing relatively good circulation.

FORCED VENTILATION

As the extent of mine workings increase, it is necessary to introduce a forced ventilation system. A fan blows air through a duct, as shown in D,a. The inlet to the fan is located in fresh air and blows it into the face of the working heading. The fan can also suck air from the heading, as shown in B. The advantage of the exhaust fan is that blasting gases can be more quickly removed from the face. Both systems have certain advantages depending on conditions.

Deep mines often have ventilation systems, as shown in E. A large fan or series of fans moves air through the mine. It is necessary to regulate the air on various levels so it can be directed to where it is needed. This is usually done by regulator doors or air locks. In E, air flows down the hoisting shaft which is then referred to as a downcast shaft. In some mines the main shaft may have air flowing up so it is called an upcast shaft.

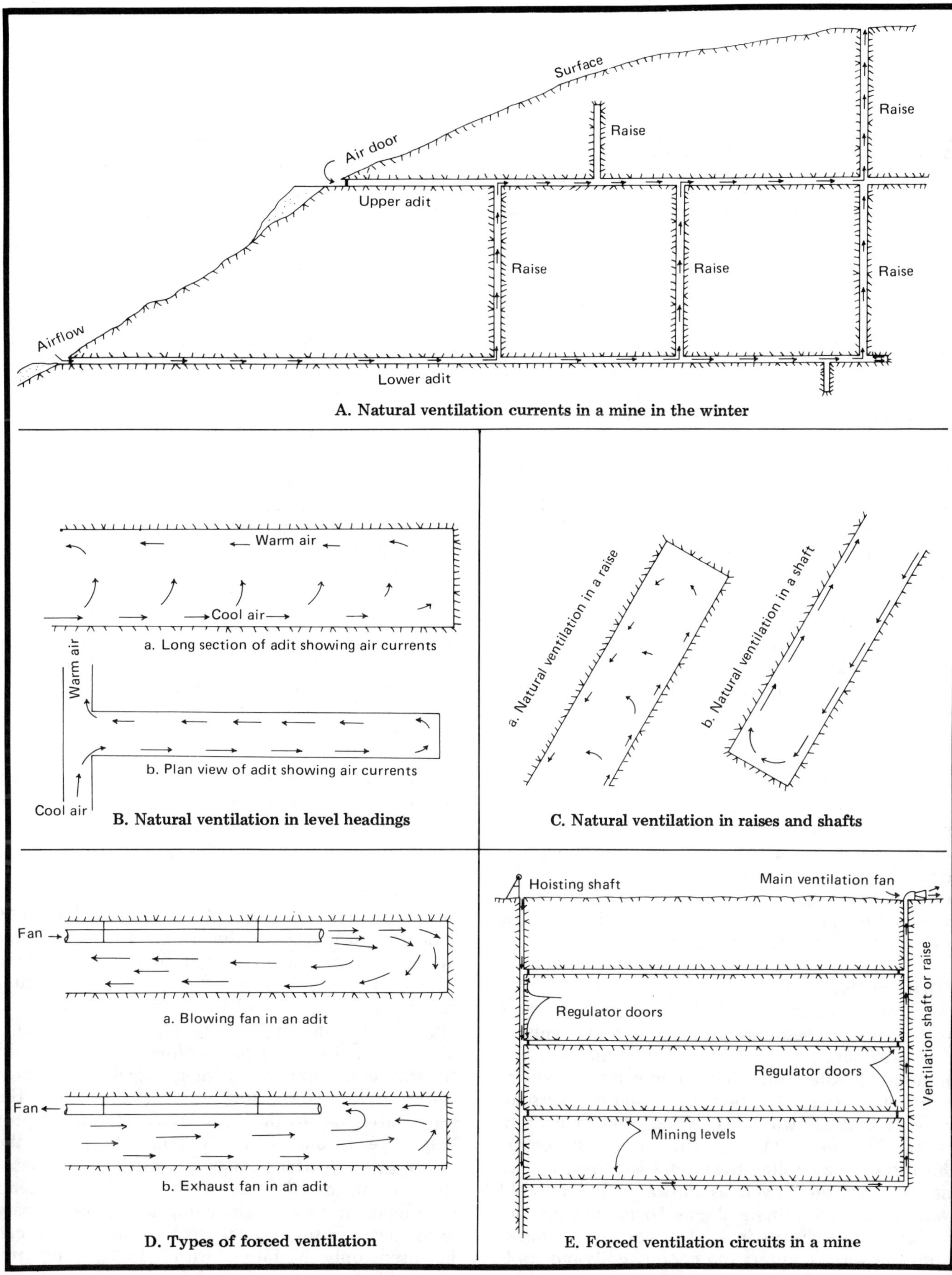

A. Natural ventilation currents in a mine in the winter

B. Natural ventilation in level headings

C. Natural ventilation in raises and shafts

D. Types of forced ventilation

E. Forced ventilation circuits in a mine

AIR CONDITIONING IN DEEP MINES

Personnel are most productive if they can work in comfortable temperature ranges, as shown in A. Both extremes of temperature cause worker inefficiency. The effective temperature shown in A is a combination of dry bulb and wet bulb temperatures. The dry bulb temperature is the one ordinarily read on the thermometer. The wet bulb temperature is the one recorded when a thermometer wrapped in a wet wick is swung through the air. This is called the evaporation temperature. The comfort temperature is measured in a large part by the wet bulb temperature, but the dry bulb termperature and the air velocity are also contributing factors.

Some mines are very cold in the winter, especially in higher elevations and higher latitudes so the intake air is often heated by specially designed heaters. If a surface plant is nearby waste heat is often utilized to warm the intake air.

AIR CONDITIONING

As mines get deeper the temperature usually increases. Eventually air brought in from the surface is not adequate to cool the mine enough to maintain a work efficient termperature at the working headings. When this occurs, it is necessary to introduce some type of air conditioning to allow the air flow to the heading to be cooled.

COOLING WATER

Cool water is one of the best and cheapest heat transfer mediums for underground mines. If a large supply of cold water is available it can be used in the mine for cooling air. However, water must sometimes be cooled, and one system, such as that shown in B, works well in dry climates. The water in a continuous circuit enters the top of the cooling tower and is cooled by the air rising up the tower, as shown in B. The cold water enters the mine and goes to an underground cooling plant, shown in C. The cold water piped from the surface enters the cooling coils and cools the mine air as it flows through. After it enters the fan, the mine air is washed with a spray and run through dust collectors to remove contaminants. The cold air from the plant is directed to mine workings by fans and air ducts. Often the water is piped from the surface cooling tower to the underground cooling plant and the warmed water is piped up the shaft to the tower. This gives very good pump efficiency, but the water pressure in a deep mine gets very high. High-pressure underground cooling plants are very expensive and difficult to move.

MECHANICAL REFRIGERATOR

Another system used in mines is to install a mechanical refrigerator underground, as shown in D. The refrigerator chills the water on the mine level where it is pumped to cooling plants shown in C, and E. Because this water is at low pressure it can be used in small air cooling plants, shown in E, which can be located near the working face. These light air cooling units can be readily moved to more advantageous positions. The drawback to this system is that water or air must be used to cool the condenser in the water cooling plant on the level, and this warm air or water must be pumped out of the mine.

A schematic system incorporating many air cooling features is shown in F. No known mine uses all of the features shown in F, but equipment exists to install such a system.

The water is cooled in the tower, and if necessary, a mechanical refrigerator can be put in series to further cool the water. The cold water under high pressure is piped to a water turbine which produces energy to help turn the water return pump. The cold water from the turbine is pumped to an air cooling plant. The warmed discharge water from the cooling plant is pumped to the condenser of a mechanical refrigerator water chiller. The heated water is returned to the pump attached to the turbine. The electric motor helps the turbine turn the pump to return the water to the surface. Such a system has some of the advantages of both of the previous systems described, however, air conditioning is a very expensive addition to the cost of mining and can be used only in mines with good paying ore.

PLATE 5-24 AIR CONDITIONING IN DEEP MINES

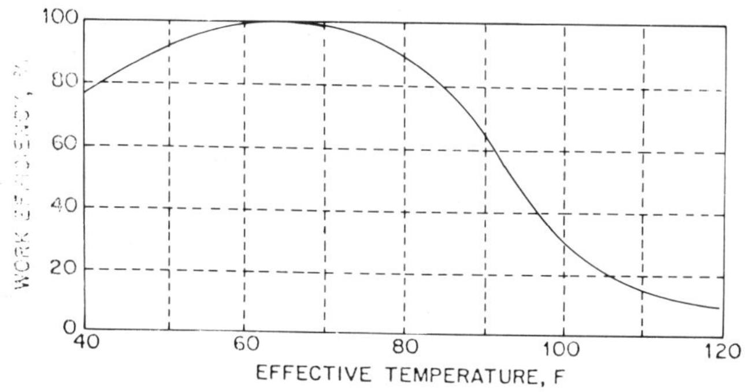

A. Effect of temperature on work efficiency

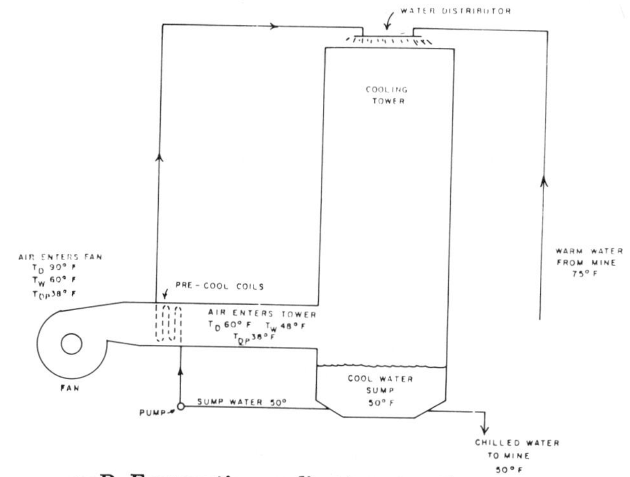

B. Evaporative cooling-tower system

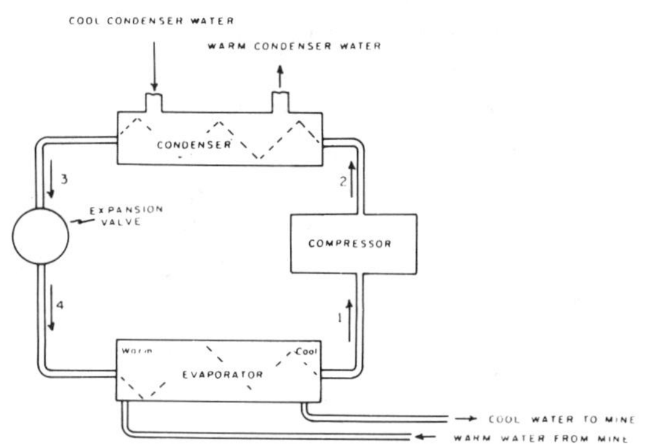

D. Vapor-mechanical refrigerator
with water chill evaporator for mine use

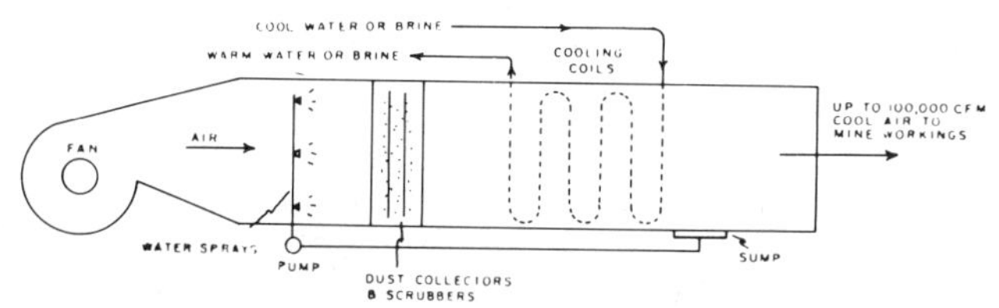

C. Underground cooling plant

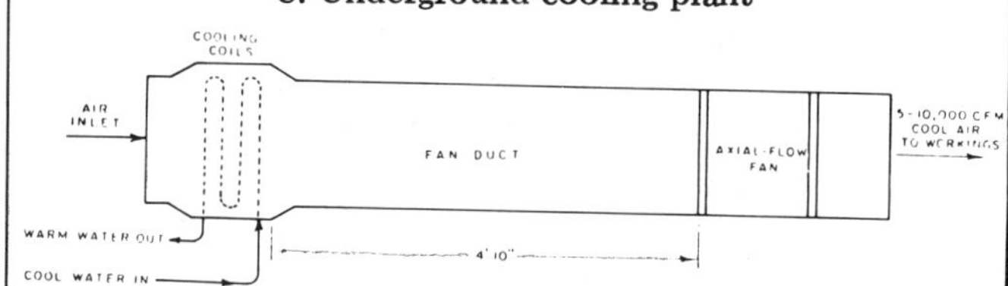

E. Underground cooling plant

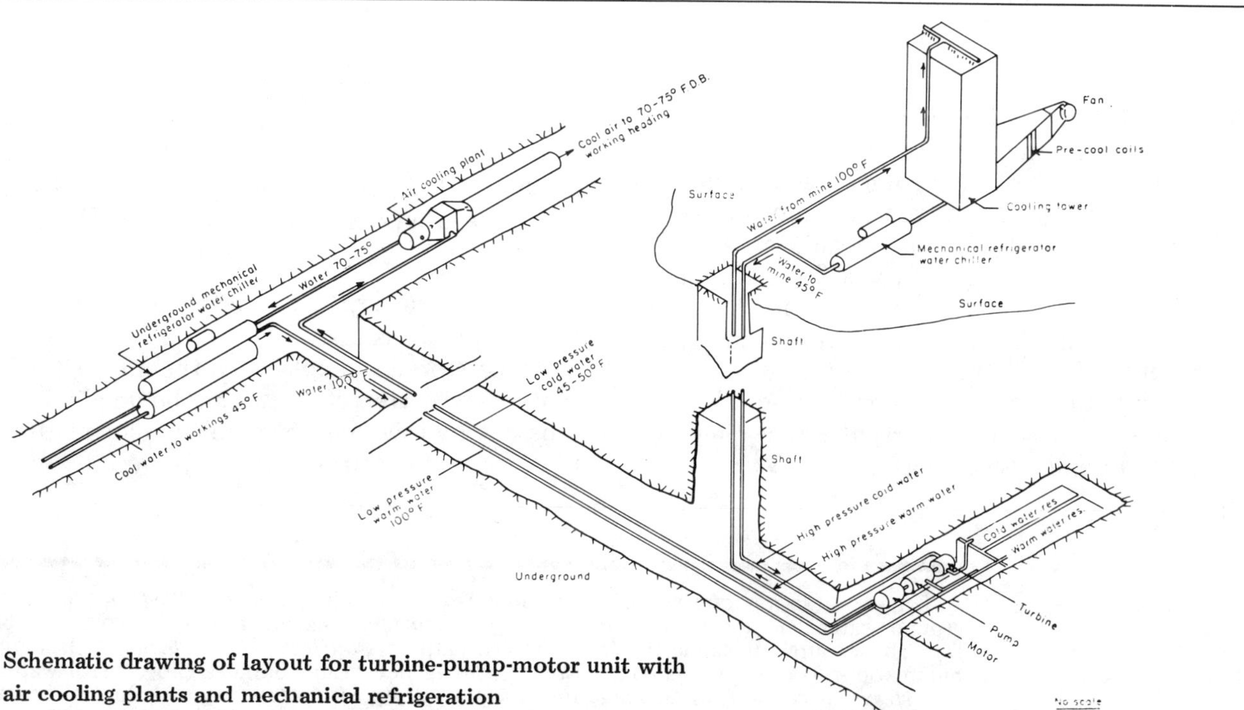

F. Schematic drawing of layout for turbine-pump-motor unit with air cooling plants and mechanical refrigeration

VENTILATION AND PUMPING EQUIPMENT

Ventilation air is frequently moved through pipes or ducts underground so it can be directed to the needed place. A large ventilation pipe or duct is shown in Plate 5-25A. Ventilation ducts are used in conjunction with fans to produce forced ventilation. Ventilation requirements are generally increasing in mines because of the increased use of diesel equipment so the size of ducts and fans are also increasing. Long, small-diameter ducts offer large resistance to air flow. To decrease the loss caused by resistance larger diameter ducts are used.

A large electric driven ventilation fan is shown in D. A fan this size is one of the main ventilation fans shown in Plate 5-23E. These fans are often located on the surface, but they can be placed underground if necessary for efficient air movement.

COMPRESSED AIR VENTURI TUBE

Short development headings driven off an underground air course are sometimes difficult to ventilate. A compressed air venturi tube, shown in C, is used to ventilate short headings. A compressed air stream is directed toward the throat of the venturi, which creates a vacuum in the back end of the venturi. Ventilation air rushes into this vacuum and is forced through the throat and into a duct with the expanded compressed air. Although not as efficient as a regular fan, they are light in weight and easy to place.

CENTRIFUGAL WATER PUMPS

Water must be controlled and pumped from the mine, and a modern mine pump room is shown in E, and G. The pump in G will pump about one thousand gallons per minute under a head of over 4,000 feet. The mine water where the pump is installed is very corrosive to ordinary metal, therefore these pumps are made from corrosion resistant metal. The pumps are centrifugal type and are built in stages to increase the pressure of the water.

USE OF CHECK VALVES

The check valve arrangement shown in E, and G is necessary because when the pump is shut down, the water in the vertical column tends to run back through the pump. The pump will then act as a turbine and could cause overspeeding which may injure the pump or motor. Thus the check valves keep the water from flowing back through the pump.

BULKHEADS AND DAMS

A concrete bulkhead or dam is shown in B. The large diameter pipe coming from the bulkhead goes to the inlet side of the pump. This is called a head pump feed system as explained in Plate 5-22F.

An emergency dam and gate is shown in F. In case power is disrupted to the pumps, the water must be stored someplace. In the mine shown in F, temporary water storage can be made in a long crosscut by lowering the gate and closing the valves in pipes running through the bulkheads.

A. Large ventilation duct with diesel powered LHD unit. *(Courtesy Wagner Co.)* **B.** Underground water dam and water feed to pumps. *(Courtesy Anaconda Co.)* **C.** Venturi air blower. **D.** High pressure adjustable fan for mine ventilation. *(Courtesy Fairchild Inc.)* **E.** Underground pump room. *(Courtesy Anaconda Co.)* **F.** Emergency, water bulkhead to control water flow. *(Courtesy Anaconda Co.)* **G.** Large underground pump with a capacity of 1,000 gallons/minute at over 4,000 ft head of water. *(Courtesy Anaconda Co.)*

PLATE 5-25 VENTILATION AND PUMPING EQUIPMENT

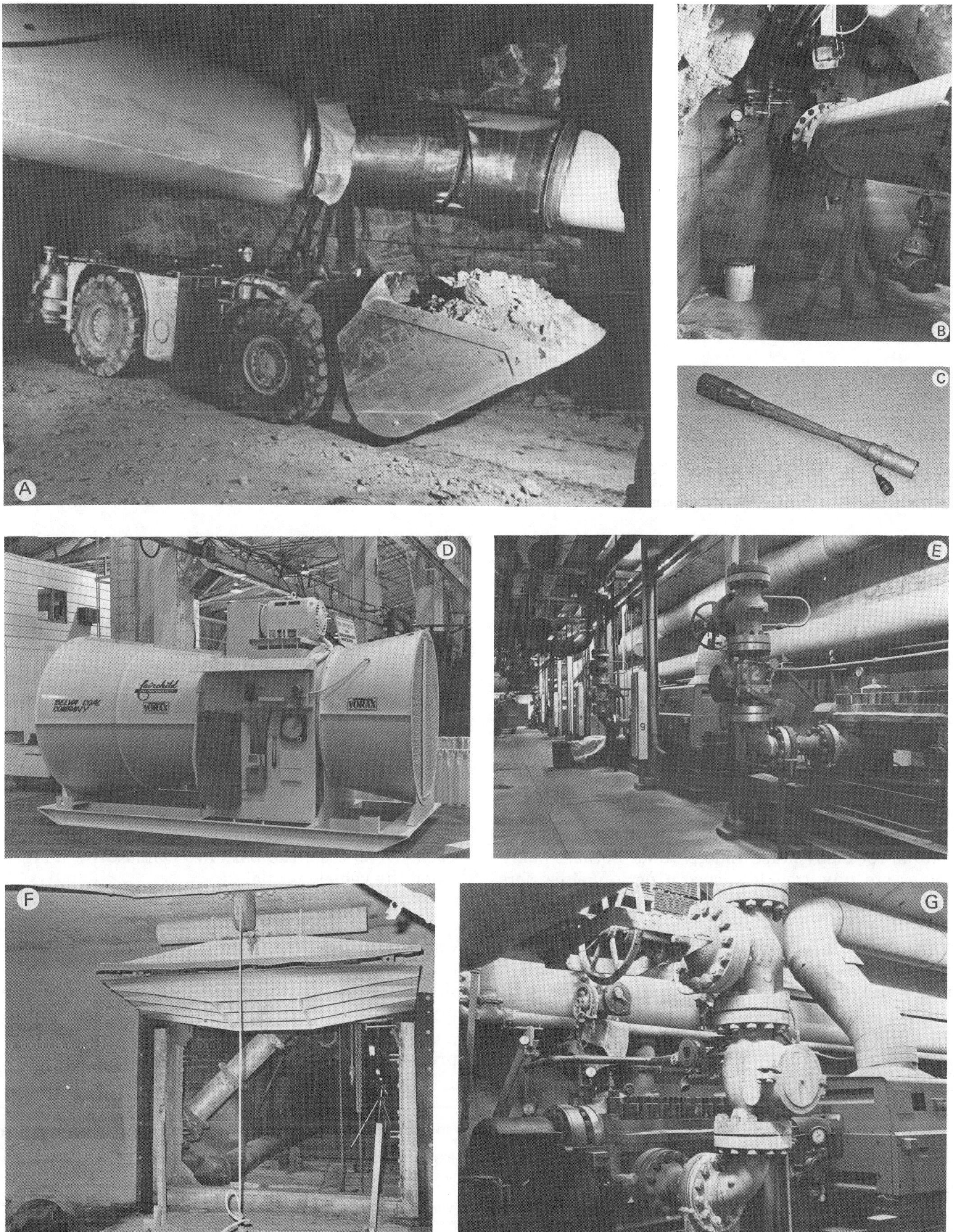

SERVICE TO MINES

Plate 5-26 shows a variety of service equipment that is used in mines. One of the potential hazards in a coal mine is fine coal dust getting air borne and being ignited. This will cause an explosion which in turn will kick up more coal dust thereby propagating the explosion and fire. To reduce this hazard, inert rock dust such as limestone is sprayed on the exposed coal faces to minimize the chance of fine coal dust becoming airborne. In A, a rock dusting machine is shown with a miner dusting the face.

PERSONNEL CARRIERS

Modern mines in bedded deposits such as coal extend for many miles underground. Because many mines are now trackless, special personnel carriers are used to transport people, equipment and supplies underground from the central shaft. One such vehicle is shown in B. One large trona mine uses electrified golf carts to move supervisors and engineering personnel underground. Where permitted, diesel equipped four-wheel drive trucks are also frequently used for moving people and supplies.

Much underground equipment is now hydraulically powered. A portable hydraulic power pack unit is shown in C with a control console on the right side. Hydraulic motors and pistons are light weight for their power output and are used in many applications.

DIESEL-POWERED CENTRIFUGAL PUMPS

A large diesel-powered centrifugal pump is shown in D. Although water may be a hinderance in an underground mine, it is needed in many operations in mining, particularly to settle dust in drilling and mucking operations. Water is one of the best dust control mediums used on mining machinery. Good clean water must be supplied to the mine for many purposes. Usually the mine water is so dirty, corrosive, and contaminated that it can't be used for ordinary mine service. A pump such as shown in D is often used to supply fresh water to a mine from some surface water source.

FORK LIFT TRUCKS

Fork lift trucks are used a great deal underground to handle equipment and supplies. A specially designed battery-operated fork lift truck is shown in E. This machine is used as a longwall support vehicle.

RAPID-CHARGE BATTERY SYSTEMS

A rapid-charge battery system is shown in F for the vehicle of the type shown in E. Batteries are made up in packs or units, and when they are discharged, the whole pack is exchanged for a charged one. The discharged pack can then be charged while the vehicle uses the replaced charged pack. Of course, the batteries can also be charged when the vehicle is idle.

Various types and styles of special track type cars and trucks are shown in G, and H. The timber cars shown in G are generally used only around the shafts because they are too high to take into the mine. The two cars in the center of G are underground toilet cars. The small trucks in H are used primarily to transport timber into the mine.

A. Rock dusting a coal face. *(Courtesy Long Airdox Co.)* **B.** Personnel carrier. *(Courtesy Long Airdox Co.)* **C.** Hydraulic power pack unit. *(Courtesy Gardner-Denver Co.)* **D.** Diesel driven centrifugal pump. *(Courtesy Ingersoll Rand Co.)* **E.** Underground fork lift, battery operated. *(Courtesy Fairchild Co. Charles Ryan Associates)* **F.** Rapid battery change unit. *(Courtesy Fairchild Co. Charles Ryan Associates)* **G.** Miscellaneous rail-type service cars and trucks. *(Courtesy World Museum of Mining)*

PLATE 5-26 SERVICE TO MINES

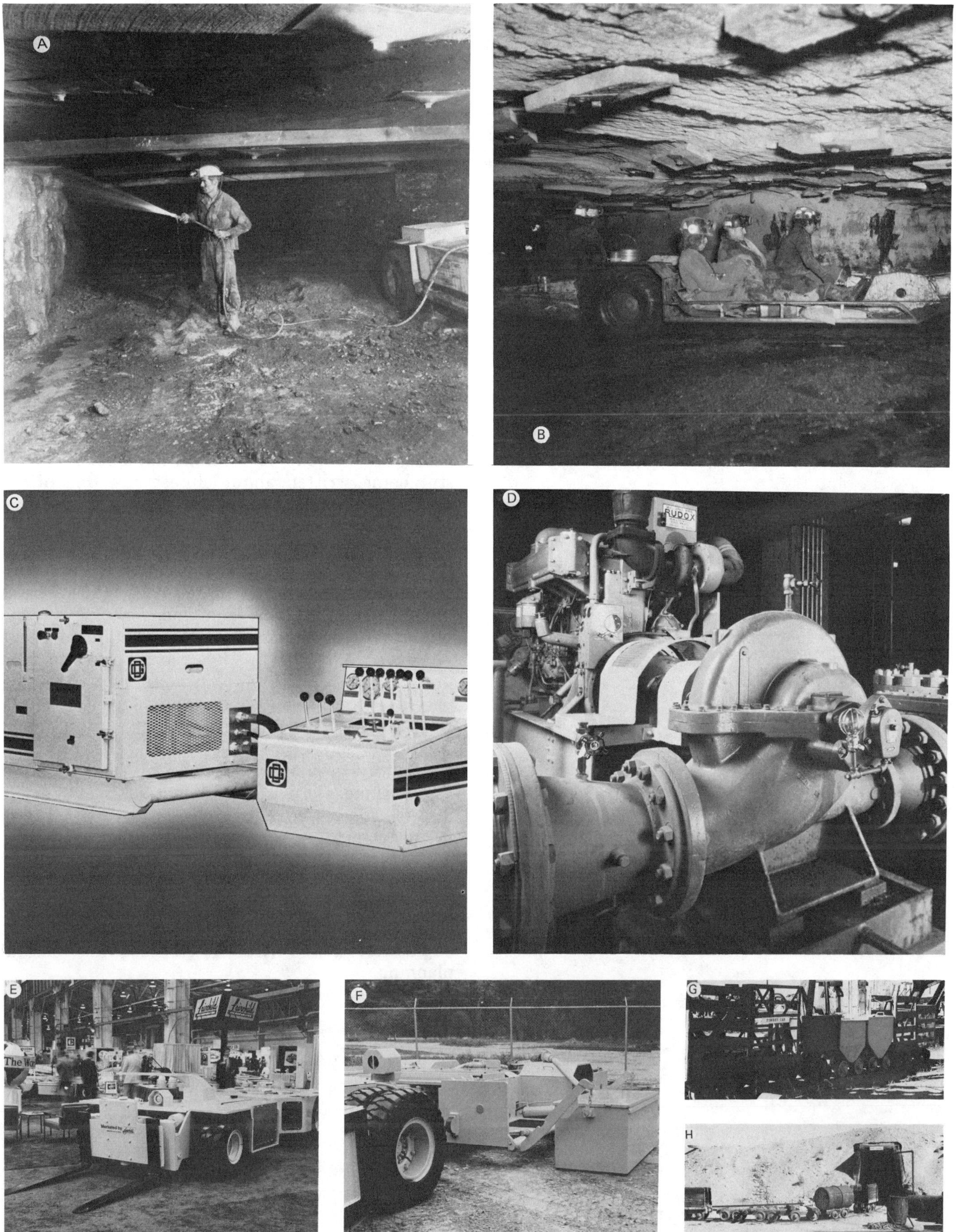

PART 5
CONCLUDING REMARKS

The various operations of mining have been examined in this section. In the few short pages alloted to this subject, only the bare essentials could be covered. Text books have been and are still being written about almost any one of the operations discussed in this section.

Much engineering and management expertise goes into the successful building and use of this equipment. Until one really studies mining, he is unimpressed with the equipment necessary, and the proper arrangment of this equipment for a successful operation. The versatility, size and productivity of the equipment is impressive, especially considering the equipment frequently works in a very hostile environment in most mines.

Supplying power, water and necessary tools to the working place is challenging. Many mines are big and cover a large area. Power lines with appropriate transformers must be installed in places where room is limited. Air and water must be supplied with adequate pressure and capacity. Because working places are scattered over a broad area, supplying tools and other necessaries such as explosives, pipe fittings, etc. requires careful planning.

PART 6

DEVELOPMENT

Few ore bodies are so situated within the earth that mining can be started immediately. The deposit must be prepared for mining. This step is called development. Even in open-pit mines, some overburden or waste must first be removed, roads must be built and dumps established. This work may be thought of as development, although it may not be referred to that way. The U.S. Bureau of Mines classifies development as "work done in a mine to open up ore bodies, as sinking shafts and driving levels." This, of course, refers to underground workings. Raises are often classed as development headings. This section describes the general methods and procedures for driving development headings, but details are omitted.

CYCLE OF OPERATIONS

The descriptions in this section will start at various steps in the cycle of operations. The cycle is generally planned so blasting is done at the end of a shift to allow time for smoke, fumes, and dust to clear out of the heading before the next shift. In sinking shafts or driving large timbered raises, blasting during every shift may be impractical if not impossible. In those cases, special measures are taken to provide adequate ventilation to clear the heading as soon after blasting as possible. In some high speed tunnel projects blasts may occur more than once in a shift, so adequate ventilation becomes especially important.

CONTINUOUS VERSUS CYCLIC OPERATIONS

Most mining operations have historically been cyclic in nature. For example, making the area safe when entering, mucking, timber or rock bolting, drilling, loading and blasting. It has been an established principle in production that continuous operations are usually more productive than cyclic operations.

In recent years in the mining industry, equipment manufacturers have produced continuous mining equipment. The coal industry was first to adopt this principle with the development of the continuous miner. The principle has also been followed in the hard rock mines but development has been much slower. The hard rock has posed problems to the cutting action of most continuous miners now developed.

In this section on development, the classical methods as well as the newer continuous innovations will be presented.

SHAFT-SINKING SEQUENCE

Shafts are of various sizes and shapes. The horizontal cross section can be square, rectangular, circular, elliptical, or some special shape. For small mines, the timbered rectangular shaft is the most popular. The shaft may be inclined or vertical. Plate 6-1 shows a bench or sump type of blasting round in a vertical shaft. Note that the holes are slanted so they will break toward a free face. The sump round is popular because it provides a low place in the shaft for water to collect so that the drill holes can be started or collared in the high rock bench above the water level. Thus mud and dirt are kept out of the shot holes.

ELECTRIC BLASTING

Because electric blasting provides greater safety, the explosives in wet shaft bottoms often are detonated by electric blasting caps. B shows the shaft bottom drilled, loaded, wired, and ready for blasting. The blast usually piles the broken rock as shown in C. After the fumes from the blast have been removed, the round is mucked out. In one method, a small clamshell loads the rock into a sinking bucket D. Notice that a small sump pump is installed in the lowest part of the shaft to remove the water so that it will not interfere with operations.

SUPPORTING THE SHAFT

Supporting the shaft is the next problem. After the round is mucked out, or even before it is all mucked out, the next set of timber is installed. The wall plates, horizontal pieces along the long sides of the shaft, are brought down and suspended by hanger bolts from the shaft set above E. The other pieces to the set are then lowered and put in their respective places; the whole timber set is aligned or plumbed and then blocked or wedged into position F. After the timber is in position, drilling is resumed and the cycle is repeated. A hand-held sinker drill is shown in F drilling the cut holes for the next sump round.

Not all operators use the sump type of round when sinking shafts. The whole bottom may be drilled with a V-cut or burn-cut type of round, see Plate 5-4G.

PLATE 6-1 SHAFT-SINKING SEQUENCE

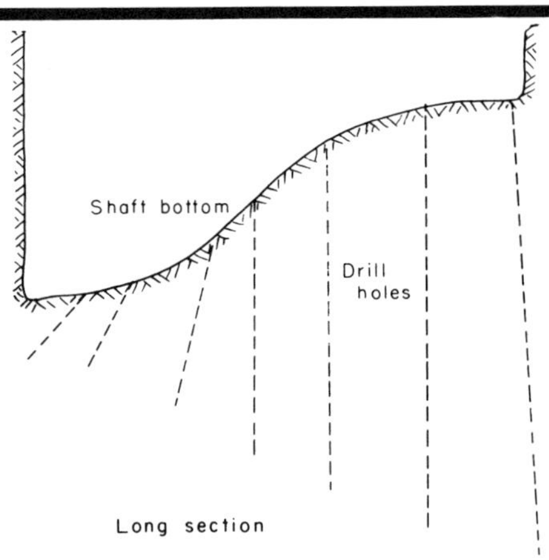

A. Shaft bottom drilled out with a sump or bench type of round

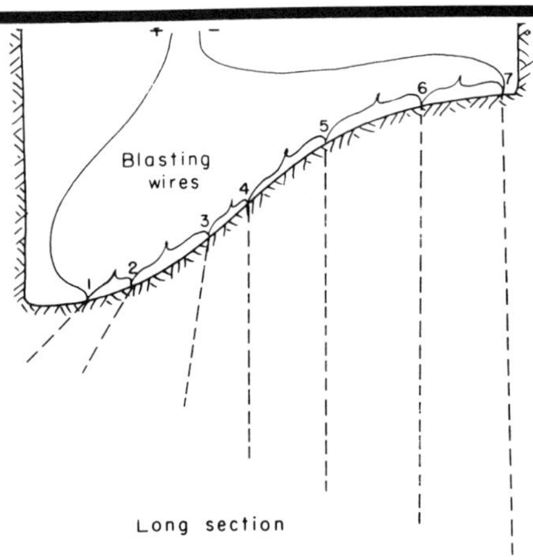

B. Shaft bottom loaded with explosive and electric blasting caps. (Numbering sequence shows order of detonation)

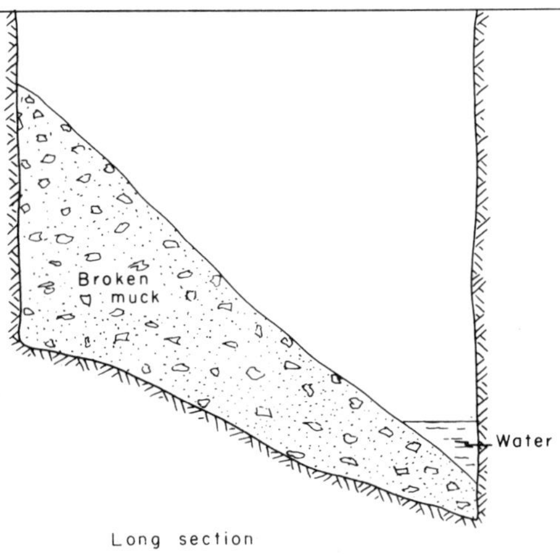

C. Round after blasting, showing muck piled up at one end of the shaft

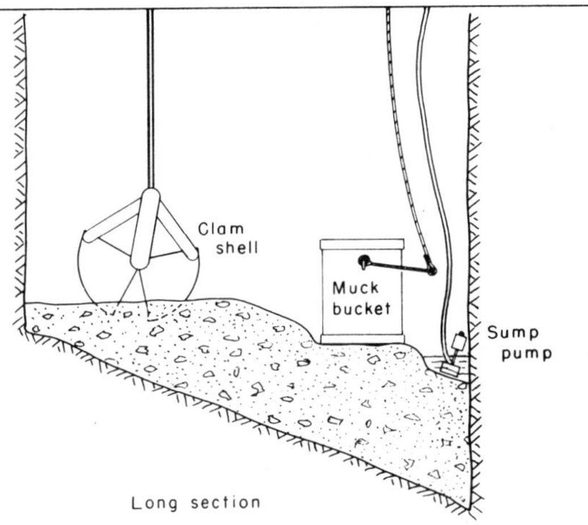

D. Mucking in the shaft with a clamshell bucket

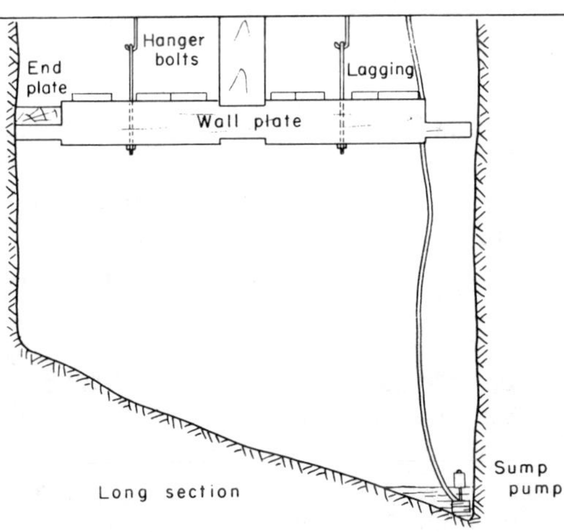

E. Timbering in shaft. Wall plate is suspended from timber set above

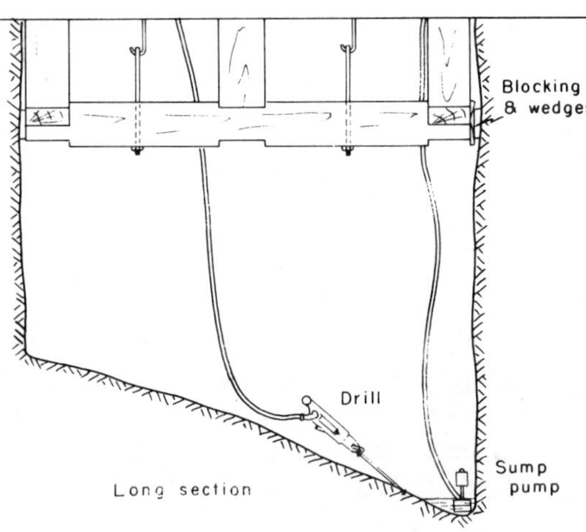

F. Timber set blocked, ready for the start of another drill round

MUCKING AND CONCRETING SHAFTS

Hand mucking is still practiced in sinking shafts at small mines because mechanical shaft-mucking equipment is costly to purchase, expensive to operate, and inefficient in cramped quarters. To use mechanical shaft muckers effectively, the supporting equipment such as sinking buckets and surface facilities must be designed to handle the large amount of muck that the mechanical equipment is capable of moving. Most small mines are not thus equipped, nor is money available to purchase such equipment.

HAND MUCKING

Two methods of hand mucking are in general use. One is direct loading of the bucket A; the other consists of first filling a low container called a sinking pan, which is then hoisted and dumped into the bucket. Shoveling into a low pan is easier than shoveling into a high bucket. The sinking pan is popular, although it may require a separate hoist.

MECHANICAL MUCKING

Two types of mechanical shaft muckers commonly used are the Riddell and the Cryderman muckers. The Riddell C uses an ordinary cable-actuated clamshell bucket. The two hoists necessary for operation of the clam are mounted on a traveling carriage, which runs back and forth in the shaft on a specially designed track suspended in the shaft. The dotted position in C shows the clamshell in the dumping position. The Cryderman D also uses a clamshell bucket, but the bucket is closed and opened by air cylinders rather than cables. The clam is attached to a telescoping air-operated arm, which is positioned in the shaft by air cylinders attached to the cage. The dotted line in D shows the bucket in the dumping position.

CONCRETE SUPPORTS

Concrete is now a popular shaft support because of its strength, permanence, and fire resistance. Forms are needed to hold the wet concrete in the desired shape until it hardens, but they must not interfere with or restrict the other shaft-sinking operations. Therefore steel forms of the desired shaft cross-section and as much as 20 feet in vertical length are commonly used. They are lowered into position and may rest on the broken muck. After the forms are carefully aligned and blocked into position, wet concrete is brought down by bucket or pipe line to fill the space behind the forms, which are left in place while the concrete hardens and shaft sinking continues. They are lowered when needed again.

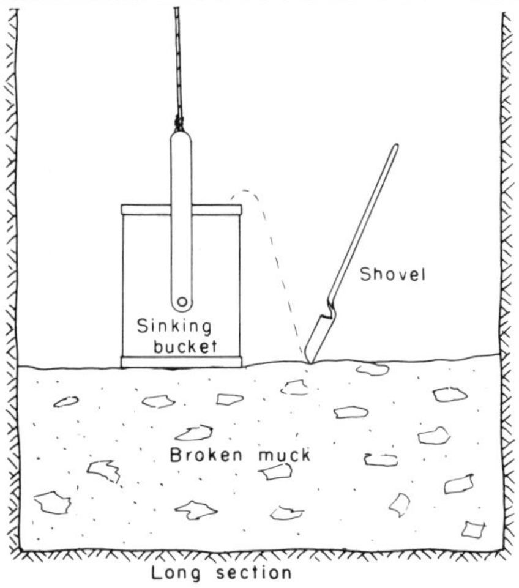

A. Shaft bottom mucking by shoveling directly into sinking bucket

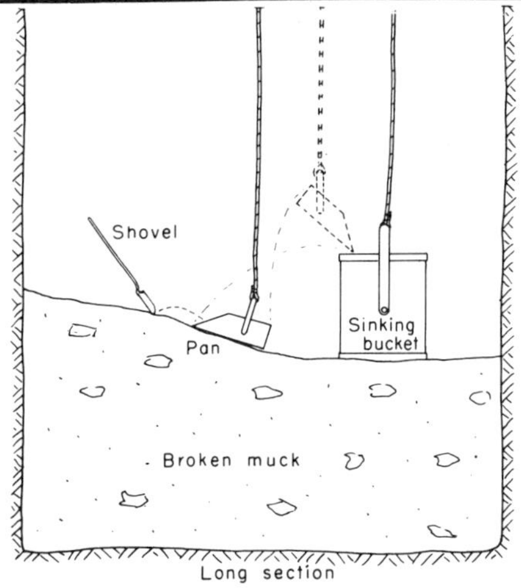

B. Sinking pan used in a shaft bottom to fill the sinking bucket

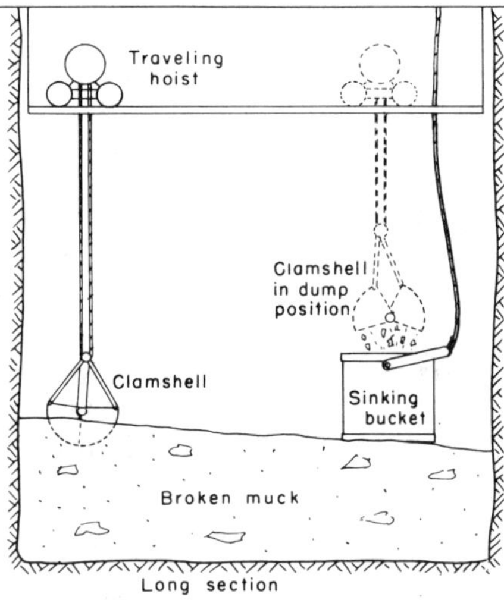

C. A Riddell shaft-mucking machine mucking into a sinking bucket

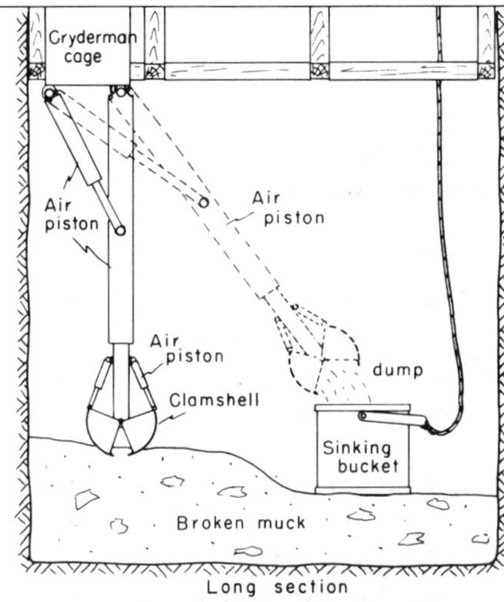

D. A Cryderman air-operated shaft mucking machine in use in a timbered shaft

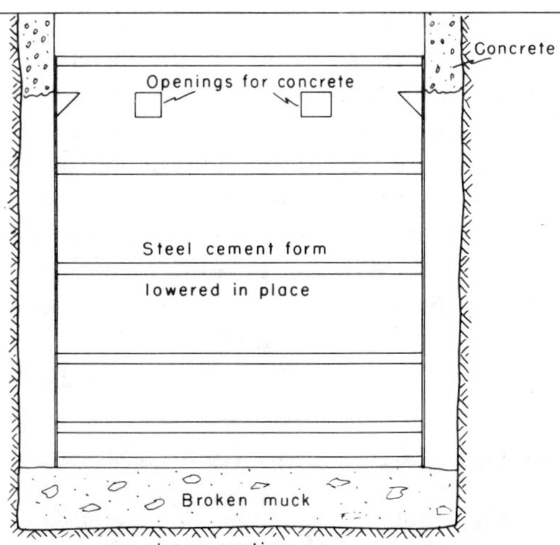

E. Steel form for concrete in position in a shaft bottom

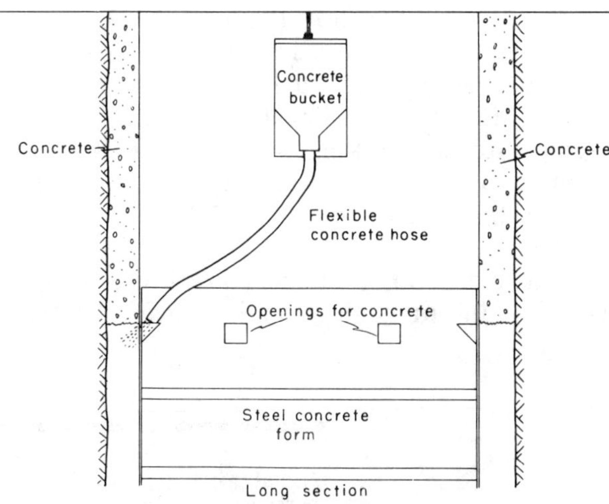

F. Concrete bucket in shaft supplying wet concrete to steel forms

SHAFT-SINKING EQUIPMENT

In Plate 6-2 the use of mechanical shaft sinking equipment was explained; this plate illustrates some of the equipment. Because of the small space underground, the full view of a large piece of equipment may be impossible to obtain.

CRYDERMAN MUCKER

The cage or supporting frame of a Cryderman mucker is shown in A. The whole machine is raised and lowered in the shaft. The telescoping air leg and positioning cylinders attached to the bottom of the supporting frame are seen from below in B. (This machine is not the one shown in A.) The bucket end of the telescoping air leg is seen in C, which shows the bucket digging and filling at the shaft bottom. The Cryderman bucket is dumping into the sinking bucket in D. Note that the air cylinders have opened the Cryderman bucket in this view.

When the sinking buckets are hoisted to the surface for dumping, they tend to swing back and forth in the shaft. To overcome this, the buckets slide into a crosshead, which guides them up and down the shaft and prevents swinging.

RIDDELL SHAFT MUCKER

The carriage and hoists of a Riddell shaft mucker are shown in E. Note the carriage wheels, which run on the track positioned so that the whole carriage can move across the long section of the shaft. Because the carriage can move in the shaft, the bucket can dig or dump in almost any position in the shaft. (Plate 6-2C shows a long-section sketch of this machine.)

Steel forms for concrete are shown in F. The bracing on the forms gives them strength so they will stay in alignment while the concrete is being poured into place. The open space within the forms allows the sinking buckets to go past the forms freely, so that the sinking cycle is not delayed. A section of the forms can be removed to decrease the perimeter distance and allow the form to come loose from the hardened concrete so that it can be lowered to a new position. The concrete used underground is high early strength, which means that it takes its initial set within a few hours after placement

A. Supporting frame for a Cryderman shaft-mucking machine. The frame is in the left compartment. **B.** Cryderman machine viewed from below. *(Courtesy Anaconda Co.)* **C.** Cryderman clam bucket picking up a load of broken rock. *(Courtesy Anaconda Co.)* **D.** Cryderman clam dumping into sinking bucket. *(Courtesy Anaconda Co.)* **E.** Riddell shaft mucker showing the traveling carriage. *(Courtesy Anaconda Co.)* **F.** Steel forms hung in shaft ready for concrete. *(Courtesy Anaconda Co.)*

PLATE 6-3 SHAFT-SINKING EQUIPMENT

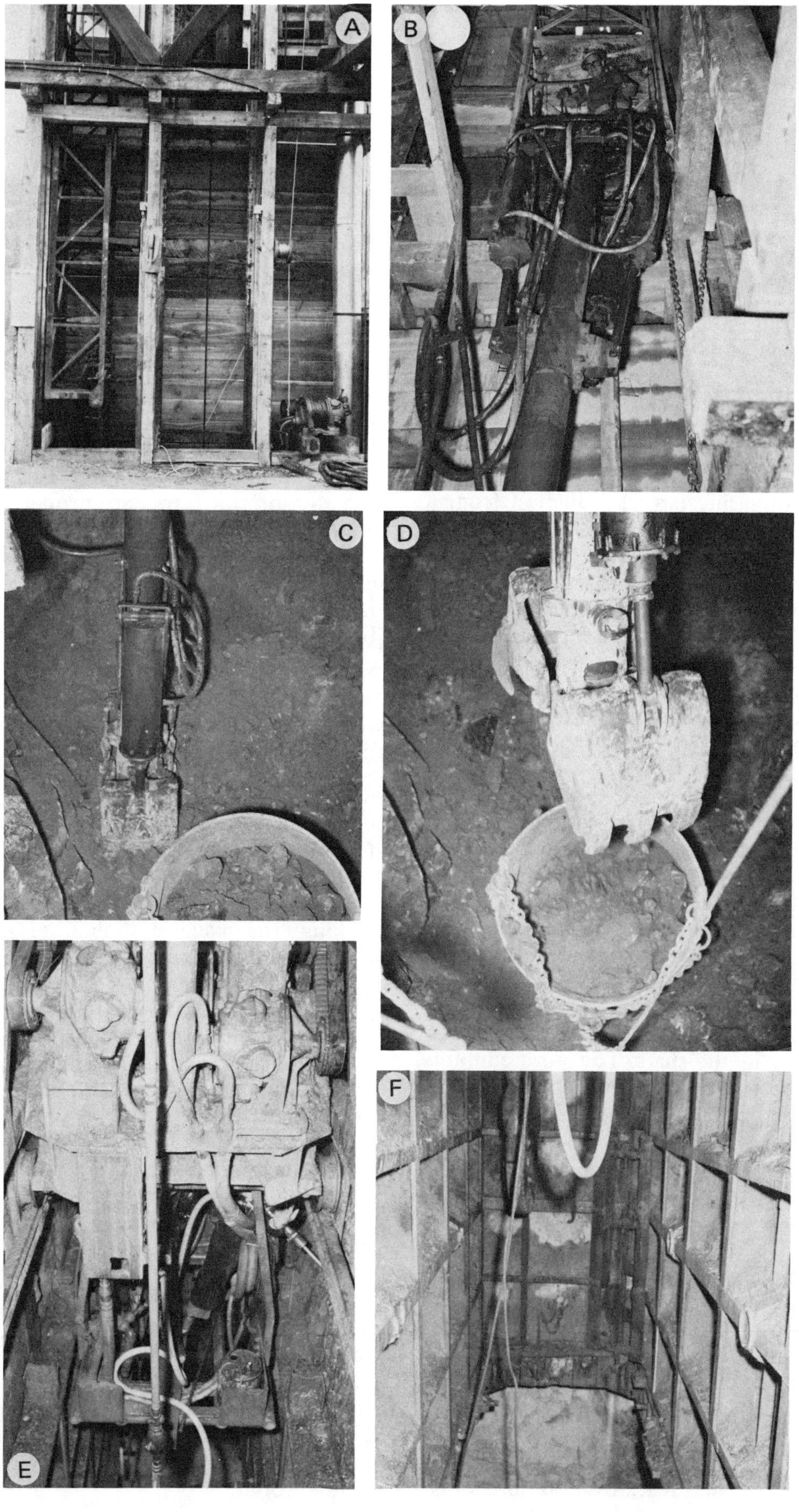

SHAFT-SINKING STAGE AND SHAFT CONFIGURATIONS

Mines contemplating large production from a deep underground deposit usually sink vertical shafts. As a rule, the shafts have big cross sectional areas so large quantities of air can be moved underground, as well as providing a cage with enough area to carry large equipment into and out of the mine. The ore skips are usually large and travel at high speeds in the shaft. To accomplish these requirements, the shafts are often circular in cross section and are up to 25 to 30 ft in diameter.

SINKING STAGE

To facilitate sinking shafts with such large diameters, sinking stages are commonly used. A rather simple 3-deck sinking stage is shown in A. The purpose of the stage is to provide working platforms and space so the shaft sinking operation will continue at a rapid pace and several functions can be accomplished at the same time. The stage is suspended from the headframe by steel ropes attached to one or more hoists. The stage can be lowered or raised in the shaft. The stage contains one or more shaft mucking devices, and a drill jumbo or drill manifold so when needed, the drills are readily available.

The shafts are frequently concreted, therefore traveling concrete forms are also a part of the stage assembly. The stage has holes on each deck or platform which allows the sinking bucket to pass through the stage and sit on the bottom of the shaft, see A,a. In large shafts, these buckets are of large capacity and often the main mine hoist is used to lift the buckets.

CONCRETING

Concreting is part of the shaft cycle and is conducted when the shaft has advanced far enough to insert the forms. After the shaft round has been blasted, the muck is leveled off and the curb ring is lowered into position and carefully placed and pinned to the walls so the shaft will be vertical. Early setting cement is poured into the ring and if everything checks out after this curb ring pour, the rest of the forms are properly placed and the concrete is poured. After pouring the cement, the round is mucked, drilled and blasted, leaving the curb ring and forms in place. Depending on the length of the forms, several rounds may be drilled, blasted and mucked before concreting is again conducted.

These large shafts are often sunk three shifts a day, seven days a week until they are completed. The shaft is usually completed to depth before any dividers are installed in the cross section. It is common practice to hoist the stage some distance from the bottom when blasting to make certain the equipment on the stage will not be damaged by the debris thrown by the blast. After completion of sinking, the stage is then used as a platform to place the steel dividers in the shaft.

SHAFT CONFIGURATIONS

The configurations shown in the circular shafts in A, and B are possible, but none is standard. The cage in B is large enough to lower and hoist most large, heavy equipment without disassembling. The cage and skips are held in position in shafts by guides. These guides can be made of wood or steel fastened to the shaft divider, but rope guides, as shown in C, are becoming more popular.

As shown in D, long rectangular shafts are still being used and may be the most desirable under certain types of ground stresses. Square sections have also been used, as shown in E. Special sinking stages can be constructed for configurations of the type shown in E, but stages are not ususally made for long rectangular shafts, shown in D because of limited room.

PLATE 6-4 SHAFT-SINKING STAGE AND SHAFT CONFIGURATIONS

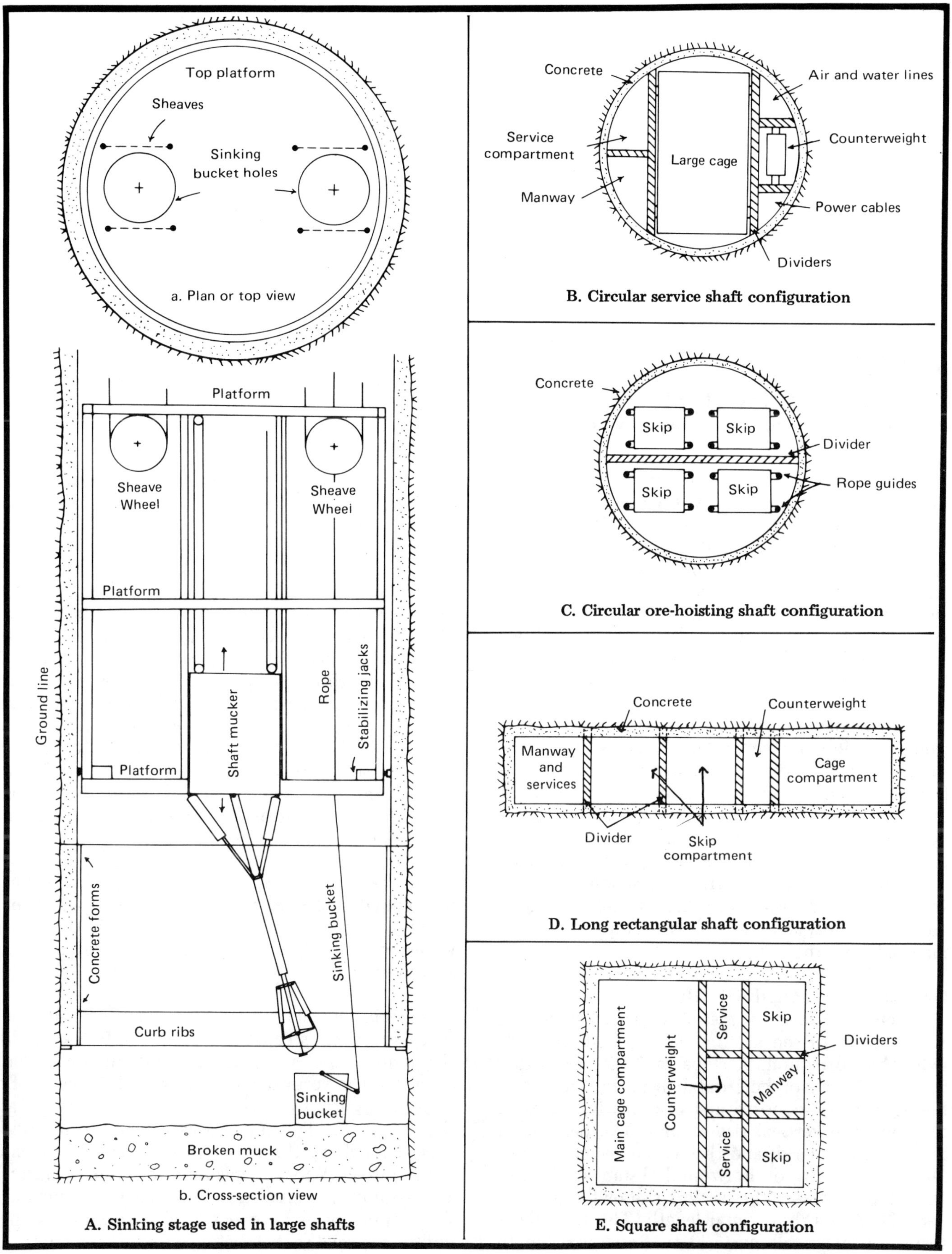

A. Sinking stage used in large shafts

B. Circular service shaft configuration

C. Circular ore-hoisting shaft configuration

D. Long rectangular shaft configuration

E. Square shaft configuration

SINKING INCLINED SHAFTS

Diesel powered load-haul-dump (LHD) equipment has increased the popularity of shallow inclined shafts. These are often referred to as declines. Shallow ore bodies may be developed entirely by declines and mined by diesel powered equipment which makes it unnecessary to sink shafts or invest in a large surface plant. Sometimes conveyor belts are installed in these declines, and ore is hoisted from the mine on belts rather than in skips or trucks.

Often, load-haul-dump units will perform fairly well in declines as steep as 20%, as shown in A. The rubber-tired jumbos can also work satisfactorily at this grade, and B shows a jumbo drilling out the face of the decline. The driving rate in these declines compares favorably with level driving if the haul distance doesn't get too great. Load-haul-dump units start losing efficiency at some distance such as 1,000 ft; therefore, they sometimes dump into larger capacity diesel trucks to carry the material from the mine.

DEGREE OR GRADE OF DIP

Inclined shafts, sometimes called slopes or declines, are driven at some slope, grade or angle from the horizontal. A slope of 10:1 means that for each 10 ft. of horizontal distance, the rise or fall is 1 ft. The horizontal distance is expressed first and the vertical rise second. The percent of grade is often used, and that is the vertical rise or fall per 100 ft. of horizontal distance. Thus, a 10% slope is where the vertical rise or fall is 10 ft. per 100 ft. of horizontal length. The formula is vertical rise or fall per 100 ft. horizontal length multiplied by 100 to put the answer in percent. The degree of grade or dip is defined as follows:

$$\text{Tan}\ \theta = \frac{\text{rise or fall}}{\text{per 100 ft. of length}}$$

where θ is the angle of dip. Using the example above tan θ = .10 or θ 5.7°. A slope of 1:1 is equal to 100% of grade or an angle of 45°.

STEEPER INCLINED SHAFTS

In sinking steeper inclined shafts, one of the problems has been mucking out the face after blasting. Hand mucking can be used but it is slow and costly. Slusher ramps with three drum slushers, as shown in C, have been used with success in inclined shafts up to 25° in dip. Overcast muckers have also been especially fitted with cables attached to reels on the back wheels to work in shafts that steep. One of the problems in mucking in inclined shafts is the limited room available to maneuver equipment when a skip is near the bottom.

Air-leg type drills are often used in drilling in inclined shafts, as shown in D. Small air-driven sump pumps usually provide satisfactory dewatering at the face unless a great deal of water is encountered.

Special machines have been designed and used successfully in steeper inclined shafts. A modification of the air-operated Cryderman mucker, often used in vertical shaft mucking, is shown in E. It is necessary to have a double track shaft, as shown in F. This facilitates lowering the skip close enough to the muck pile so the loader can dump into it. With the ingenuity of mine operators and equipment suppliers, other types of inclined shaft mucking equipment will undoubtedly be built for special shaft jobs.

PLATE 6-5 SINKING INCLINED SHAFTS

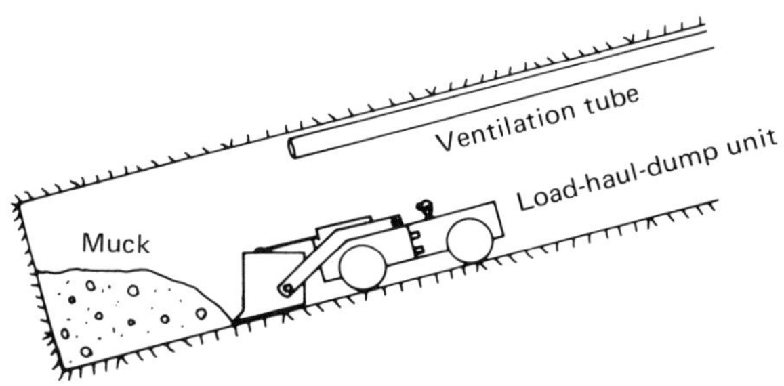

A. Mucking a shallow inclined shaft with an L-H-D unit

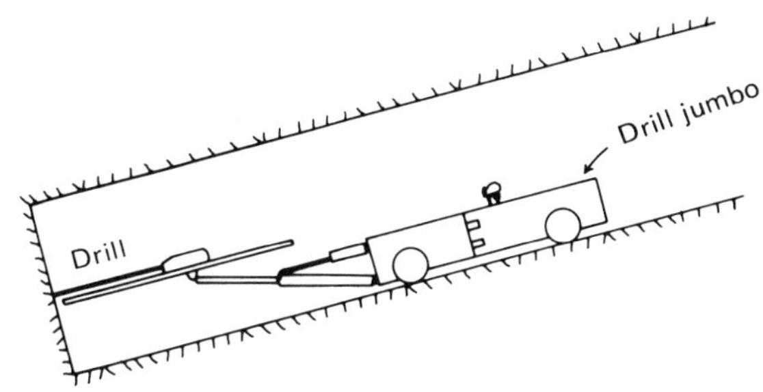

B. Drilling a shallow inclined shaft with a jumbo drill

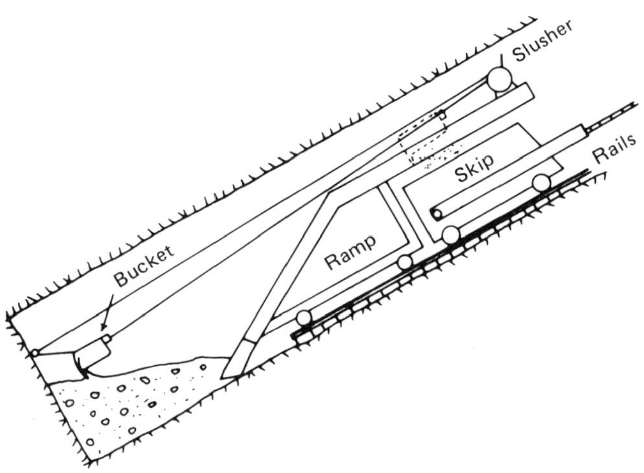

C. Mucking with a slusher ramp

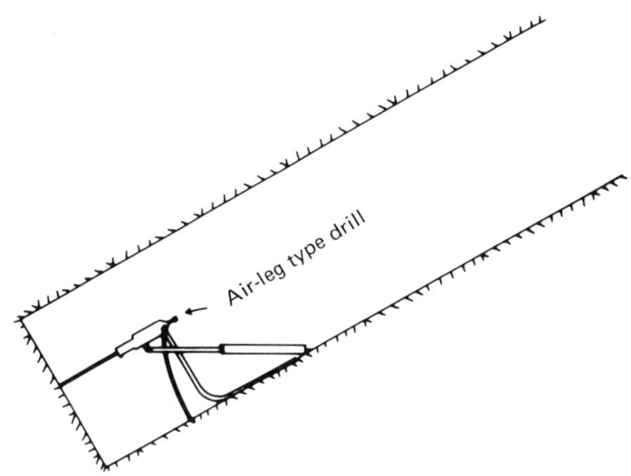

D. Drilling in a shaft with an air-leg drill

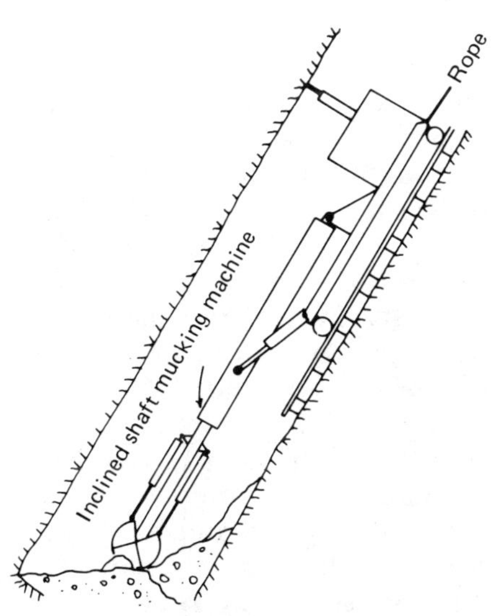

E. Special design clamshell loader for steep inclined shafts

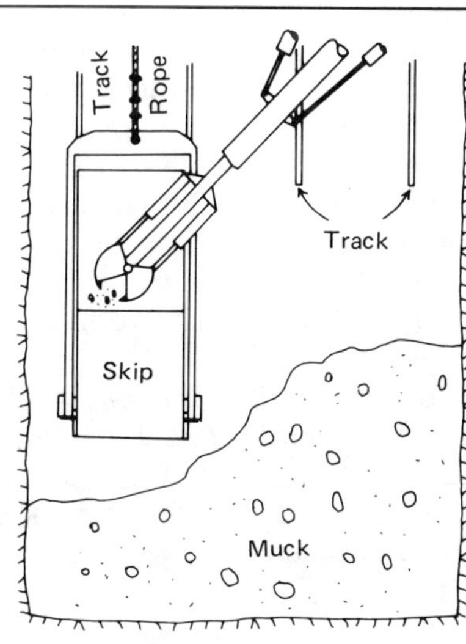

F. Plan of shaft view of machine in E loading a skip

DRIFT CYCLE

In a freshly blasted heading, the first thing to be done is to scale or bar down loose rocks, then water the walls and muck pile to control dust A. In other words, to make the area safe for working. Mucking the blasted mound comes next; B shows the mucking machine filling the car. After all of the muck is removed, the heading is rock bolted C, and D or timbered. If the ground needs no support, then drilling of the round is next. An air-leg drill is shown being used for drilling E; jumbo drills may be used in this operation. A burn-cut round for breaking the rocks is shown F, and G. Loading the round comes next. The first stick of explosive that goes into the hole should be the detonator stick, which contains the cap and fuse. Putting one or two sticks of powder into the hole ahead of the detonator stick is regarded as poor practice. The rest of the hole is then filled with explosive.

AN/FO

Ammonium nitrate, like that used for fertilizer, if treated with ordinary diesel fuel can be used as an explosive with great success, both in open-pit and underground mines. The nitrate is in bulk form and can either be poured into a vertical hole or blown into a horizontal, or up, hole with a compressed air nozzle type of loader.

FUSE LENGTH

Fuses should originally all be of the same length. Fuse burns at a fairly constant rate, so the timing of the blasting sequence of the holes is set by the length of the fuse. After the round is loaded, the fuse to the hole that is to be shot first should be shortened most. The fuse for the next hole has a smaller length of fuse cut from it, and so on until the round is timed. Assume all fuses are originally 12 feet in length and that the fuse burns at a rate of 1 foot in 30 seconds. If a 4-foot length were cut off the 12-foot fuse, leaving 8 feet, it would take 4 minutes after lighting until the explosion. If the next fuse is shortened by 3½ feet, it would take 4 minutes 15 seconds for the next hole, and so on J, and K. There are now timing devices on the market which delay the fuses in the proper sequence. They are easy to use, and reduce the chance of error in timing the round.

PLATE 6-6 DRIFT CYCLE

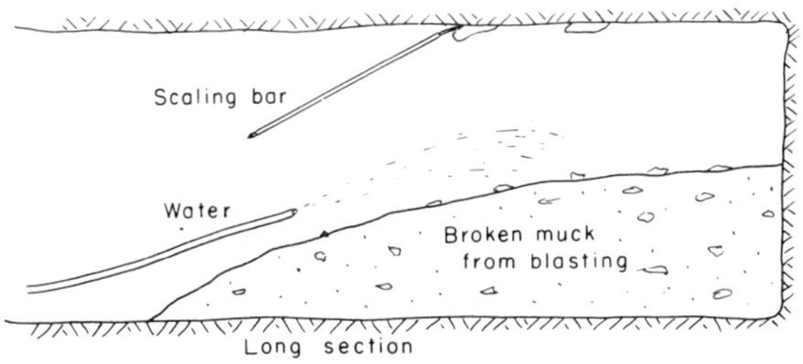

A. Entering blasted drift (Scaling or barring loose rock and wetting of muck pile required for safety operations)

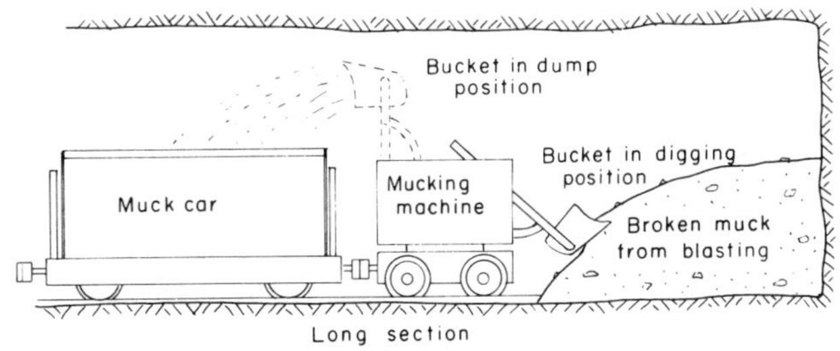

B. Mucking broken rock into a mine car with an overcast mechanical loader

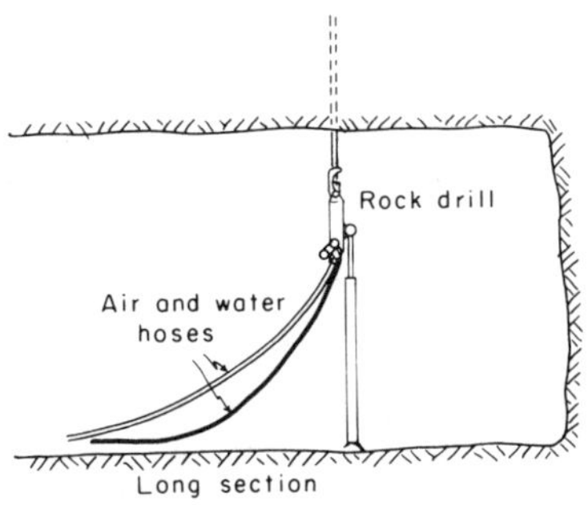

C. Drilling holes in back of drift for installation of rock bolts

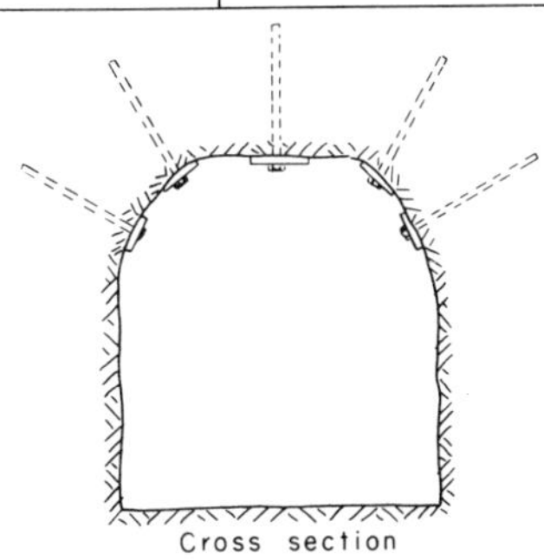

D. Rock bolts installed in section of drift

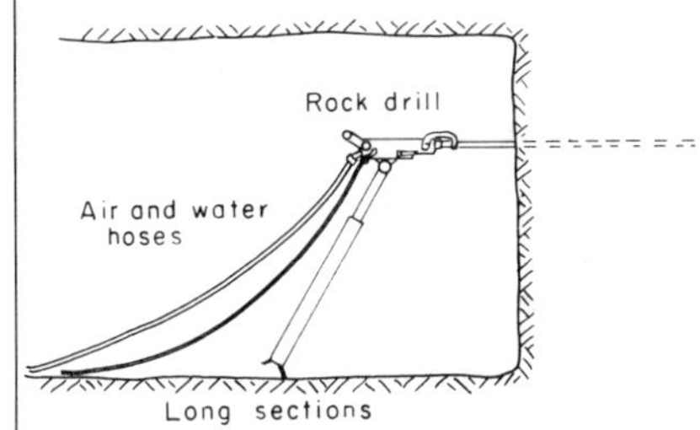

E. Drill set up to drill holes, preparatory to blasting

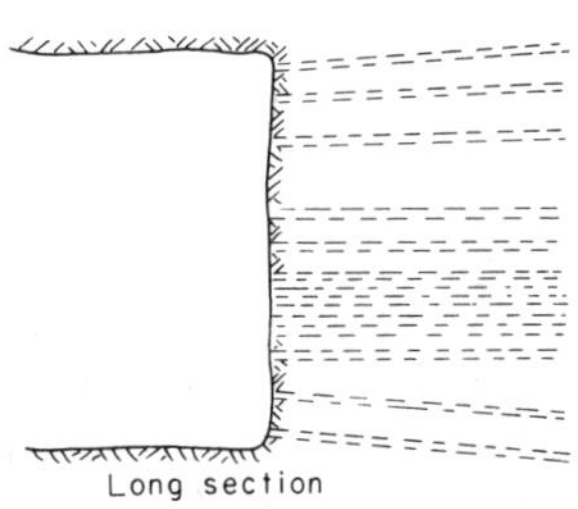

F. Pattern of drill holes in a drift heading

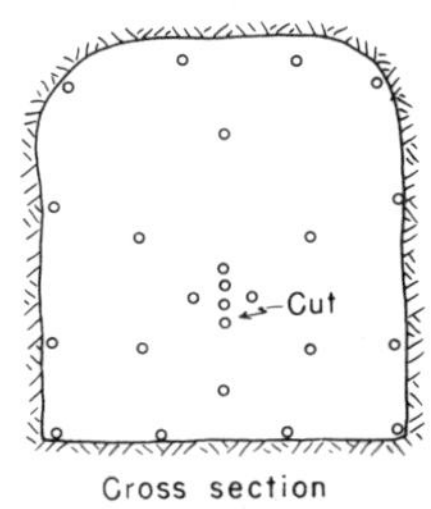

G. Front view of drill holes in a drift heading (burn cut round)

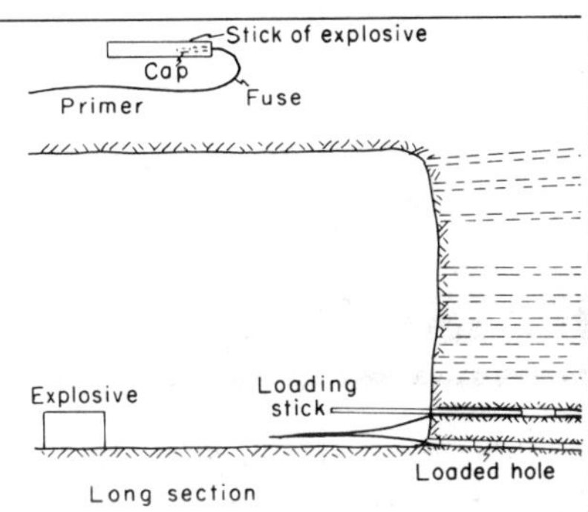

H. Loading the drill round with explosives

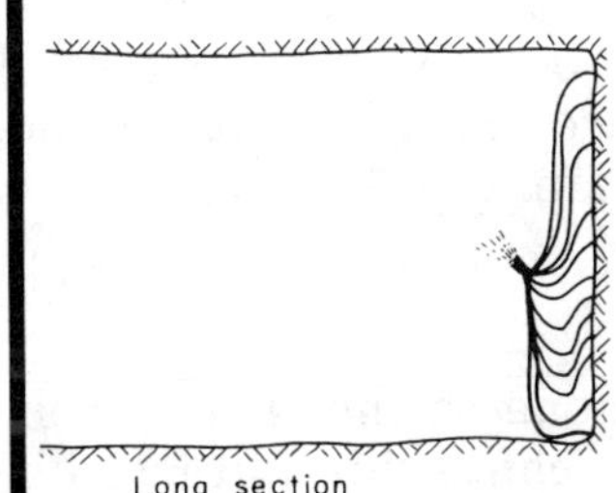

I. Lighting the fuses (Fuse and caps are used here instead of electric detonators)

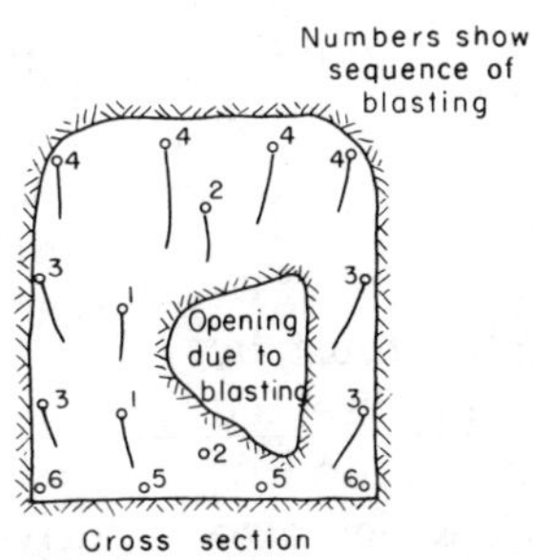

J. Front view of round after the cut and two relievers have been detonated

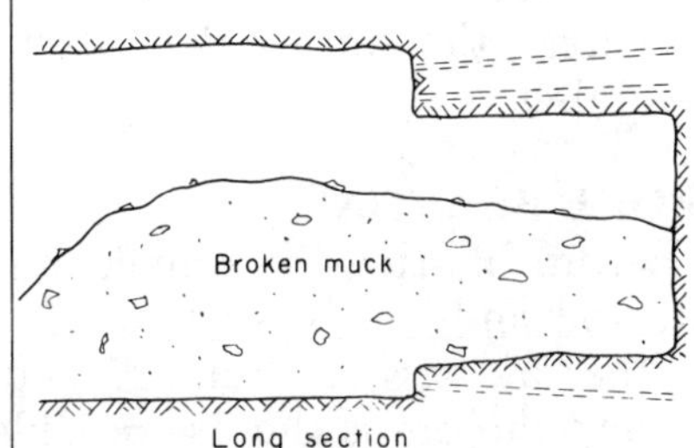

K. View of round after all holes are detonated except the back holes and lifters

RAISE CYCLE

The top of a 6-post timbered raise, as prepared for the drilling sequence, is shown in A. Only three posts can be seen in this long-section sketch, but there are three more posts behind the three shown, hence the term 6-post raise. A stoper drill is commonly used for drilling in a raise. In timbered raises, the V-cut type of round is popular B, whereas the burn cut is often used in untimbered raises.

BEFORE BLASTING

Before blasting the round, the timber must be protected and the blasted rocks must be directed into the ore pass or chute. A common method of directing the ore is the use of slides C, which also keep broken rock from dropping down the manway. The timbers that will be directly exposed to the blast should be covered with lagging. The round is then loaded and blasted.

AFTER BLASTING

After the blast, it is necessary to clean up loose broken rock that may have fallen into the manway and to water the whole area to reduce dust. The loose broken rock on the back and sides must be scaled down, usually starting from one corner, as shown in D.

After the raise has been made safe, the manway slide is removed, a floor is built on top of the timber set, and the timber for the next set is hoisted and set in place E, then aligned and blocked tightly in position F. The cycle is then ready to start again.

Jobs that must be done as the raise is being driven include raising the chute side and installing ladders, lengths of ventilation hose or bag, and air and water pipes. These necessary miscellaneous jobs are performed whenever the miners can get them done during the cycle.

PLATE 6-7 RAISE CYCLE

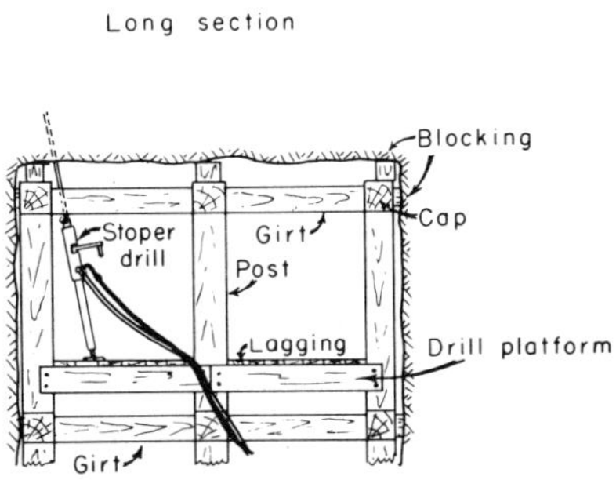

A. View of a 6-post timber raise (being drilled with a stoper-type drill)

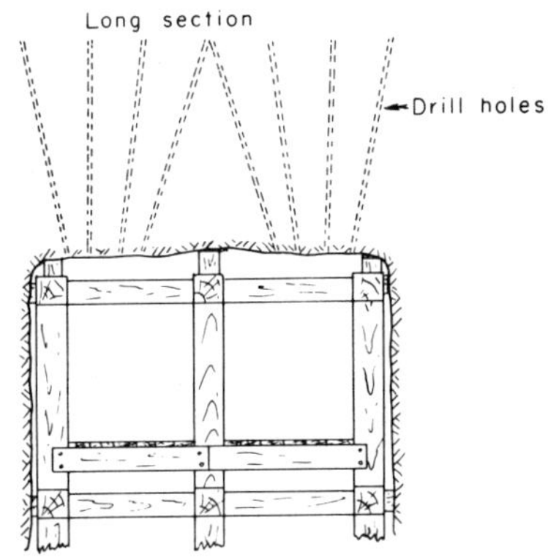

B. Long-section of a V-cut type of raise round

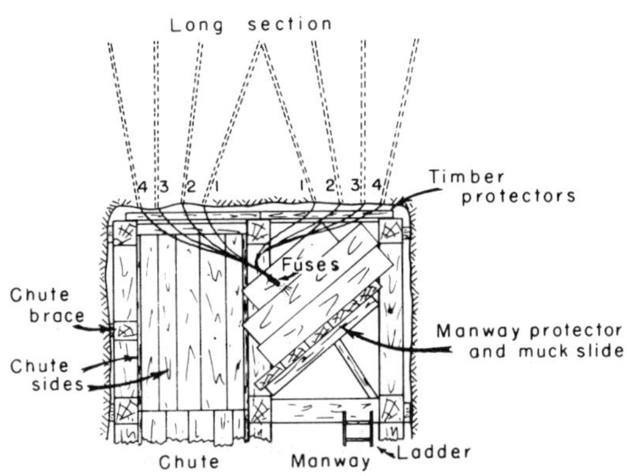

C. Raise prepared for blasting; chutes, slides, and timber protectors are in place

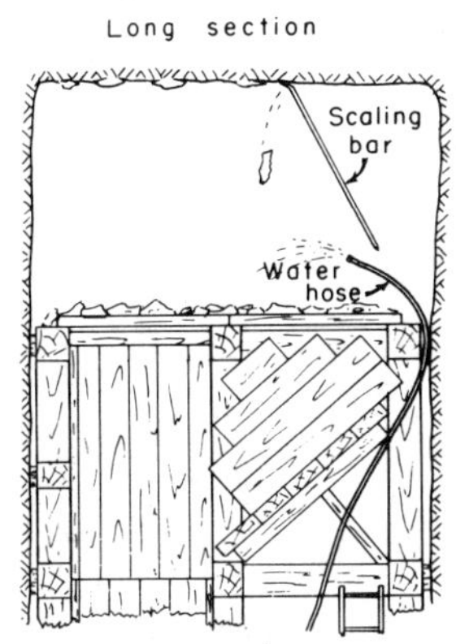

D. Raise after blasting; scaling or barring down, and wetting entire area

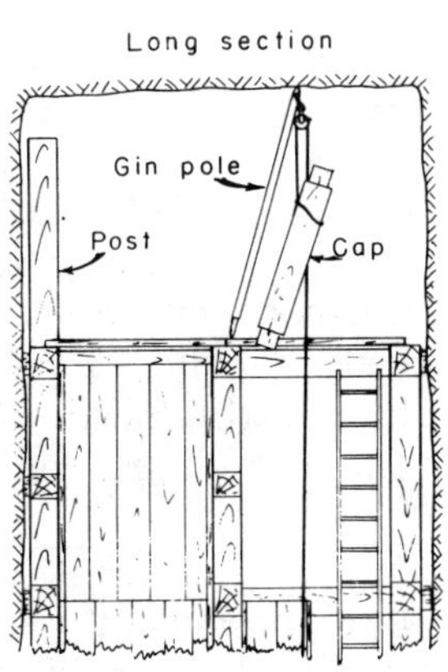

E. Hoisting timber in scaled-down raise

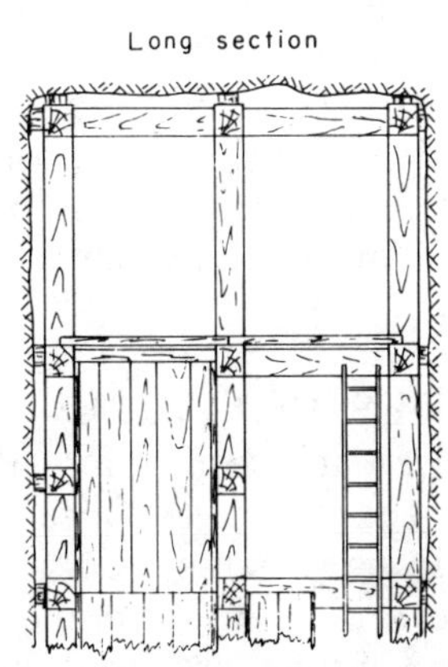

F. Next floor of timber in place in raise (cycle is ready to start)

TYPES OF RAISES

The raise illustrated in Plate 6-7 is a 6-post timbered raise containing one ore-pass or chute compartment and one manway compartment. Another popular style of timbered raise is the 8-post raise Plate 6-8A, which has two ore-pass or chute compartments and one manway compartment. The advantage of an 8-post raise is that stopes can be mined from both sides of the raise at the same time more conveniently than with a 6-post raise.

TIMBERED RAISES AND STULLS

Timbered raises are used where the ground is heavy or where the timbered raise is an integral part of the mining system. Untimbered raises are driven, however, and the sketches in Plate 6-8 show some types. Stulls are used in many one-compartment raises B. The stulls give support to the back and also provide a base for a drilling platform. If the ground does not need support, then stulls may be horizontal C as steps and drill platform. The perpendicular stull system B may be used in steeply dipping raises, but the horizontal system C is used only in moderately inclined raises.

In hard ground, stulls may be broken or knocked out by the blast. Two systems are designed to overcome this. The pin raise D requires steel pins about 14 inches long. A short sinker type of drill is used to drill holes in the footwall so the pins can be installed. Mud from drilling is usually enough to cement the pins in place firmly. They provide means of support when climbing the raise. Alternatively, anchor pins are inserted in holes in the side of the raise E to support the drill platform and the chain, rope, and chain ladder.

A recent development in raise driving is the mechanical raise-climbing machine, which runs on a track fastened to the sides of the raise F. It includes a drilling platform and a means of hauling supplies up the raise.

PLATE 6-8 TYPES OF RAISES

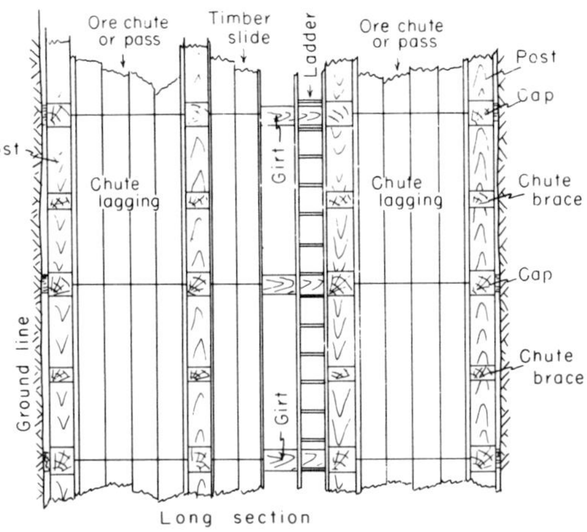

A. **Eight-post raise with ore passes or chutes on each side of the manway**

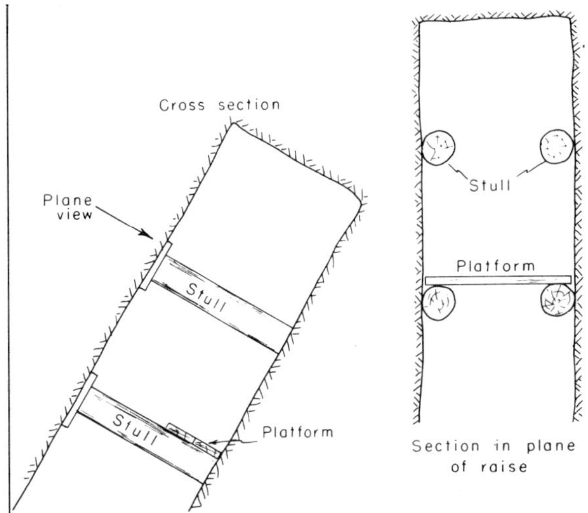

B. **Stull type of raise, single compartment with stulls perpendicular**

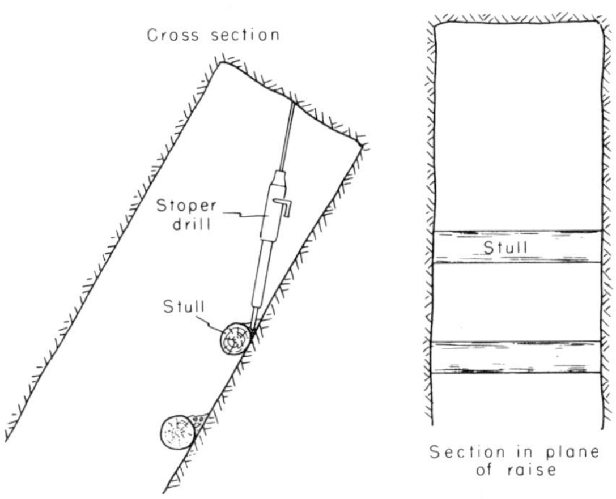

C. **Stull type of raise, single compartment with stulls horizontal**

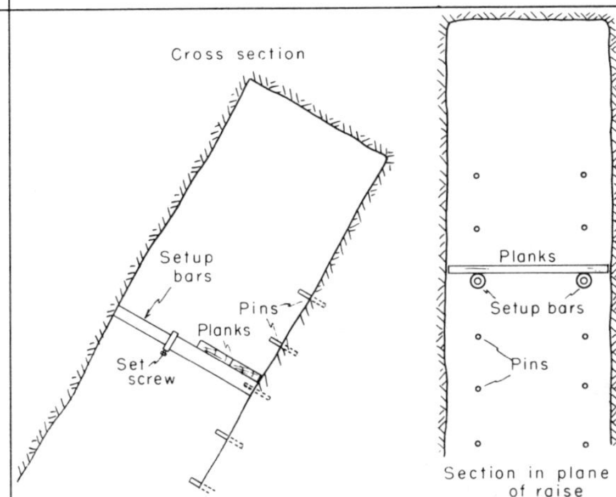

D. **Pin type of raise, single compartment showing setup bars and drill planks**

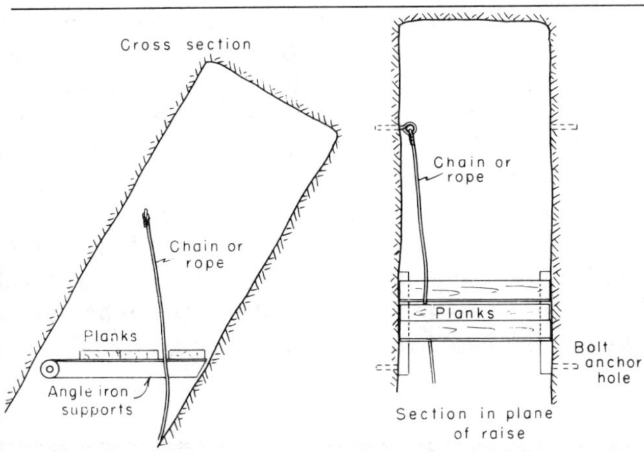

E. **Single-compartment raise with chain, rope, or chain ladder for access**

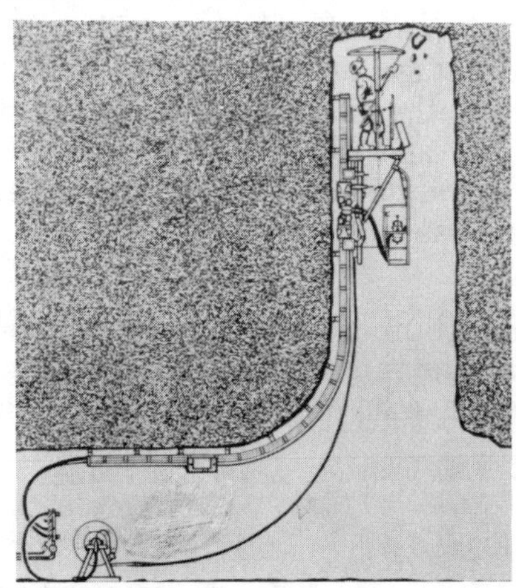

F. **Sketch of raise-climbing machine (Courtesy Anaconda Company)**

RAISES AND RAISE MACHINES

Pin raises were sketched on Plate 6-8. The major disadvantage of such a raise is that everything including the machine must be carried up the raise. A block and cable may be carried up for hoisting the equipment, but this practice is generally regarded as impractical.

The raise climber was also sketched on Plate 6-8. The machine runs on a track, which is securely fastened to the wall of the raise by rock bolts A. On this track is a row of teeth, which meshes with the gear on the raise climber. Turning the gear directs the climber up or down the raise. The climber is driven by compressed air. The miners can ride up and down the raise in the man carrier just below the platform B, and C. When the climber gets to the top of the raise, the miners climb onto the deck, scale down the loose rock, and are ready to drill D. When drilling is finished, another section of track is installed and the climber is run down the raise so the explosives can be brought up and the round loaded. The track sections contain the air and water lines, and a ventilation tube can be installed behind the track, which protects it from the blast. This machine has greatly increased the rate of raise driving.

A. Track and pipe assembly on which a raise climber runs. *(Courtesy Anaconda Co.)* **B.** Raise climber starting to climb the raise *(Courtesy Anaconda Co.)* **C.** Raise climber viewed from the side; the drilling platform is at top. *(Courtesy Anaconda Co.)* **D.** A drill on the raise-climber platform. *(Courtesy Anaconda Co.)*

PLATE 6-9 RAISES AND RAISE MACHINES

DRILLING OR BORING DEVELOPMENT HEADINGS

Drilling or boring development headings has increased greatly in the past decade. Driving raises has often been a problem in many mines using the conventional methods described in Plates 6-7, and 6-8. Ventilation, supplies, safety of personnel, and cost are problems in conventional raises. Hard rock can now be cut with various types of rotary bits so drilling or boring large diameter openings has become practical.

RAISE BORING MACHINES

A raise boring machine is shown in A. A hole, large enough in diameter to hold the drill rod shown in B, and D is drilled to the level below. The drill rod or shaft is extended through the hole to the level below where the large diameter bit is fastened on, as shown in B, and D. The machine in A lifts the bit and creates a high force or thrust of the bit against the face while rotating the bit. The cutting action is seen in C.

The advantages of drilled or bored raises are that they can usually be drilled faster, they are cost competitive in many cases, and personnel are not exposed to the hazards of raise driving. The finished section is often smooth which offers less resistance to the flow of ventilation air and the drilled raise may not require support. Often, a large cross section raise is not needed, and a small diameter raise is easier to drill than it would be to drive by conventional methods.

SHAFT BORING

Boring shafts has also gained in popularity in recent years. The bits are similar to the ones shown in B, C, and D, but it is necessary to have a center section. The bit is rotated by a machine similar to the one in A. However, the weight of the bit often provides enough drilling force, and the cuttings are carried away by circulating heavy mud. The mud is pumped down the drill rod to the cutters, and the cuttings are carried up the hole by the ascending mud. As experience is gained in shaft drilling, larger diameter shafts are now being bored. Some shafts similar to raises are drilled; that is, they have a hole drilled to the level below. In this case, the bit is designed to discharge the cuttings to the center hole and mud is not necessary to carry the cuttings away.

TUNNEL BORER

The machine shown in E is a tunnel borer, and it is equipped with two rotating cutting heads, only one of which can be seen clearly in E. The two cutters mesh so that they do not interfere with one another, and a cutting chain operating between the two cutters trims off the top and bottom of the bore which is left between the two rotating cutting heads. These heads are not equipped with roller bits as are the raise borers, but the heads have cutting teeth that gouge out the rock. These machines work well in coal and softer rock such as salt, trona or potash. As of this date, they have not performed well in harder rocks or ore.

A. Raise boring machine set on the level above the raise. *(Courtesy Ingersoll-Rand Co.)* **B.** Large diameter raise boring bit. *(Courtesy Teton Drilling Co.)* **C.** Raise boring bit pulled away from face. *(Courtesy Ingersoll Rand Co.)* **D.** Raise boring bit cutting. *(Courtesy Ingersoll-Rand Co.)* **E.** Large tunnel borer. *(Courtesy FMC Corp-Trona Mine)*

A

B

C

D

E

DEVELOPMENT OPERATIONS

Pictures of drilling in raises are few, not only because of the difficulty of getting the camera in a suitable position, but also because the miners working in a raise are often paid an incentive bonus and they do not want to waste time posing for pictures.

STOPER-TYPE DRILL IN RAISE

Plate 6-11 A shows the operation of a stoper-type drill in a raise. Water is used in drilling to control dust, and a spray of water can be seen issuing from the hole where the steel is rotating. The compressed air exhaust often forms a fog behind the machine.

AIR-LEG MACHINE IN DRIFT HEADING

Plate 6-11 B shows an air-leg type machine in a drift heading. The machine is pivoted on the leg so that the angle between the leg and the machine can be adjusted. The dark streak below the hole is caused by water running down the side of the drift. The large hose carries compressed air to the machine, and the smaller hose carries water.

A cutting machine is sometimes used in softer rock mines such as a coal mine to cut a slot or kerf near the floor or near the roof so that explosive charges have a face to break toward. This cut usually stops the action of the explosive, and a smooth floor or roof results. This often helps in maintaining the integrity of the opening. In other mines, the cut is placed vertically in the center of the opening to act as a free face for the explosive charges in the round. Cutting a vertical slot is shown in C. A close up view of the cutter in action is shown in D. Note the roof bolts supporting the roof in both C, and F.

A load-haul-dump unit (LHD) is shown in E mucking out an adit heading. As the load is carried out and dumped, the bucket is lowered as the machine enters the heading to clean up spilled and blasted thrown rock. The muck gathered is then dumped in the heading face with the remaining muck as shown in E. The machine can then get a full load of muck from the pile in the face. A very clean heading results when the mucking operation is finished.

The (LHD) unit is shown dumping in F. Usually the muck is dumped on a grizzly or into a breaker to allow the large chunks to be broken before passing onto the next phase of transport in the mine.

A. Drilling in a raise with a stoper drill. *(Courtesy Gardner-Denver Co.)* **B.** Drilling in a drift with an airleg pusher drill. *(Courtesy Gardner Denver Co.)* **C.** Cutting machine in operation. *(Courtesy FMC Corp Trona Mines)* **D.** Close up of cutter bar. *(Courtesy FMC Corp Trona Mines)* **E.** Mucking with an LHD unit. *(Courtesy Wagner Mining Equipment Co.)* **F.** Dumping a LHD unit *(Courtesy Wagner Mining Equipment Co.)*

PLATE 6-11 DEVELOPMENT OPERATIONS

PART 6 CONCLUDING REMARKS

The development process in a mine is a time consuming task. In a new underground mine, it is necessary to sink the shaft and drive the levels before any actual mining can take place.

In a sense, the development of a mine is similar to building a factory, i.e. the buildings and machinery must be installed before the factory can start to produce. However, in a factory, the building process can often be speeded up by using more personnel and by conducting several operations at the same time.

In a mine, however, levels cannot commence until the shaft reaches that elevation. Usually in high-speed shaft sinking operations, the shaft is completed before any levels are driven. The level stations are cut in the shaft sinking process, but because shaft sinking requires use of all of the shaft equipment, level driving is delayed until the completion of the shaft.

A shaft progresses just so fast, and only so many people can work there, so adding workers is not the answer. There is no room or equipment for them to work. The progress of the shaft determines when the next phase of the operation can begin. After the shaft is finished, it still takes considerable time to drive levels, raises, and other development openings.

Understandably an underground mine can't suddenly start to produce ore. As previously explained, the necessary development work must be done before any ore can be mined. To develop a large underground mine often takes several years after discovery before it starts to produce ore. An open-pit mine, especially if the ore outcrops on the surface, can be put into production much sooner than an underground mine.

The mining industry is continually seeking new ways to speed up the development process. Even though the cost per foot of development work may be higher for some of the newer and faster methods, the fact that the work can be done faster may result in an overall cost savings. The higher production rate usually results in lower overhead and investment costs assigned to the development work.

PART 7

SURFACE MINING

The lure of placer gold brought thousands of miners into the unexplored areas of the world in the late 1800's. Although the easily obtained placer gold has long since been removed, the soaring price of gold, increasing 10 or more times in the past decade, is again luring prospectors to placer prospects. Gold panning is becoming a popular sport again with participants hoping to find prize gold nuggets to add to their jewelry collections. Although very strict reclamation laws are now being enforced on placer mining it still is being done. The usual placer miner methods are discussed in this section.

If an ore body is large enough and occurs close enough to the surface, open-pitting is one technique for mining. The rapidly increasing cost, the improvements in milling and processing in handling lower grade ores, and the length of time involved in putting an underground mine into production are all factors favoring surface mining. However, not all ore bodies are amenable to surface mining. Frequently it is necessary to switch from surface to underground mining because of the increased costs of surface mining as the pit gets deeper. The big factor in cost in an open-pit is overburden or barren rock removal. When the ratio of waste-to-ore, called stripping ratio, gets too large, underground methods may be required to recover the rest of the ore. Open-pit methods are described in this section.

The Western part of the United States has large reserves of coal lying close to the surface which makes open-cut or strip mining feasible. Large equipment is used in open-cut coal operations thereby lowering the cost of producing a ton of coal. Open-cut or strip coal mining disturbs the ground overlying the coal seam. Land reclamation laws are being enforced requiring that the disturbed land be redeposited to permit revegetation of the land. Open-cut or strip coal mining is briefly discussed in this part, while reclamation practices are covered in more detail in Part 9.

PLACER MINING METHODS

The gold pan is seldom used in production placer mining, because it has very small capacity, but it is used extensively in testing potential placer ground. The pan is filled with gravel A and immersed in water, the clay broken by hand, and the large rocks removed. The pan is then rotated under water to settle the gold and heavy minerals to the bottom, then tilted and gyrated so that the lighter material is washed away. The pan should be jarred occasionally to settle the gold. As the quantity of material decreases, the washing rate must be reduced to avoid washing away the gold.

SLUICING AND ROCKING

A sluice box B will wash much more gravel than a pan, but an ample supply of water must be available for sluicing. If water is scarce, a rocker may be used C. The washing action is provided by rocking the rocker back and forth. The coarse gravel is caught in the screen and periodically dumped, so that only fine material goes to the riffles. Although the rocker has less capacity than a sluice, it is faster than a pan.

HYDRAULIC GIANT OR GROUND-SLUICE METHOD

If water is plentiful and if disposal of tailings is not a problem, the hydraulic giant or ground-sluice method may be used. The hydraulic giant is a pivotable water nozzle attached to the end of a large pipe. If the water comes from an elevation much higher than the nozzle, water issues from the nozzle with tremendous force and will wash boulders and gravel through the sluice box D. In the ground-sluice method, the force of a stream of water directed into the cut washes the gravel into the sluice E. In ground sluicing, the water upstream is dammed and a large volume of water is released at one time. The ground that the water is to wash may be loosened by blasting. Mechanical washing plants are also popular F. The gravel is fed into the hopper by a power shovel, dragline, scraper, or other device. The operation of the plant is described in the next section. Abundant water is essential, and many operators catch the used water in a pond downstream and pump it back to the plant.

PLATE 7-1 PLACER MINING METHODS

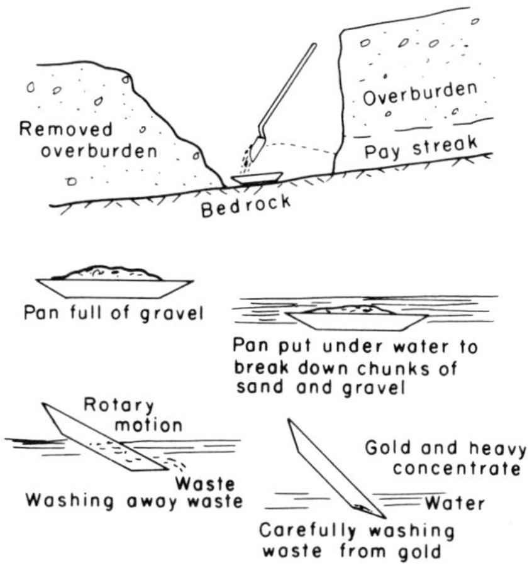

A. Steps in gold panning

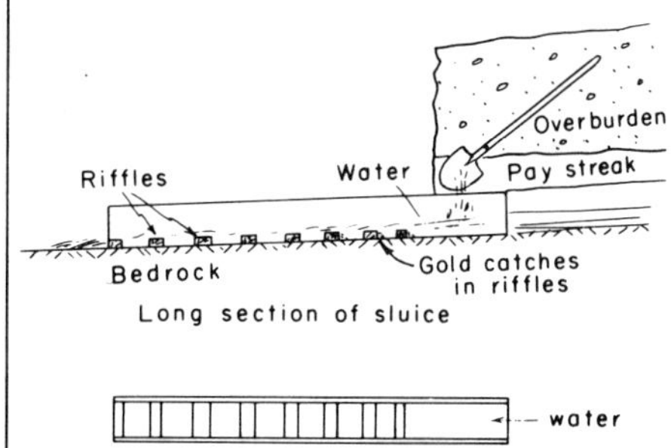

B. Washing gravel in a sluice box filled by hand shoveling

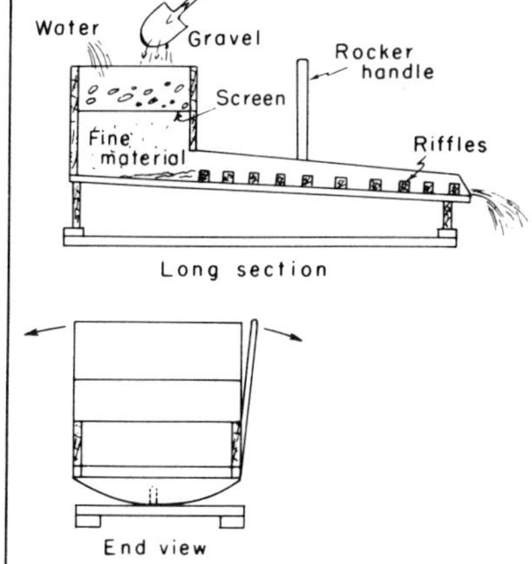

C. Washing gravel in rocker

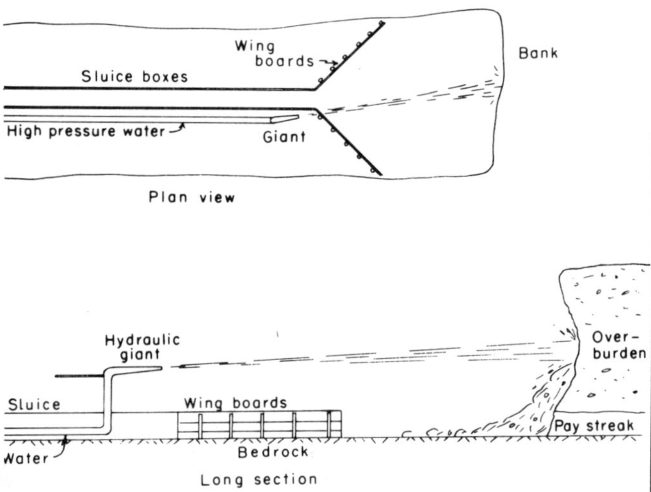

D. Views of placer mining by hydraulic giant

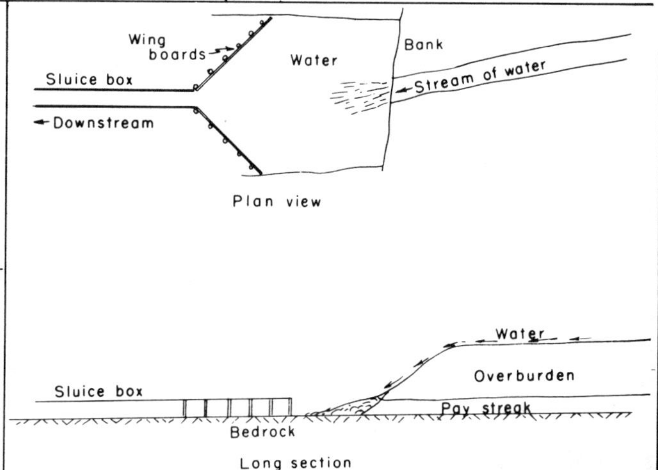

E. Views of placer mining by the ground-sluice method

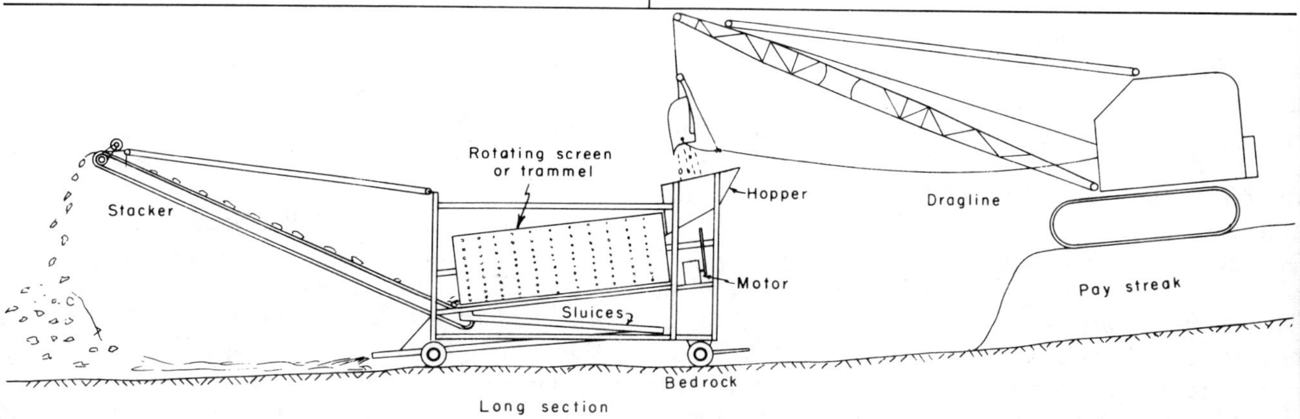

F. Sketch of a dry-land dredge or washing plant. A dragline or power shovel fills the hopper with gravel. The rotating trommel separates the coarse and fine gravel, only the fine gravel is sluiced. Water must be brought to the plant

DREDGING

A dredge floats in water and digs the gravel by an endless string of buckets A, and B. The gravel is deposited in a hopper that feeds it to the washing plant, which is similar to that of a dry-land dredge Plate 7-1F, Plate 7-3 C, D, G, and H. The material from the hopper flows into a rotating screen. The hole size in the screen ranges from 1/4 inch to 5/8 inch. Water is sprayed into the screen to break up clay and wash large rocks. The coarse material from the screen falls on a conveyor belt called a stacker, which transports it away from the dredge. The fine material passes through the screen and into a series of sluices equipped with riffles where the gold is recovered. The fine material coming out of the sluices is deposited at the back of the boat and forms a dam to retain water in the pond. The large quantities of water needed are pumped from the pond that contains the dredge. Although the water may be dirty, the dirt has no effect on gold recovery.

TRAP FOR FINE GOLD

On some dredges, other devices in addition to sluices are used to recover gold or other valuable material. A copper plate covered with mercury makes a good trap for fine gold. Jigs, which are concentrating devices used in many mills, may be used to recover gold and other heavy minerals.

ANCHORING FLOATING DREDGE

To make a floating dredge dig effectively, it must be held in position. This is done by cables attached to winches on the boat and to anchors on the bank D, and E. By proper placement of the anchors and proper tension on the cables, the boat can be held at any place in the pond. To aid in digging into the corners of the pond, pivots or spuds are driven into the bottom of the pond so that the boat can rotate about them as an anchor C. The series of sketches C, D, and E shows how the boat moves across the pond.

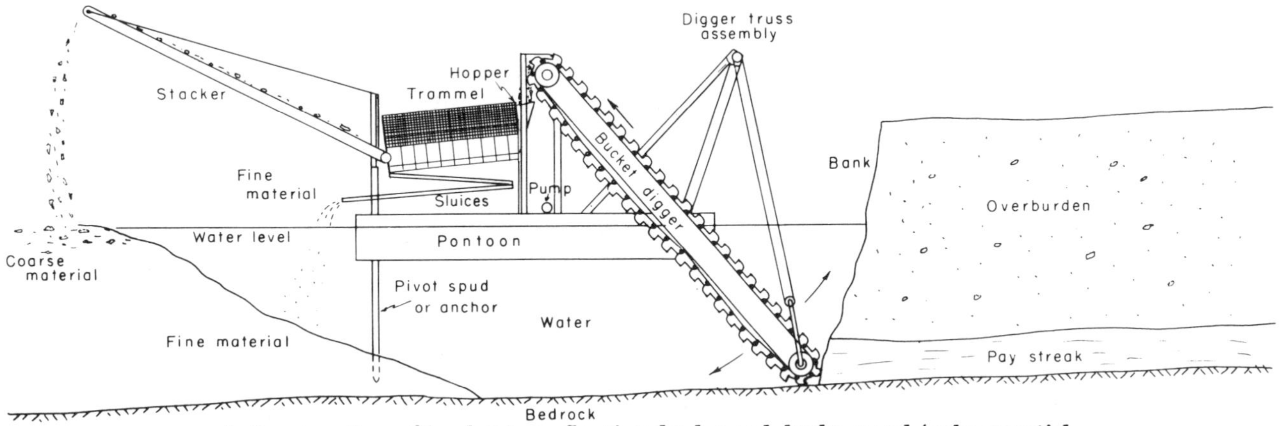

A. Long section of bucket-type floating dredge and dredge pond (only essential features of digging, washing, and stacking parts of dredge are shown)

B. Floating dredge

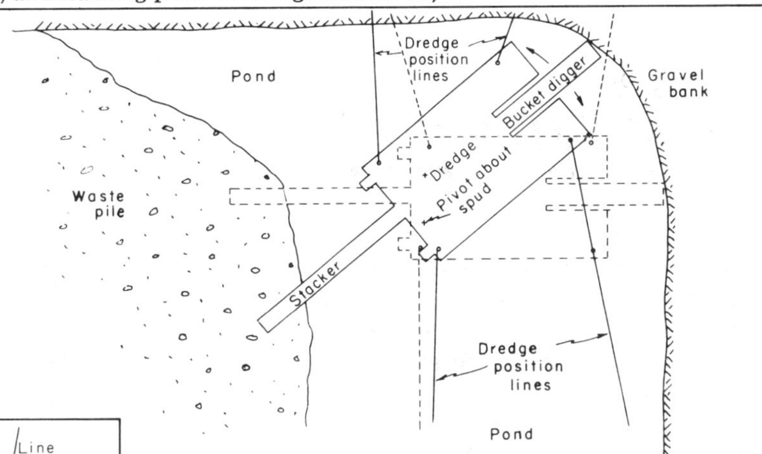

C. Plan view of dredge digging against corner of pond (solid and dotted positions show how dredge swings about pivot)

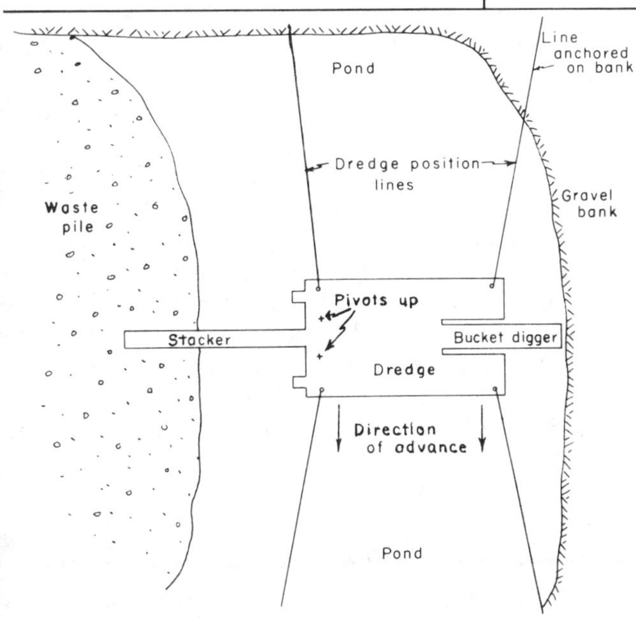

D. Plan view of dredge traversing or moving across the pond

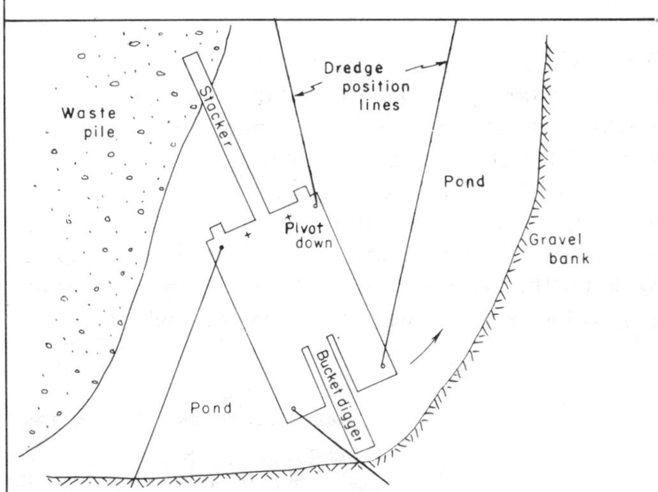

E. Plan view of dredge at other corner of dredge pond

PLACER MINING EQUIPMENT

The ordinary hand placer equipment is shown in A, and B. An assembled rocker A, upper, includes the screen, right, and the riffles, left. The sluice box A, middle, is actually a model of a box showing many different riffle arrangements that can be used. Almost every prospector chooses the particular type riffle personally felt to be best for gold recovery. A gold pan A, lower right, is shown with a prospector's pick B for size comparison. The pick is about the same length as a hammer.

DRY-LAND WASHING PLANT

A dry-land washing plant C includes the hopper at the left, the rotary screen or trommel to the right of the hopper, and the sluice underneath the trommel. The back view of the plant D shows the rock stacker. To provide water for the plant, a dam is installed downstream E, and water is pumped back to the washing plant. Since this mine was in a wooded area, it was necessary to clear off the brush by bulldozer F before the dragline could strip off the overburden lying above the pay streak, which is near bedrock. The unproductive overburden can be stripped off and piled to one side, whereas a dredge, Plate 7-2, must wash everything, including the overburden, unless it is stripped away by other types of equipment.

One of the problems with an ordinary gold washing plant and dredge is that the fine material collects in the bottom of the cut, and the coarse rocks are piled on top. In fact, in a floating dredge, this is necessary because the fine material holds the water in the dredge pond. The large coarse rock pile is not conducive to vegetation growth. The gold washing plant shown in G, and H has a sand stacking wheel attached. The wheel has a series of buckets that catch the sand as it emerges from the sluices and carries it above the level of the stacker belt. The coarse rocks and finer material are then blended together on the stacker pile. When leveled by a bulldozer, this type of soil will support soil growth much better than will bare boulders. The stripped overburden dragline which loads the washing plant and the plant itself are shown in H.

A. Rocker, in background, sluice box with various styles of riffles, in middle, and gold pan, in front. **B.** Gold pan partly filled with gravel. **C.** Side view of a dry-land dredge or gold-washing outfit. Loading hopper on belt, rotating screen in middle, and sluice on lower right. **D.** The same washing plant viewed from the rear. **E.** Dump and pond that supply water to the washing plant. **F.** Clearing brush and stripping overburden preparatory to mining. **G.** Gold washing plant with a sand stacking wheel attached. **H.** View of placer operation.

PLATE 7-3 PLACER MINING EQUIPMENT

OPEN-PIT MINING SEQUENCE

Before an ore body A can be mined, it is necessary to expose the ore by stripping off the overburden B. As this is waste material, it is dumped far enough from the mining area so that it will not interfere with future operations. A system of benches is used, B through F. Bench height ranges from 25 to 60 feet, depending on the depth that holes can be drilled and the convenient bank height that a power shovel or other loading device can work. The width of the bench depends on the size of the equipment used to haul the ore and on the room needed to load the equipment with the shovel. Waste stripping continues after the ore production has started C.

STRIPPING RATIO

In order that the sides do not become too steep, it is necessary to continue to strip back the waste as the pit is deepened and ore is produced D. Mining and stripping continue at the same time E. In an open-pit operation a time finally comes when the economic limit is reached; for each foot of depth several thousand cubic yards of waste must be stripped and carried away F. The usual stripping ratio for waste to ore for most copper and iron mines is about 2 to 1; i. e., two tons of waste can be removed for each ton of ore removed, but this ratio depends on the value of the ore.

If the walls could be steepened, more ore could be removed. Judging the maximum safe steepness of open-pit walls, sides, or slopes is one of the biggest problems facing operators of open pits, because a difference of a few degrees can mean the recovery of many thousands or even millions of tons of additional ore. Unfortunately, the answer is not easy, but much work is being done on the problem by mining companies, colleges and universities, and government agencies to attempt to arrive at the maximum safe angle at which a given pit wall will stand.

PLATE 7-4 OPEN-PIT MINING SEQUENCE

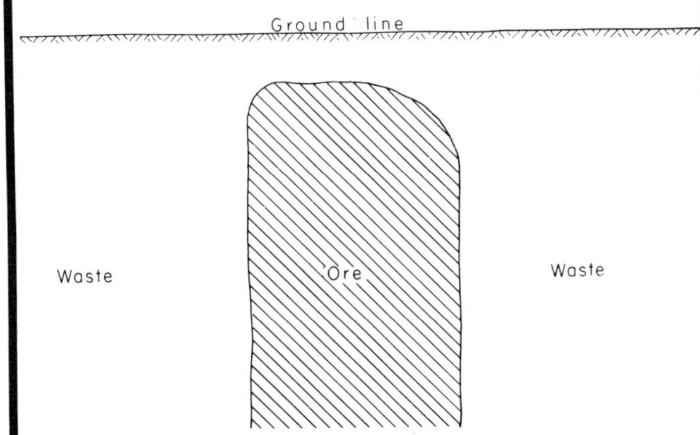

A. Section through an ore body before open-pit operations start

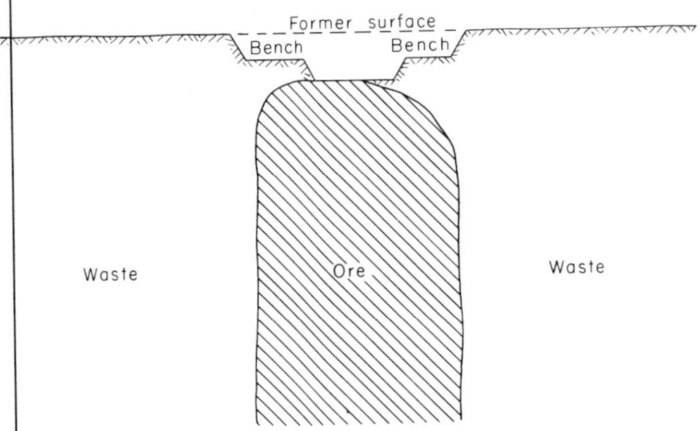

B. Section through ore body just as ore is exposed

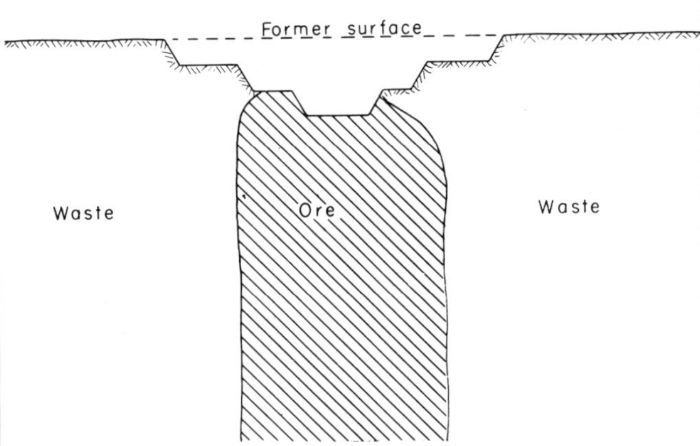

C. Section through ore body as ore production commences

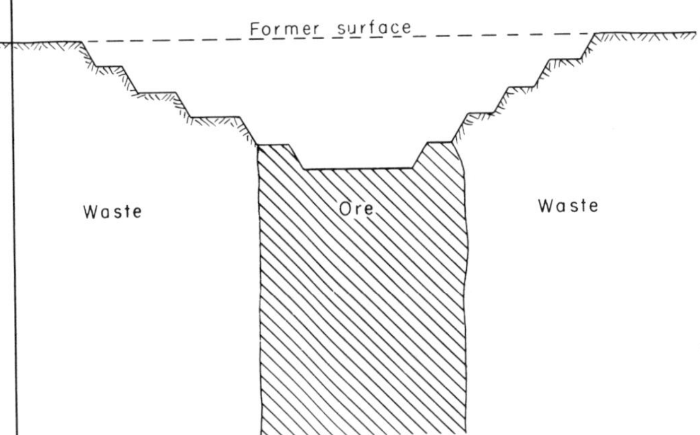

D. Section through ore body as waste stripping and ore production proceed

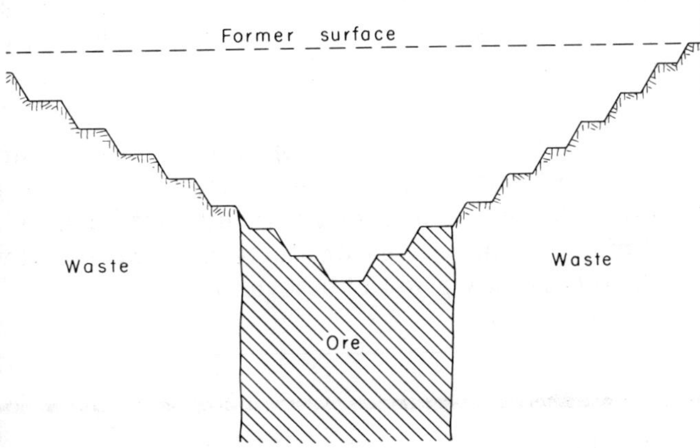

E. Section through ore body as mining continues

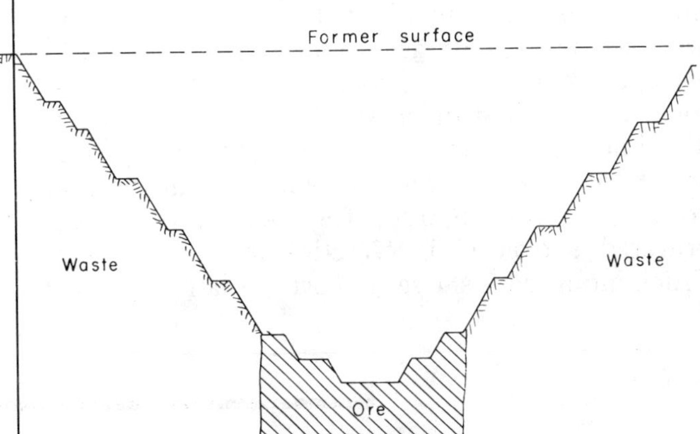

F. Section through ore body as economic limit of pit operations is reached

LOADING IN OPEN PITS

Not all operators use equipment as large as that in the Berkeley pit Plate 7-6. Power shovels are used in most large production pits. A 6-yard electric shovel is suitable for loading blasted rock into a 35-ton diesel truck A. A 2½-yard diesel-powered shovel is adequate for loading an 18-ton truck B. The pick-up truck in the foreground is a service truck for the shovel. A diesel-powered shovel C is effective in remote areas. Truck-mounted shovels have relatively small capacity D. On the next bench above the shovel a tractor loader is loading a truck.

Another type of loader E has the same action as an underground loader. This machine does not have to turn around like most tractor loaders, but throws the load directly up and over the back of the machine into a truck F.

Tractor or wheel loaders are popular in small open pits because they are versatile, can operate in small areas, and they are readily available. When a pit starts large-scale production, however, power shovels usually give the best service.

A. A 6-yard electric shovel loading a 30-ton truck in the Berkeley pit. *(Courtesy Anaconda Co.)* **B.** 2½-cu. yd. diesel shovel loading 18-ton truck. **C.** A diesel shovel loading waste rock at tungsten mine in southwestern Montana. **D.** A small truck-mounted shovel loading barite in a pit in southwestern Montana. **E.** An overcast tractor loader removing waste rock at the tungsten mine shown in C. **F.** A tractor loader loading tungsten ore at the mine shown in C.

PLATE 7-5 LOADING IN OPEN PITS

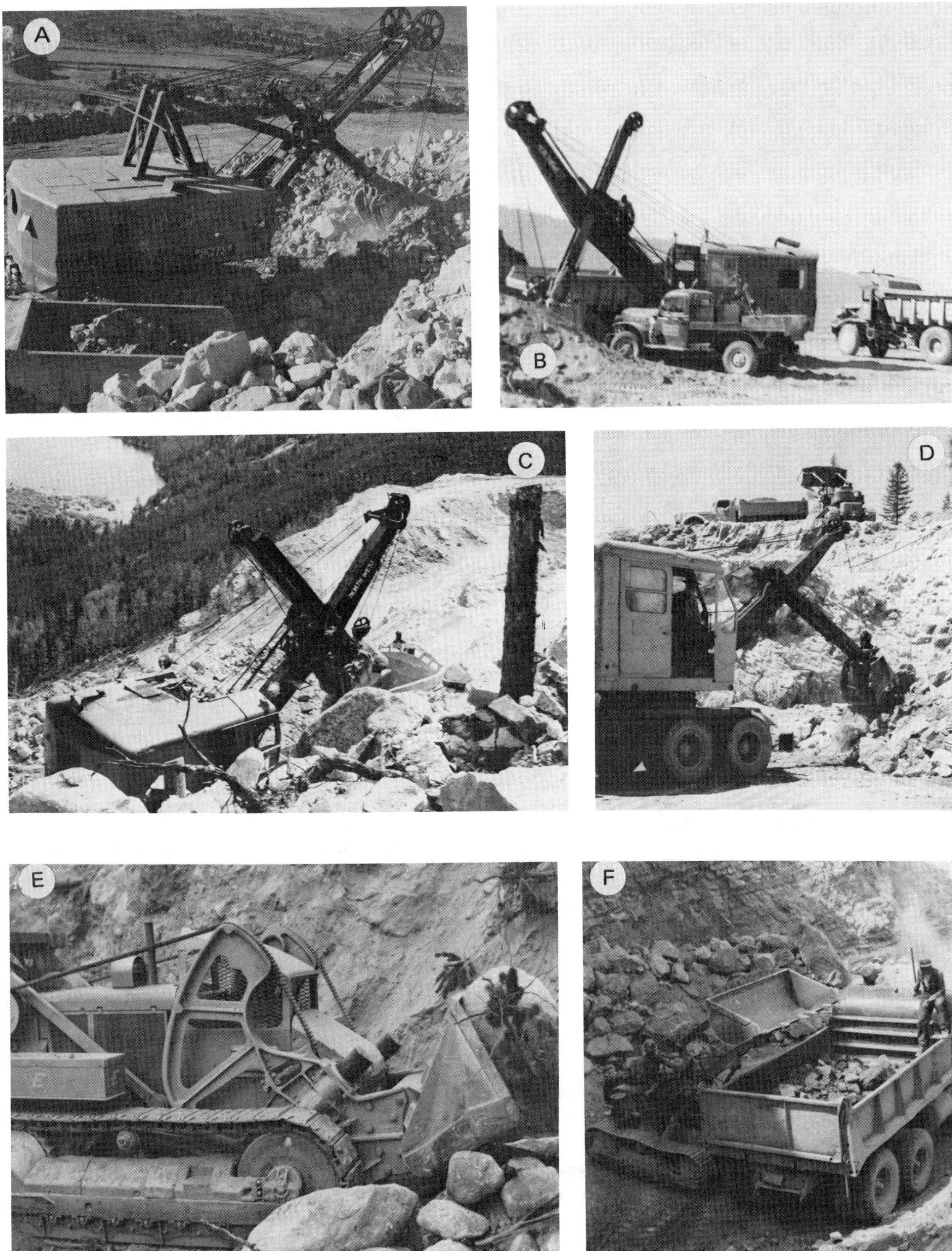

THE BERKELEY PIT

The largest open-pit copper mine in Montana is at Butte. All pictures in this plate are published through the courtesy of the Communication Department of the Anaconda Company.

A general view of the pit A shows benches and haul roads. A power shovel, trucks, and drills are shown in the foreground. The white piles in the lower center, between the truck and the drill, are sacks of explosives, which will be loaded into the drill holes.

The large electric-powered rotary drills B drill holes of 9-inch diameter as much as 42 feet deep in just a few minutes. The first part of a blast is shown in C, and D was taken just a fraction of a second later. A blast is not designed to throw broken rock into the air, but rather to break the rock just enough that the power shovels can move it. The white cloud in these pictures is mostly dust. The shovels C are 6-cubic-yard electric shovels.

A 13-cubic-yard electric-driven power shovel loads a 65-ton-capacity diesel truck E. The bucket on this shovel could hold an ordinary automobile and will load the truck with three or four scoops in a couple of minutes. The bulldozer in the foreground is used to clean up loose spilled rock around the shovel so that the tires on the trucks will not be damaged.

Most rock from the pit is hauled to an area well away from the ore deposit and dumped over the bank F by 65-ton-capacity trucks.

A. The Berkeley pit in Butte showing haulage roads, benches, and equipment. **B.** Rotary type blast-hole drill used in the Berkeley pit. **C.** Blast at the Berkeley pit. **D.** The same blast a fraction of a second later. **E.** Power shovel loading a 60-ton-capacity truck. The bulldozer cleans up around the shovel. **F.** Waste from the pit is dumped over the end of the waste pile.

PLATE 7-6 THE BERKELEY PIT

OPEN-CUT STRIPPING

Most bedded deposits lying within a reasonable distance below the surface are recovered by stripping operations. The overlying rock or waste material is first blasted and removed. The valuable mineral thus exposed is then mined. Much coal at Colstrip, Montana, has been mined in this manner, as are some placer deposits, i. e., the worthless overburden is thrown off to the side and only the paydirt is sluiced.

A long section of a bedded deposit is shown in A; B shows this same bedded deposit after part of the waste is removed and cast to the side. The waste pile is beyond the exposed coal; compare with E.

Cross sections of the cut show the overburden removed C and the coal removed D. This sequence is shown in more detail in E. The waste rock is cast to the side by the stripping shovel, usually into the area that has already been mined. Draglines may be used instead of stripping shovels to remove the waste rock. As a rule, a dragline has a longer reach than a stripping shovel, and hence can pile the waste higher.

SEQUENCE OF STRIPPING OPERATION

The sequence of stripping operation is shown in F. The stripping shovel started at the left end of the picture and is moving toward the right, casting the waste toward the upper side of the sketch. It is followed by the loading shovel mining the coal. Whenever the shovel turns a corner, a problem arises as to where to cast the waste. It cannot be cast into the mined-out portion until the loading shovel has mined out the coal. On corners it may be necessary to bring in a dragline to move some of the waste out of the way of both the stripping shovel and the loading shovel. After the corner is turned, the going is good—as far as the next corner. Different layouts or arrangements of strip mining are used, depending upon the size and shape of the property, the size and type of excavation equipment, and the type of haulage equipment. Trucks and railroad cars are used for haulage.

PLATE 7-7 OPEN-CUT STRIPPING

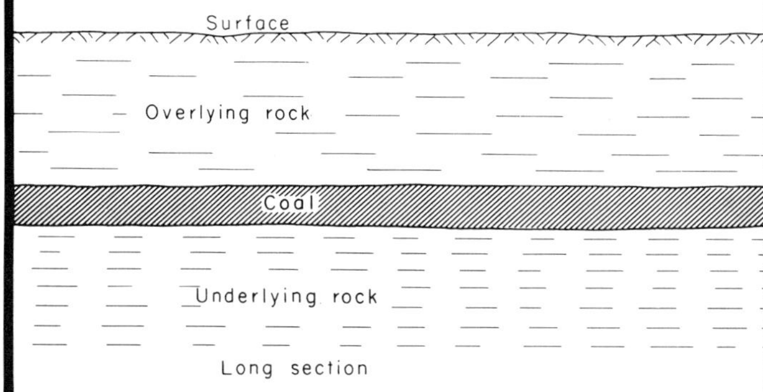

A. Coal seam lying close enough to the surface for stripping operations

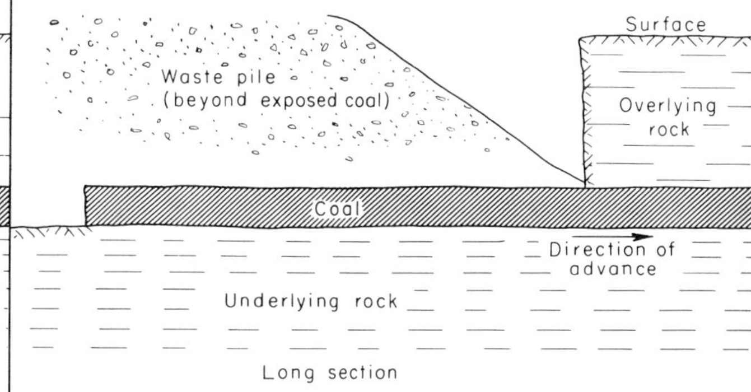

B. Overlying rock is stripped off and cast to one side. Exposed coal is then loaded

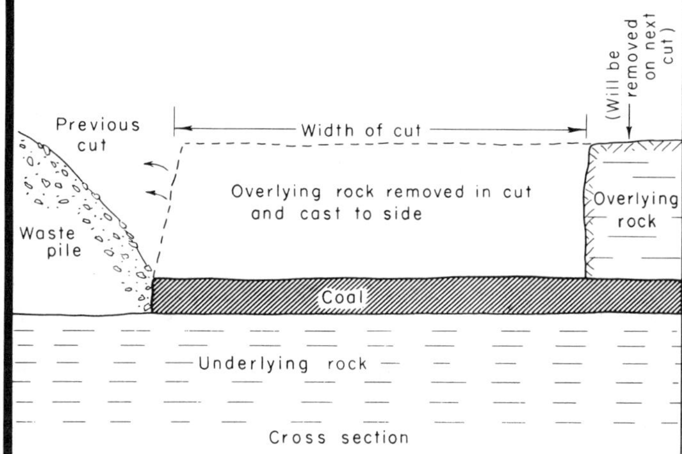

C. Coal seam exposed after overlying rock has been removed

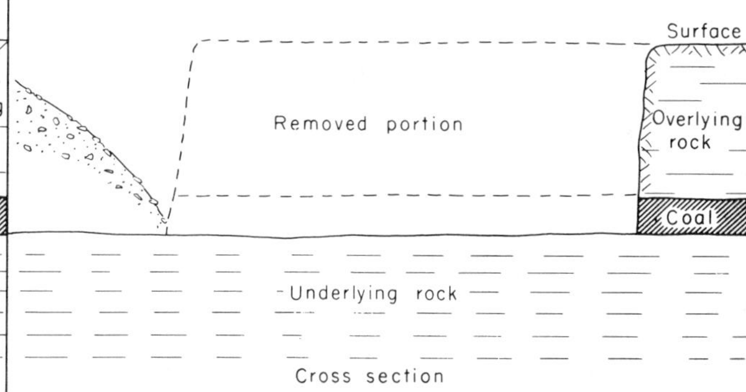

D. Cut after the coal has been removed

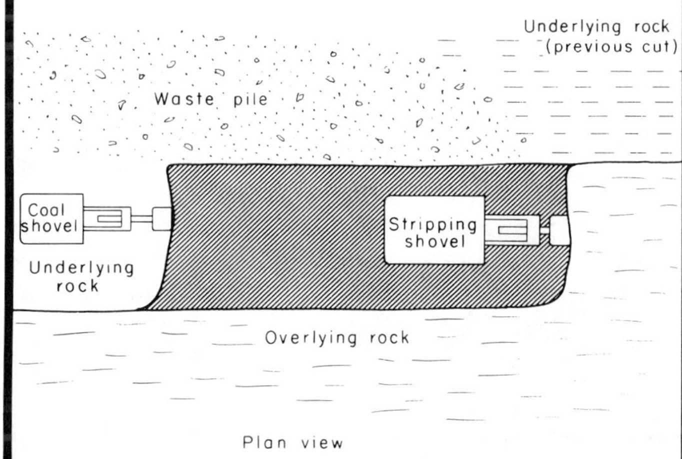

E. Plan of cut showing stripping shovel working to expose coal seam

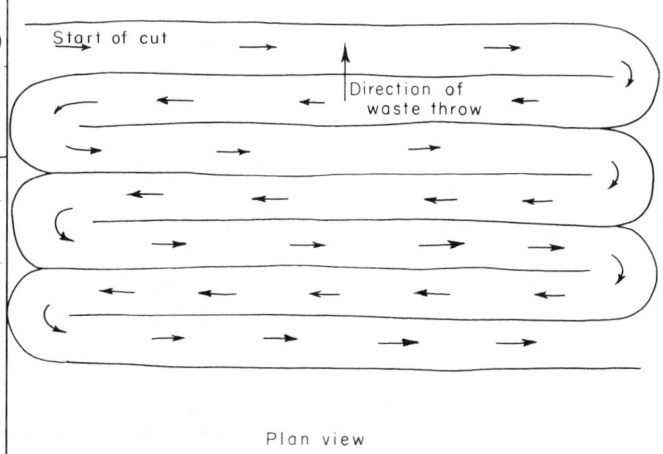

F. Plan of cut sequence to work out a tract of coal land

COAL STRIPPING AND LOADING

The photographs in Plates 7-8 were taken about 25 years ago at Colstrip, Montana. At that time, the equipment in use was large compared with that at other mines. The equipment is still in use at mining operations in the area. A 20-cubic-yard stripping shovel is shown in the foreground, and a 25-cubic-yard dragline in the left background A. The stripping shovel is casting a load of waste on the waste pile B. The machine just to the left of the telephone post is a churn drill which was used to make blast holes in the overlying waste rock.

LOADING SEQUENCE

The next four pictures show the loading sequence of the 13-cubic-yard loading shovel. The bucket is digging into coal, C, and D. The bucket full of coal is swinging to the railroad car E, and the load is being dumped into the car F. It required only three or four scoops of the coal-loading shovel to fill a 50 ton capacity railroad car completely.

As can be seen, the shovel loaded directly into the rail-cars which were used extensively in pits and strip mines in the past. There are pits where rail service is still utilized, but with less frequency since the increase in size of diesel powered trucks. The diesel powered truck has flexibility, it can go wherever there is a road, it can climb steeper grades than railroad locomotives, and it does not require relocation of track.

The picture may change in the future because diesel fuel is very expensive, and hauling ore and waste by truck may become uneconomical.

Railroads are still very cost efficient for long hauls and may find their place in some pit and open strip mines in the future. However, conveyor belts are also being used more in pits and may be a keen competitor to railroad equipped mines. See Plate 7-11.

A. Stripping shovel and dragline used to uncover a coal deposit. The dragline bucket capacity is 25 cu. yd. and the shovel 20 cu. yd. **B.** The shovel casting overburden to the side. A churn drill used for drilling blast holes is at left of pole. **C.** A 13-cu. yd. shovel digging into a coal seam. **D.** The shovel preparing to dig into coal. **E.** Swinging a scoop of coal toward a railroad car. **F.** Dumping coal into the car.

PLATE 7-8 COAL STRIPPING AND LOADING

MODERN STRIP MINING

The pictures in Plate 7-9 are all the courtesy of the Western Energy Company in Colstrip, Montana. The pictures in Plate 7-8 were also taken at Colstrip, about 25 years ago, but the coal then was loaded directly into rail cars, and the mine was much smaller at that time.

An aerial view of the cut is shown in A. The large machine near the top center is a dragline. The coal loading shovels and trucks are shown near the center of A, and the blast hole drill is to the left. The overburden being stripped and stockpiled is to the left, and the leveled spoil piles on the right.

The blast hole drill is shown in B, and a series of drilled holes is shown by the drill in A. The holes are loaded with Amonium Nitrate mixed with diesel fuel to form an explosive mixture. Because the holes require a large bulk of explosives, a powder truck with feeder is shown loading the holes in C.

DRAGLINES AND SHOVELS

A stripping dragline, coal loading shovel and coal hauler are shown in D. The dragline has a 60 cubic yard capacity bucket. As most broken rock weighs from 2,500 up to 3,000 pounds per cubic yard, the tonnage moved in each bucket is from 75 to 90 tons. Some coal stripping draglines and shovels have bucket capacities of over 100 cubic yards. The coal loading shovel shown in D has a bucket capacity of 17 cubic yards and is filling a coal hauler with a 120 ton capacity.

COAL HAULERS

A close-up view of a diesel coal hauler being loaded by the coal shovel is shown in E. The dragline and shovels are powered by electricity, while the primary source of power for the coal hauler is a diesel engine.

The coal haulers take the coal from the mine to a central loading point where it is stockpiled. From the stockpile the coal is moved by conveyor belt to a unit train loading facility shown in F. The train in F is just being moved into position for loading. Each coal car in a unit train has a capacity of 100 tons. These are cars in a train making a full load of 10,000 tons of coal.

A. Aerial view of an open cut coal strip mine. *(Courtesy of Western Energy Co.)* **B.** Close-up view of blast hole drill. *(Courtesy of Western Energy Co.)* **C.** Loading blast holes with explosives. *(Courtesy of Western Electric Energy Co.)* **D.** Stripping overburden and loading coal. *(Courtesy of Western Energy Co.)* **E.** Close-up view of coal hauling truck and trailer. *(Courtesy of Western Energy Co.)* **F.** A unit train pulling into position for loading. *(Courtesy of Western Energy Co.)*

PLATE 7-9 MODERN STRIP MINING

STRIPPING OPERATIONS

As mentioned previously, it is necessary to strip off and stockpile the surface soil in a strip mining operation as well as level the spoil piles produced by draglines. A bulldozer is shown in Plate 7-10 A leveling a spoil pile. The machine shown to the left in A is a scraper, frequently used in highway construction, which loads and hauls material. Scrapers are often used to scrape off, load, and transport the topsoil either to stockpiles or to spread it over the leveled spoil piles.

STRIPPING DRAGLINES

A stripping dragline dumping a load of material is shown in B. Long-boom and long-reach shovels are also used for stripping. One such machine is shown in C. The size of a dragline bucket compared with that of a man is shown in E. Not only is the bucket large, but the size of the chain and cables is huge.

BUCKET WHEEL EXCAVATORS

Continuous bucket wheel excavators are popular in Europe and other parts of the world. They are now being introduced and used in the United States. The bucket wheel digs into the face at a continuous rate. The muck from the wheel is loaded on conveyor belts and removed from the mining area. These machines work very well where the overburden is soft, and where blasting is not necessary to break the material. Note in D, the conveyor belt from the machine goes to another conveyor belt removing the material from the mine. These huge machines are crawler mounted and can move when necessary. The bucket wheel excavator in F is loading railroad cars while the one shown in G is piling or spoiling the overburden.

These machines dig at a constant rate, and no time is lost in swinging and dumping the material. The dug material loads onto a conveyor belt system which moves the material rapidly. Capacities of some of the larger bucket wheel excavators are in excess of 5,000 cubic yards per hour.

The large draglines and stripping shovels are designed for high capacity output, and are tremendous in size, highly productive, and, of course, very expensive.

A. Bulldozer and scraper used in strip mines. *(Courtesy Western Energy Co.)* B. Dragline used in stripping. *(Courtesy Western Energy Co.)* C. Coal stripping shovel. *(Courtesy Montana Bureau of Mines and Geology)* D. Hydraulic bucket wheel excavator. *(Courtesy O & K, Orenstein and Koppel)* E. Large capacity dragline bucket. *(Courtesy Western Energy Co.)* F. Bucket wheel excavator, conveyor belt and railroad loading hopper. *(Courtesy Krupp Industries)* G. Bucket wheel excavator stripping overburden. *(Courtesy Krupp Industries)*

PLATE 7-10 STRIPPING OPERATIONS

PIT LOADING AND CONVEYOR SYSTEMS

Wheel loaders have developed into large capacity machines with great maneuverability. They are used for a variety of jobs in pit and open-cut mines. In smaller pits or open-cuts they are often the primary loading machine. With their high speed and versatility they can move rapidly from one job to another. Not only can they load, but they can carry the load for considerable distances before dumping. Machines of this type are used underground in some types of mining and are called (LHD) units.

A large wheel loader is loading a truck in an open pit in A. In B, the machine is loading railroad cars with coal from a stockpile.

Although large capacity diesel-powered trucks have been used extensively in pit and open cut mines, the recent price rise in diesel fuel may make other transportation systems more attractive. Rail systems were discussed in Plate 7-8. Conveyor belts have been and are being used in pits and strip mines, especially with bucket wheel excavators.

CONVEYOR BELTS

Most load-haul-dump units have a maximum point at which it is uneconomical to travel further. A conveyor belt is shown in C receiving muck from a scraper system of stripping. The belt then carries the material from the pit to a dump area. A transfer point and stacking belt is shown in D. Generally speaking, a counter-balance hoisting system or a conveyor system can lift material more energy efficiently than a truck running up a grade. This is because the descending skip and rope or conveyor belt partially compensates for the ascending skip and rope conveyor belt. The only power needed is to hoist the ore, and to overcome friction of the system. In a truck, energy must be used to elevate the truck and load. There is no compensation and a certain amount of energy is required to get the truck back down the hill.

A belt conveyor system hoisting material from a pit is shown in E. This is a truck pit and a truck is on the dumping point over the belt. A much more elaborate conveyor belt system is shown in F. One of the disadvantages of a belt or rail system is that it must be moved from time to time to service the pit. Belt sections and transfer points, mounted on skids so they can be easily moved, help make up for this disadvantage.

A. Loader loading trucks in a pit. *(Courtesy International Harvester Co.)* **B.** Loader loading railroad cars. *(Courtesy Clark Equipment Co.)* **C.** Conveyor belt and scrapers in a pit. *(Courtesy Goodyear Rubber Co.)* **D.** Conveyor belt and stacker. *(Courtesy Goodyear Rubber Co.)* **E.** Hauling material from a pit on a conveyor belt. *(Courtesy Goodyear Rubber Co.)* **F.** Conveyor belt system in a pit. *(Courtesy Goodyear Rubber Co.)*

PLATE 7-11 PIT LOADING AND CONVEYOR SYSTEMS

PART 7
CONCLUDING REMARKS

As can be seen in this part, the size and capacity of the equipment used in surface mining has continued to increase in the past decade. However, some mine operators feel there is an upper limit to the size of the equipment, with many reasons given for that opinion. One fear is that if one machine does all of the production work, and it breaks down or is out of service for any reason, all production will cease.

As machines get larger they generally become more cumbersome. They are more difficult and slower to move, losing desired flexibility. Although the machine has tremendous capacity when it is operating, it loses much of this advantage when it is out of service while being moved and repositioned.

The trend now is to use matched equipment. That is, the shovel and truck size are such that both are kept busy with neither being unduly idle. Draglines are matched to the loading and hauling facilities so they do not have to wait for them to catch up, or conversely so haulage equipment does not have to wait for the dragline.

Labor and investment costs are very important items in equipment selection. Large equipment costs much more, and unless it can produce significantly more than smaller units on a cost comparison basis it loses any advantage.

PART 8

UNDERGROUND MINING METHODS

Although pit and open-cut mining are used when practical, there are many situations where they can't be used. A steeply dipping narrow vein will produce very little ore by surface techniques. Even though an ore deposit may be large, and can initially be mined successfully by surface mining techniques, a point may be reached where it may be more economical to use underground methods.

Underground mining methods are classified or distinguished from each other by the type of support required in the stope. Supports required may range from almost nothing to heavy timbers combined with rock bolts supported by tailings, sand, or mine waste rock. Even in the strongest rock, large underground openings will cave eventually. If the ore is weak and caves readily, a system of stoping is sometimes used that takes advantage of this weakness. These are called caving systems. Mining methods usually fall into one or more of the following systems:

1. Open stopes—gophering, glory-hole, room pillar, sublevel.
2. Cut-and-fill—horizontal or flat back, rill, resuing, and timbered.
3. Shrinkage.
4. Timber—square set and stull.
5. Caving—topslice, sublevel cave, longwall, and block cave.
6. Combination methods.
7. Solution and insitu methods of recovering minerals.

There is some overlap. For example, in a room-and-pillar system, timber or stulls may be required to support weak areas. A stope may be started as cut-and-fill, but as the ground gets heavy or weak, timber may be required. Conversely, the timber method may be changed to cut-and-fill if the workings enter stronger ground.

The general underground systems will be described in the order listed in this part, with the exception of longwall and shortwall systems which are described after the room-and-pillar system. This is done because in some mines one may be substituted for the other. In some cases, new mining systems may be more cost effective than those in current use. For example, in a mine where square-set timber seems to be required, the less costly top slice method or sublevel cave system may be substituted under certain conditions.

GOPHERING

Gophering, used frequently in small mines, is really not a method, but consists of following the high-grade ore wherever it goes. The miner uses only the support necessary. It is very difficult to picture this system. Gophering frequently harms the ore body for further mining because it causes zones of weakness in the ore, particularly so if some systemized mining method is to be used later.

GLORY-HOLE

The glory-hole system, frequently used in the past, has been almost completely replaced by the sublevel long-hole or open-pit type of mining. A glory-hole was just a cavity in the earth continually enlarged by mining. Generally there was very little system used, and the miners often worked under large exposures of back or roof. The size of the cavity was often extended to the point where caving started, and the project was abandoned.

SYSTEMATIC ROOM-AND-PILLAR METHOD

In the United States the room-and-pillar method of mining produces more tonnage than any other underground method used. Even with continuous mining machines a room-and-pillar method is still used. In Europe, and other areas of the world, longwall mining may produce more coal than the room-and-pillar method. Longwall and shortwall systems are relatively new in the United States and generally have not yet replaced the room-and-pillar system.

In a flat bedded deposit it may be necessary to sink a shaft or drive an incline to the deposit.

VENTILATION

A double entry, or entries side by side, driven in the seam provides two passages for ventilation air. See A. In coal mines large amounts of air are required to dilute gases to levels below explosive concentration. Connections are made every so often in the entries to direct the air, see arrows in A, B, C, and D.

BLOCKING COAL

To outline a block of coal, side entries are driven B. Rooms are started in sequence B, and C from the side entry, which is extended to prepare more coal for mining. The block is mined as shown in D. Pillars are left between the rooms to support the roof or back.

RECOVERING PILLARS

In the room-and-pillar method, 30 to 60 per cent of the coal or other mineral remains in the pillars after the rooms are mined. To recover the pillars E, a pillar E,a is mined by using timber for temporary support and then allowing the area to cave. Then pillars E,b are mined and the ground is allowed to cave. As pillar robbing progresses, the whole mined-out block of ground caves F. This procedure is called retreating mining because after the pillar recovery starts no attempt is made to go back into the block. It is allowed to cave and is abandoned.

A model of a coal mine G shows a complete picture of the mine's development. The model of a large zinc and lead mine H shows a variation of the room-and-pillar system. Large capacity equipment can be used in mines of this type.

Unless the roof or back stands well, the room-and-pillar system cannot be used. Therefore, roof bolts and other means of artificial support are used extensively to keep the roof or back from caving. This system produces ore at a low cost per ton, and is used under the proper conditions. The equipment shown in Plates 5-18 E, F, and 5-16 are used extensively in this system of mining.

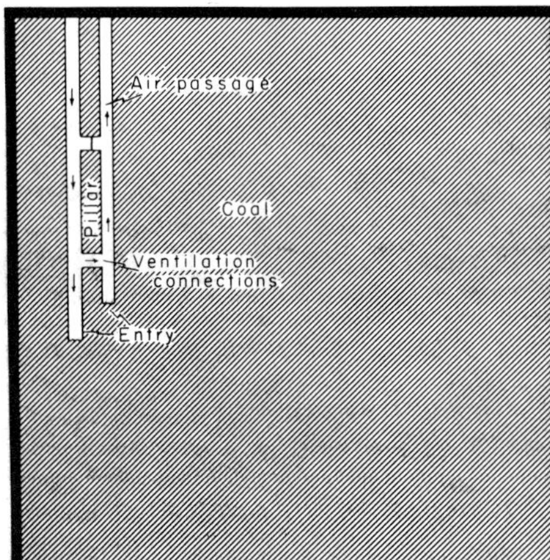

A. Initial development of a horizontal coal seam consists of driving the main entries

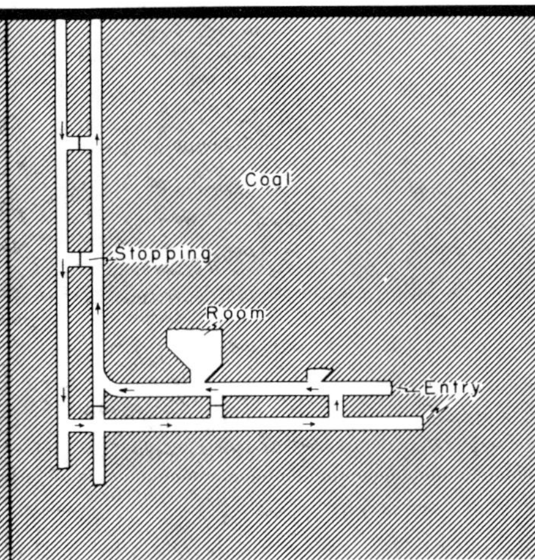

B. Development of a block of coal involves driving entries for mining block

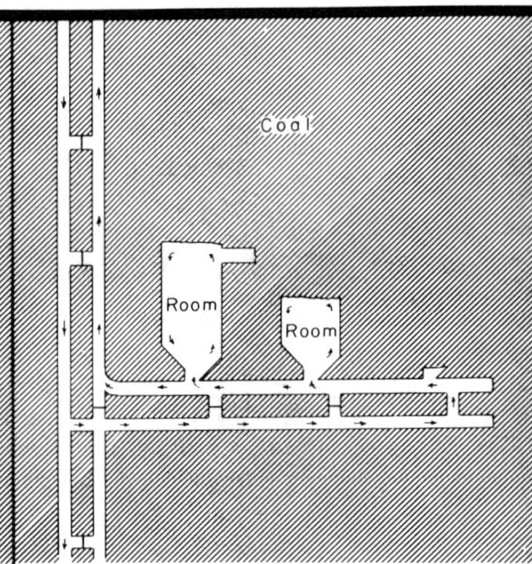

C. A series of room is started in the block of coal

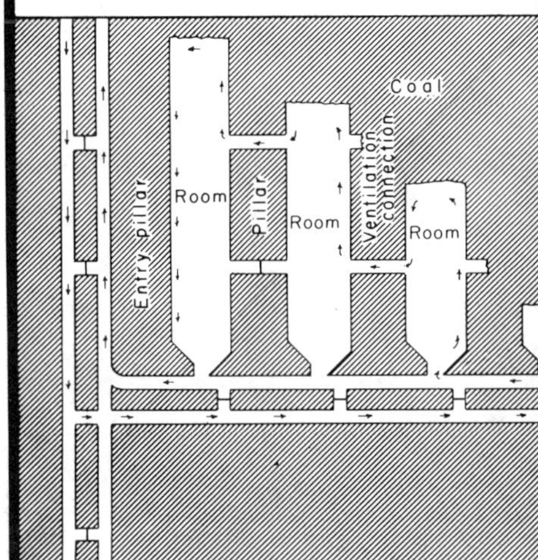

D. Mining the block of coal by driving rooms and leaving pillars for support

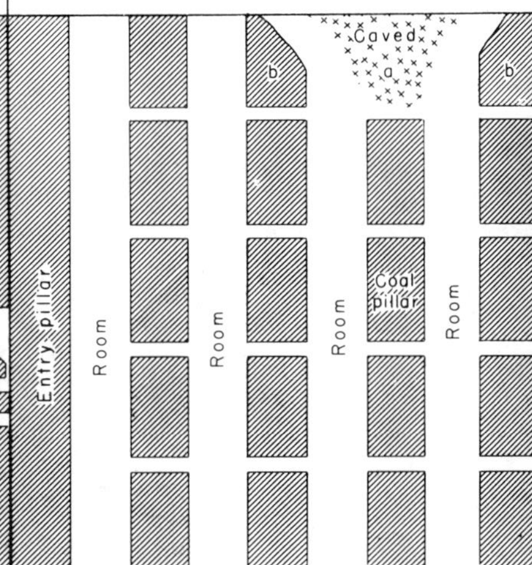

E. The block is mined out by room and pillar. Recovery of pillars is started

F. Retreating method of pillar recovery. Area caves behind mined pillars

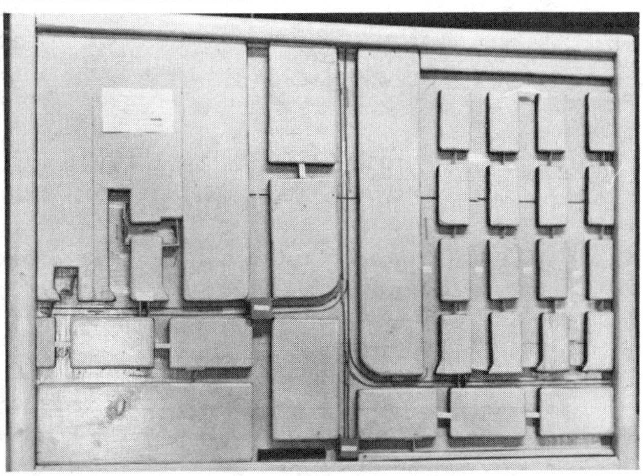

G. Model of a coal mine showing development and mining of a coal seam

H. Model of a lead mine showing a different system of pillars in a thick ore zone

CONTINUOUS MINERS AND SUPPORT EQUIPMENT

Continuous miners are being used extensively for mining coal and other flat bedded mineral deposits. Plate 8-1 shows the fundamentals of room-and-pillar mining. Continuous miners are used for driving entries as well as driving the rooms. However, some mines use the continuous miners for one job such as driving entries, and use conventional room-and-pillar equipment for driving the rooms. Many combinations and applications of equipment are in use. Continuous miners are used in the shortwall method discussed in Plate 8-7.

CONTINUOUS MINING MACHINES

All of the photographs in plate 8-2 are courtesy of the Jeffrey Mining Machinery Division of Dresser Industries Inc. The continuous mining machine shown in A is being used to cut trona which is much more difficult to cut than most coals. To hold up under the severe forces created by cutting harder minerals, these machines must be sturdily built.

The machine in B is designed for low or narrow seam coal. Note the way the cutter teeth are attached to a spiral rotation drum. The action of the spiral carries the cut coal toward the center of the cutter head where it can be picked up by the conveyor belt on the machine.

The coal pan and gathering arms, positioned slightly behind the cutting head, is shown on the machine in C. A continuous miner not only cuts and breaks the coal, but it loads it into some type of transportation conveyance as well. The particular machine in C is a narrow profile machine which will cut a narrower passage. Narrow passages are desirable in some operations. However, the machine can cut right alongside the first cut to make wider passages if desired.

CHAIN-TYPE BRIDGE CONVEYORS

Continuous miners produce minerals at a steady rate. Hence, it is desirable to remove the mineral as fast as it is cut. Conveyor belts, if properly sized and positioned, will do this, and such a conveyor system is shown in D. As the miner cuts and moves ahead at a fast rate the belts must move ahead also. One way of doing this is to use telescoping-chain-type bridge conveyors. One transfer point from one conveyor to the other is shown in D.

COAL HAULERS

Coal haulers are still used extensively with continuous miners. One battery powered unit is shown in E. This machine is articulated so that it can turn sharp corners, and is highly maneuverable. Articulated vehicles are normally more maneuverable than regular vehicles such as the shuttle car shown in Plate 5-18 F. However, shuttle cars are also used to service continuous miners.

Many mines still use rail systems, and an electric powered locomotive is shown in F. Electricity is delivered to the locomotive through a trolley line located at the top of the picture. A trolley pole and shoe make contact with the wire and locomotive.

All photographs are the courtesy of Jeffrey Mining Machinery Division of Dresser Industries, Inc.

A. HELIMINER (R) Continuous Mining machine. **B.** HELIMATIC (R) Continuous Mining machine. **C.** HELIMINER (R) Continuous Mining machine. **D.** Continuous haulage system. **E.** Battery powered RAMCAR (R). **F.** Electric powered supply locomotive.

PLATE 8-2 CONTINUOUS MINERS AND SUPPORT EQUIPMENT

LOADERS, CONTINUOUS MINERS AND SUPPORT EQUIPMENT

The (LHD) unit in A is loading and hauling ore in a large underground room-and-pillar mine. In this mine, the ore is over 40 feet thick, and as can be seen, stands well in large rooms. Note that the operator is able to fill the bucket to capacity when he has a good muck pile. Also, the unit will handle large sized boulders.

Another type of continuous mining machine employing the boring and auger principle is shown in B. It is made for low seam coal. The outboard booms extend to the width of the room, and hydraulic jacks are positioned firmly against the roof and floor. Notice the cables attached to these jacks. Tightening the cable on one side moves the machine and cutter borers toward that jack. Loosening the cable on that side and tightening the one on the other side causes the machine to swing across the face cutting the coal. The auger cutters force the coal toward the chain conveyor which picks it up and moves the coal away from the machine. C shows the machine in action with the augers rotating and cutting the coal.

The back end of the machine is shown in D with the chain conveyors in action. Note the sharp turn that can be made with two conveyors.

COAL TRACTORS

Service trucks or tractors are commonly needed in a mine. A battery operated coal tractor is shown in E. This can be attached to a coal trailer shown in F. These machines are designed for use in low coal seams. Note that the tractor in E is articulated to allow it to turn sharp corners.

The trailer shown in F is wide and long. To eject coal from the loaded trailer, it is necessary to have an ejection system. This particular tractor has an ejection system powered by a three stage hydraulic cylinder controlled from the tractor unit. These low trailer units have large carrying capacity and work well in low seam coal.

A. Loading in a large room and pillar mine. *(Courtesy Eimco Corp.)* **B.** Mark 21 continuous miner. *(Courtesy Fairchild Inc.)* **C.** Continuous miner in operation. *(Courtesy Fairchild Inc.)* **D.** Back end of continuous miner. *(Courtesy Fairchild Inc.)* **E.** Underground coal tractor. *(Courtesy Fairchild Inc.)* **F.** Coal trailer. *(Courtesy Fairchild Inc.)*

PLATE 8-3 LOADERS, CONTINUOUS MINERS AND SUPPORT EQUIPMENT

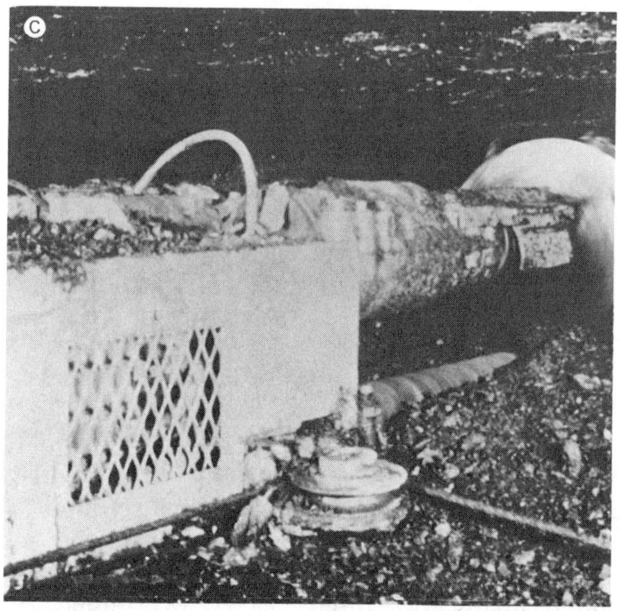

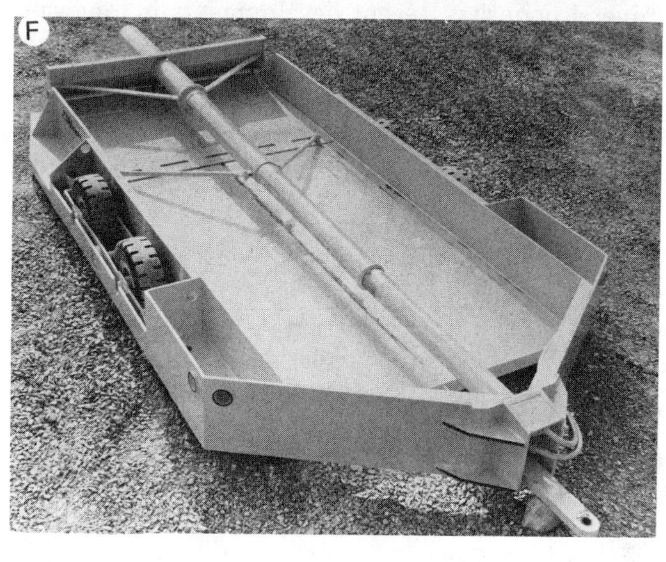

LONGWALL CAVING SYSTEM (EARLY METHODS)

As mentioned in the section on the room-and-pillar system, only 30 to 60 per cent of the coal is recovered on the first mining by that method; that is, before the pillars are robbed. To recover almost all of the coal in a single operation, the longwall system was developed. Although this system is fairly new in the United States, it is being used for mining coal and other flat-lying mineral deposits. Recent developments in longwall mining equipment have certainly increased the popularity of the system. However, the cost of equipping a modern longwall system is very high when compared with conventional mining systems. The safety record has been good in longwall blocks.

RETREATING LONGWALL SYSTEM

To develop a retreating longwall system, entries are driven in the seam A from the shaft or incline. The deposit is similar to the one shown in Plate 2-1 B. This is a triple entry heading, popular where mining faces are to be driven both ways. The entries are driven to the limit of the block and then longwall side entries are driven in both directions B. Note both the plan and long-section views in these sketches. The entries are up to and sometimes over 500 feet apart, and the longwall face extends between the entries. It is necessary to have an air circulation system in a coal mine to dilute any explosive gases with fresh air to eliminate the explosion hazard.

SUPPORTS AND CAVING

The longwall side entries are driven wider, progressing to the left in the sketch C, which produces a longwall face. The open ground is supported by wooden timbers called props. Before the use of steel chocks or shields, hydraulic actuated props were used instead of timber. As the face progresses to the left in C, and D, the supporting timbers are blasted or the hydraulic props are released and pulled from the caving area. Relieving the support by these techniques causes the unsupported roof to cave behind the protected mining face. The mining face is protected by the props or other support and by the unmined face long sections, C, and D. This method mines almost all of the coal in the first mining. As a rule, explosive cost is less in this system than in the room-and-pillar system because the caving action causes some breakage of the solid coal. Equipment used is similar to that used in the room-and-pillar system, but slushers and scrapers can be used to pull the coal along the face. Plate 5-17 B, C, and D shows a slusher and bucket in operation.

PLATE 8-4 LONGWALL CAVING SYSTEM (EARLY METHODS)

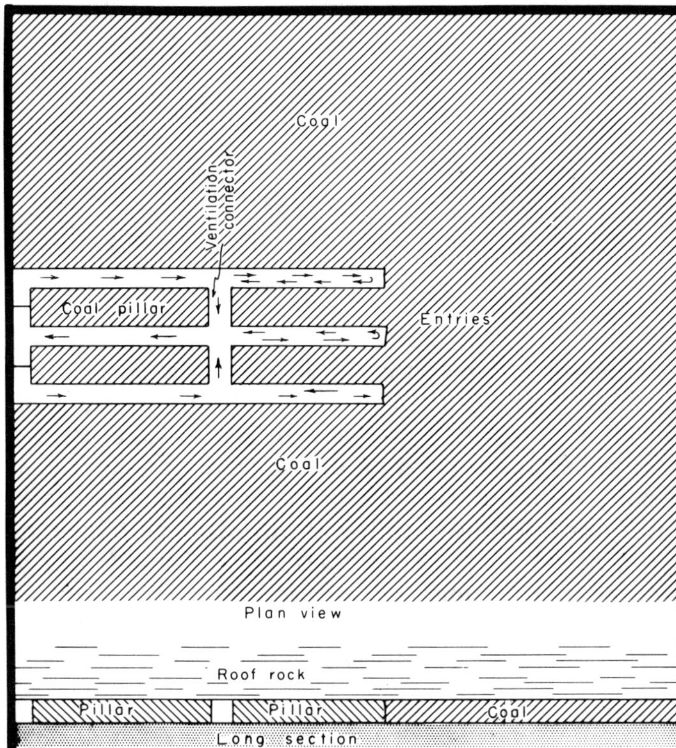

A. Developing a longwall block by driving a triple entry

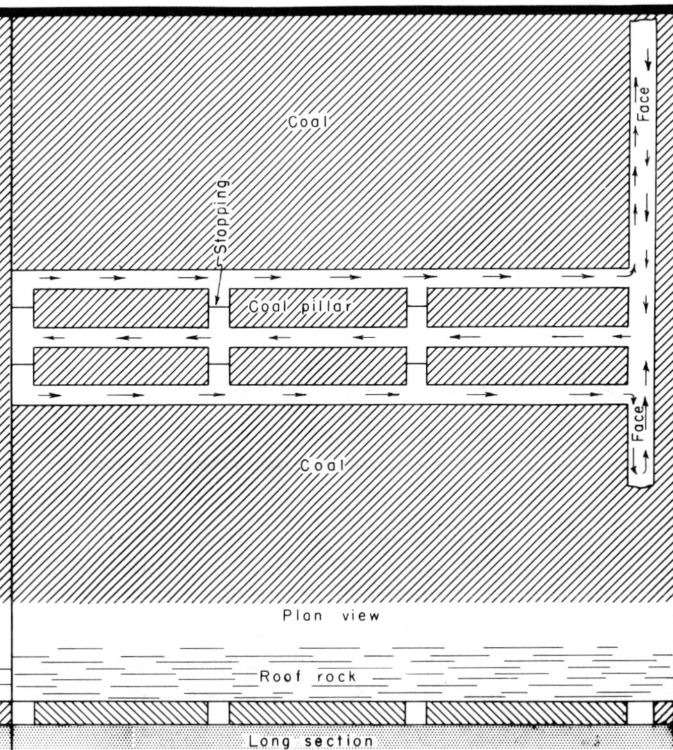

B. Developing the end of the block by driving two single entries so a longwall face can be started

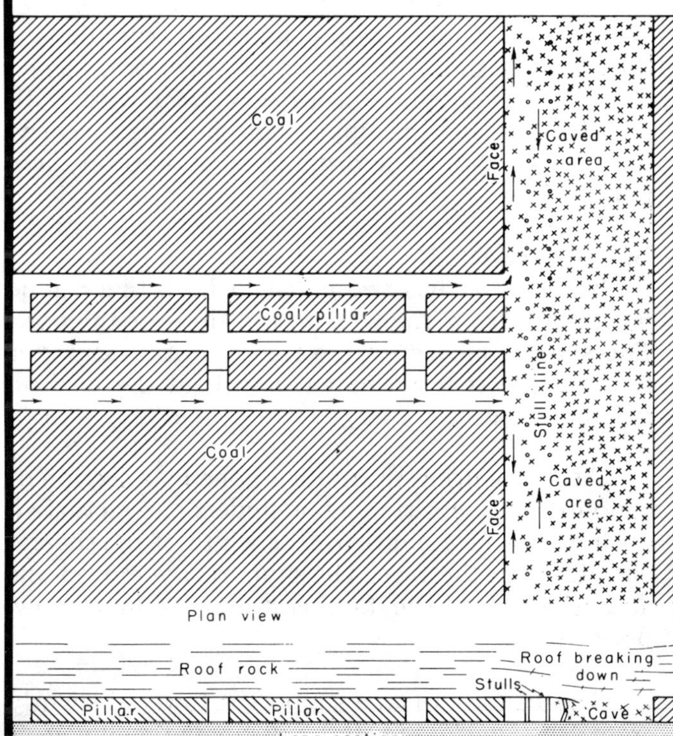

C. The longwall face retreating from the limit of the block. Caving of the roof rock has started

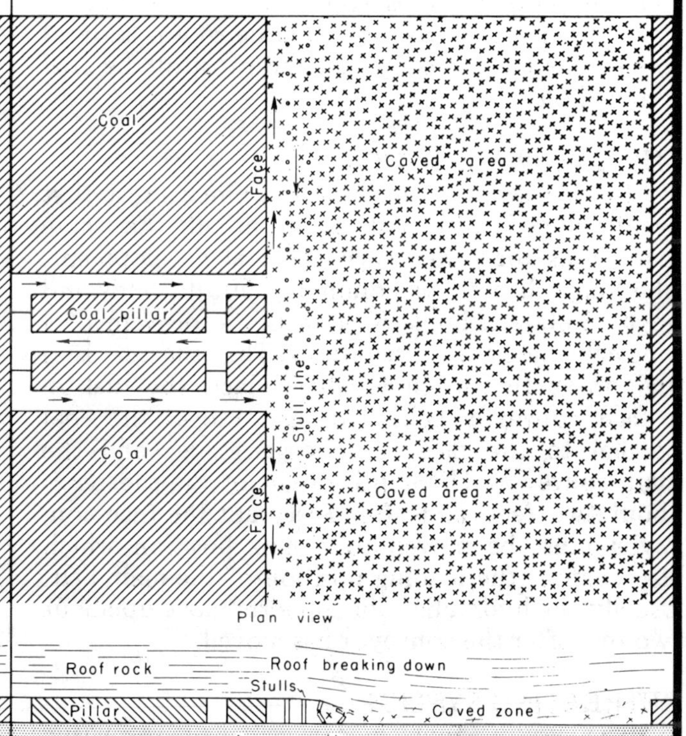

D. Further progress of the retreating longwall face. The mined-out zone is allowed to cave

MODERN LONGWALL MINING SYSTEM

Longwall mining has been greatly mechanized in recent times. Hydraulic actuated ground supports have been developed that hold the roof up while the longwall cut is made and can be moved into a new position with the use of hydraulic cylinders. Drum shearers, coal plows and other types of cutting devices have been developed that run on a conveyor belt frame which cuts the coal from the face without the necessity of drilling and blasting. The supports hold the roof in place while the coal is being mined so there is little need for roof supports such as roof bolts in the mining cycle. As the supports move up, the roof is allowed to cave behind the supports, and most of the coal is mined out.

THE LONGWALL BLOCK

In this system, the longwall block must first be developed with entries in about the same manner as for other types of coal development. The longwall block is commonly 500 ft. wide and up to 3,000 ft. long. A is an isometric vein of a longwall block in operation. The shearer is a high-production machine which places the coal on a face conveyor belt. This belt moves the coal parallel to the face to a stage loader and the main conveyor. The depth of cut of the shearer is about two feet, and after every cut the face conveyor must be moved to a new position which also changes the loading position of the main belt. Usually the hydraulic roof supports move up about two feet after the conveyor has moved.

RETREATING SYSTEM

The system shown in A is called the retreating system of mining. The development work is completed to the boundary, the longwall system is set up and the coal is mined toward the shaft. No one goes back into the area after it has been mined and it caves. Advancing longwall has been used in coal mines in other countries, but only in one or two mines in the United States. With advancing longwall, mining can start near the shaft, therefore not as much development work is required before coal production starts. However, the maingate and tailgate must be maintained through the fringes of the caved area. This sometimes is difficult to do. It requires extensive construction work to protect the passageways which may be more expensive than the retreating system.

HYDRAULIC PROPS

The longwall system consists of a long section of hydraulic actuated props called chocks or shields. Each one of these is about 5 ft. wide; consequently it would take 100 of them for a face 500 ft. long. They are attached to the face conveyor by a double acting hydraulic cylinder, as shown in B, and C. The face conveyor is usually a chain conveyor built on a heavy framework with rails fastened to the upper side and the face side of the conveyor. The drum shearer rides on these rails.

DRUM SHEARER

The shearer has two cutting wheels that can be positioned from the roof to the floor of the coal. The shearer moves on the tracks on the conveyor frame while the wheels cut and load the coal on the conveyor.

After the drum shearer has gone by a section of supports the positioning cylinders are actuated, pushing the face conveyor the width of the shearer cut closer to the face. The supports are moved one at a time after the face conveyor is in its new position.

The canopy of one support is lowered slightly from the roof, and the positioning cylinder is actuated which pulls the support toward the conveyor. When in position, the canopy is again raised and pressurized against the roof. The roof rock is allowed to cave in the space left vacant by the support that was moved.

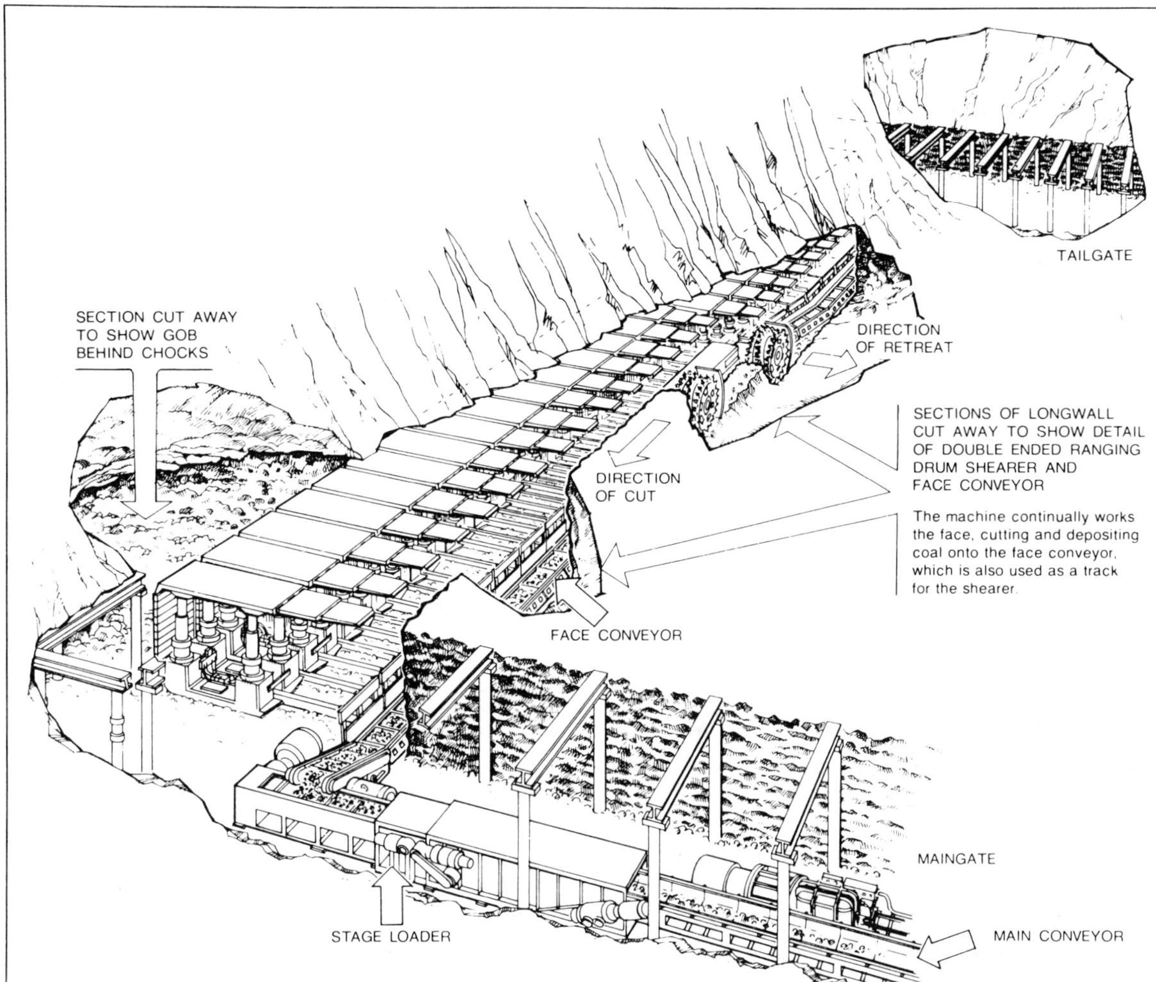

A. Isometric view of a modern longwall mining system

(courtesy of McGraw-hill, Inc.)

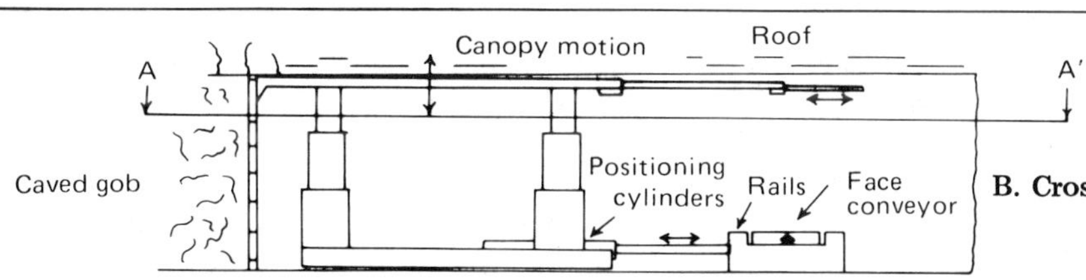

B. Cross-section through chock line

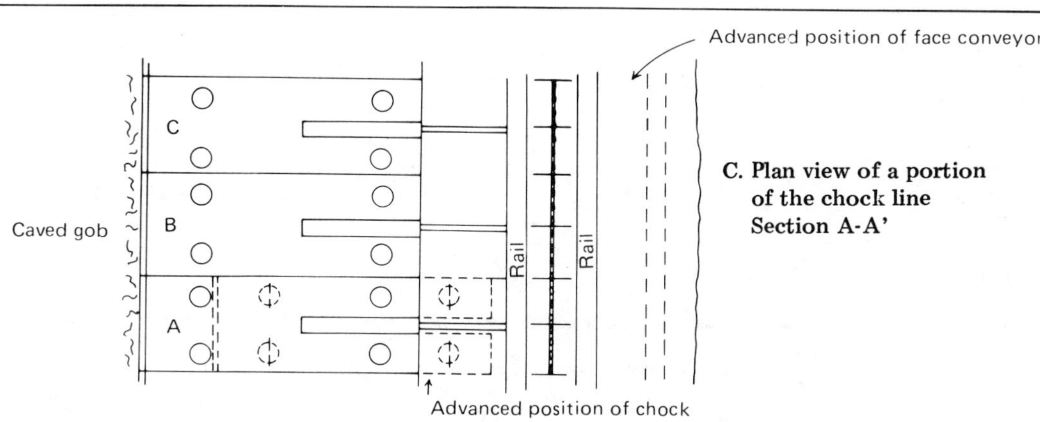

C. Plan view of a portion of the chock line Section A-A'

LONGWALL MACHINERY

A visitor to a longwall mining section is impressed with the amount of machinery present in such a small area of the mine. The present longwall support and mining system is described in Plate 8-5. This plate shows drawings of the actual machinery in operation.

The longwall cutter or shearer is shown in A, E, and F. The upper or head end of a double-ended ranging drum shearer is shown in A. The coal is being cut, and some of it tends to fall toward the face chain conveyor. The machine is moving toward the observer. A close-up of the shearer wheel is shown in E.

The face conveyor is shown in both A, and F which also acts as a track for the double ended ranging drum shearer.

The drum shearer machine must move back and forth along the longwall face. There is a toothed track shown in both A, and F where a meshing gear wheel engages the track and causes the drum shearer to move along the track. Some drum shearers use a chain system to move the drum shearer which is attached to each end of the panel.

HYDRAULIC SUPPORT SYSTEM

The hydraulic support system is shown protecting the face conveyor and track in B. The way the supports are moved up is explained in Plate 8-5. It will be noticed that there is not a great deal of room for personnel between the vertical supports and the longwall track and face conveyor.

Each hydraulic support, whether it is a chock or shield, has individual controls so it can be actuated separately from the other supports. A miner, adjusting or controlling one of the units, is shown in C. Fluid under pressure is required for the hydraulic support, and a pump and pressure storage unit is shown in D. Practically all of these longwall systems have communications with the surface or other sections of the mine. Note the telephone next to the control panel in D.

A. Longwall face showing shearer and conveyor belt. *(Courtesy Eickhoff Inc.)* **B.** Hydraulic supports in a longwall. *(Courtesy Gullick Dobson)* **C.** Controls for hydraulic supports. *(Courtesy Gullick Dobson)* **D.** Power Pac and Control system for supports. *(Courtesy Gullick Dobson)* **E.** Cutting wheel on shearer. *(Courtesy Eickhoff, Inc.)* **F.** Cutting wheel on shearer. *(Courtesy Eickhoff, Inc.)*

PLATE 8-6 LONGWALL MACHINERY

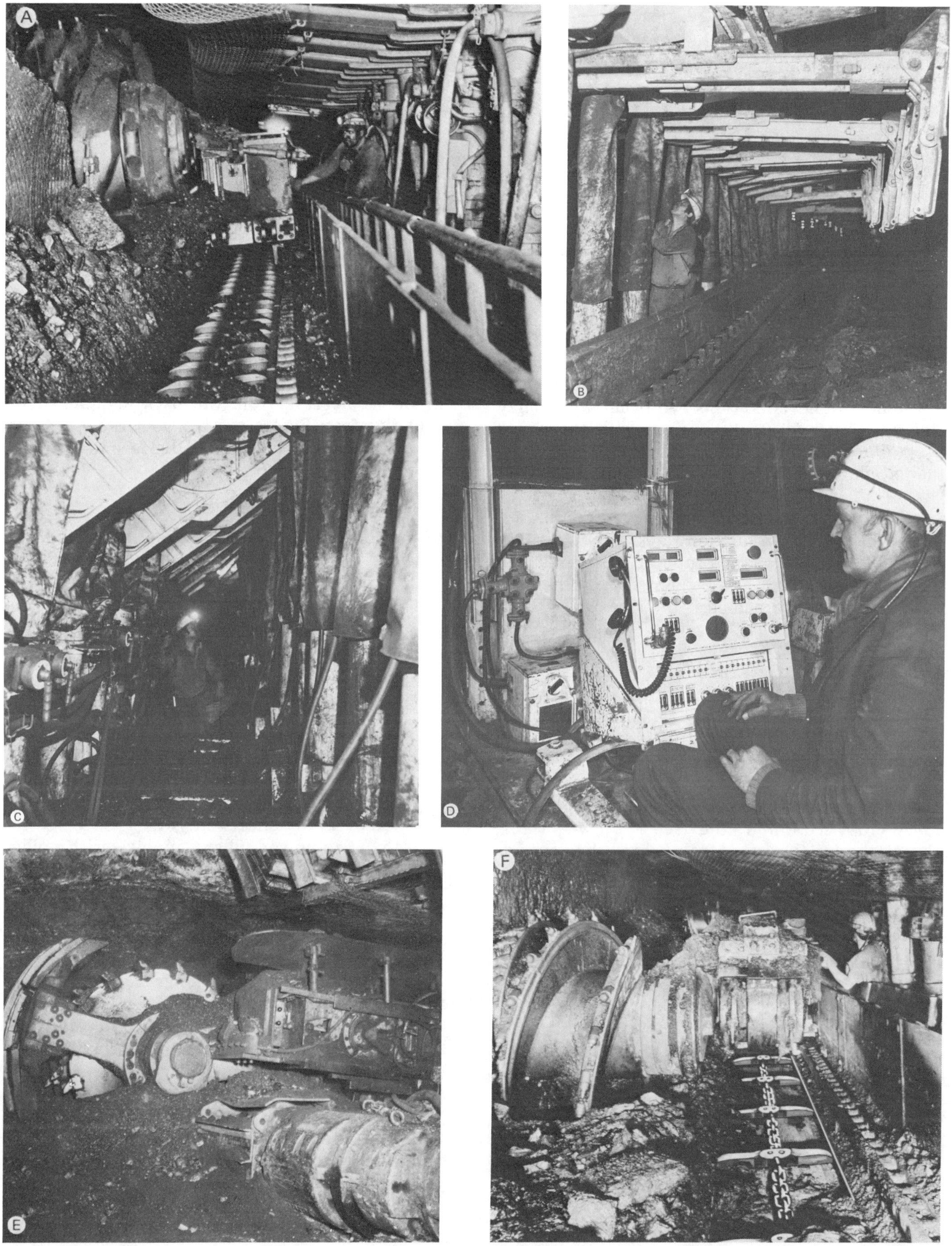

SHORTWALL SYSTEM OF MINING

One disadvantage of the longwall system is that if the face cutting machine or any other part of the system malfunctions, production of coal from the entire system stops. Too, over a 500 foot face different ground conditions can occur which may lead to mining difficulties. Even with this drawback, the system extracts most of the coal from the ground, and the safety record is excellent.

Conditions permitting, present day continuous miners have high production rates. Coal operators have combined the roof support system of the longwall method with the flexibility of the continuous miners and named it the shortwall system of mining.

SHORTWALL FACE LENGTH

The length of the mining face in the shortwall system is about one third the length of the typical longwall system or about 150 ft. The block is developed as shown in A. The initial cut along the face is made by the continuous miner, and a line of hydraulic supports is installed as shown in A. A cross section of the face is shown in E,1, except there is no caved gob behind the support.

The continuous miner takes a fairly wide cut. About ten feet wide at a time. Therefore, it starts cutting ahead of the supports as shown in B and E. The support itself, and the roof canopy of the support, are advanced and extended over the miner as it advances. This is shown in B and E,2. The miner operator is protected from the roof by the supports.

REMOVING COAL FROM THE MINER

Removing the coal from the miner is done in different ways, and a series of conveyor belts may be used. The ordinary shuttle car or some specially developed coal hauler is also frequently used. If the shuttle car or coal hauler is used, the roadway must be kept at the maximum width for passage of the conveyance until the entire length of the face is mined. After the miner advances the full length of the cut, it backs out under the supports and the spill plate and supports are moved up as shown in C, and E, 3 and 4.

ADVANCING THE SPILL PLATE

The spill plate is pushed ahead by the positioning cylinder about five feet while the supports are still in place, as shown in E,3. Then each support is released from the back and pulled up about five feet. Before the next cut is started, the spill plate is advanced toward the face as shown in E,4.

Shortwall mining is a relatively new method and proponents of the method feel that it has the following advantages over conventional mining.

1. Miners work under cover of the roof supports at all times; therefore the hazards of roof and rib fall are reduced.
2. Because the flow of air is directed along the face, the miners work in intake air all the time, reducing dust hazards and methane concentrations.
3. The continuous miners do not have to be trammed as far for the next cut so more production time is available.
4. Recovery by shortwall is as good or better than room-and-pillar mining with pillar recovery.
5. Preliminary results show the costs are lower and production is higher per man shift than in conventional room-and-pillar.

In comparison to the longwall method, the following advantages appear.

1. It is easier to move and more flexible than the longwall system.
2. It does not require the extensive development of longwall and can be more easily integrated into conventional mining.
3. It requires less capitalization than the longwall method.
4. It is not as affected by rolls, clay veins, etc. as is the longwall method.

PLATE 8-7 SHORTWALL SYSTEM OF MINING

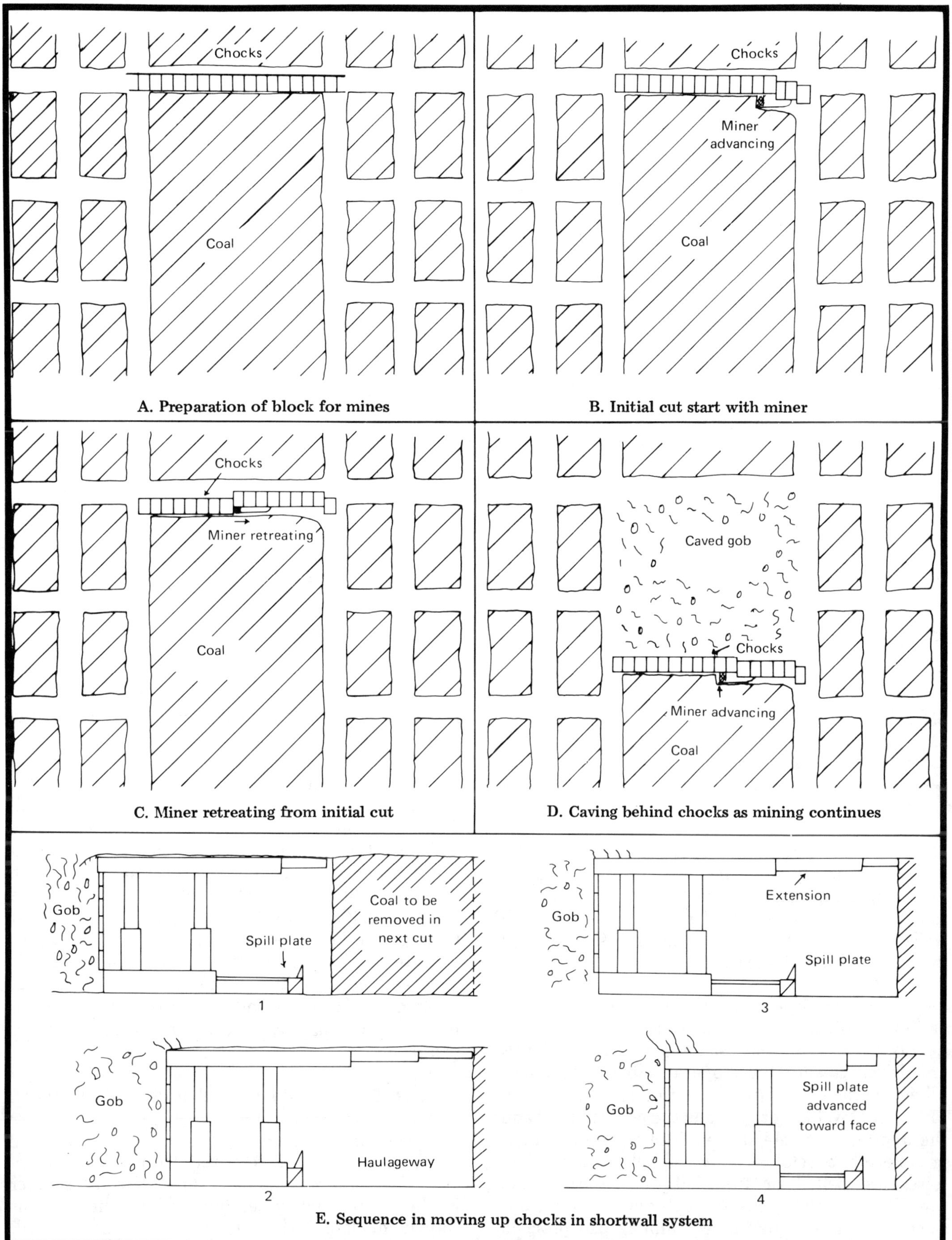

SUBLEVEL LONG-HOLE STOPES

A system resembling room-and-pillar mining can be used in dipping veins Plate 8-8 A, through E. For steeply dipping veins, 50° to 90°, the sublevel long-hole system is popular. The ground, both walls and ore, must be strong enough to stand unsupported over large spans.

A typical development for a sublevel stope is illustrated. A double raise is driven A; connections are driven between raises, and short levels are driven toward the stope at sublevel intervals. When the raise is completed B, the slusher drift and lowest sublevel are driven, and the next sublevel and the opening raise are started. The sublevel stope is completely developed C when all of the finger raises are belled or funneled out to catch the ore. This method requires much development, but once the stope is developed, mining can proceed rapidly. A three-dimension view of a model of the stope is shown in I, and J.

To start mining, it is first necessary to cut a slot across the ore, from footwall to hanging wall. This is usually done by drilling and blasting the walls of the opening raise C. Once this raise is open from footwall to hanging wall, mining can progress rapidly. The open raise provides a free face for blasting, and lines of parallel holes F or rings G can be drilled. The blast throws the rocks into the open stope. The whole mining face is advanced toward the manway raise D, and E. Broken ore can be removed from the stope as needed. Should the wall begin to fail, a pillar can be left E.

A drill H is used to drill long holes F, which may be as much as 100 feet in length.

The ore can be removed from the stope in any of several ways. The slusher drift and scraper is shown in this plate, but any of the systems shown in Plate 5-19 A, C, F, or G could be used.

PLATE 8-8 SUBLEVEL LONG-HOLE STOPES

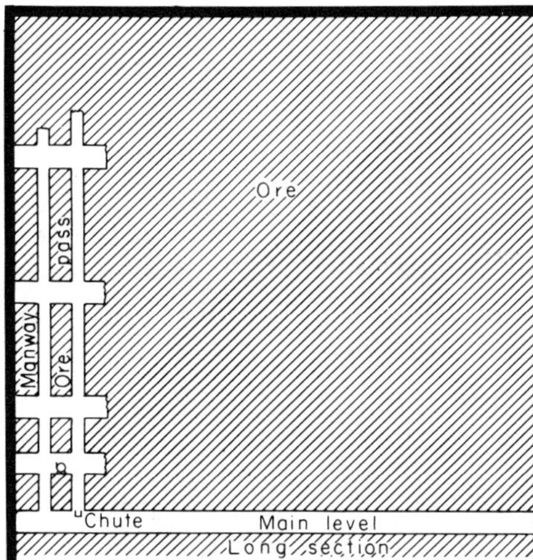

A. Initial development of a sublevel stope consists of driving raises to level above

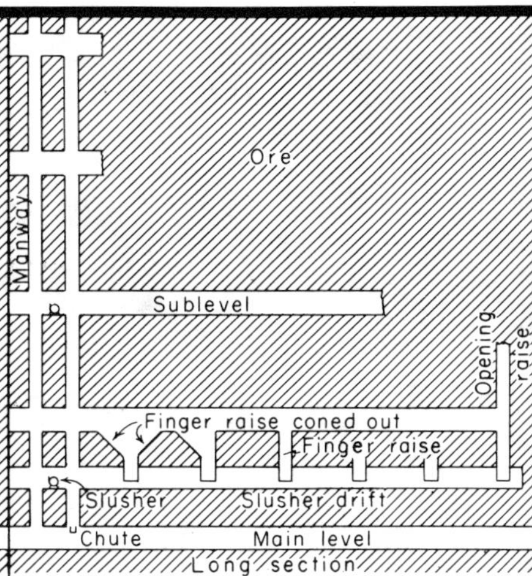

B. Raises are completed. Slusher drift, raises, and sublevel drifts are being driven

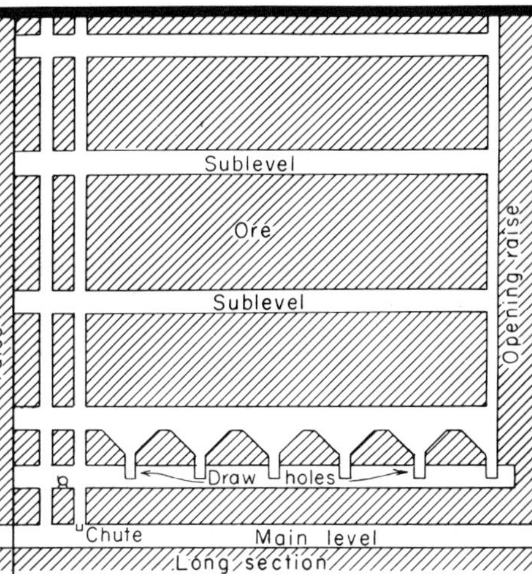

C. Stope completely developed. Opening raise is extended from foot to hanging wall

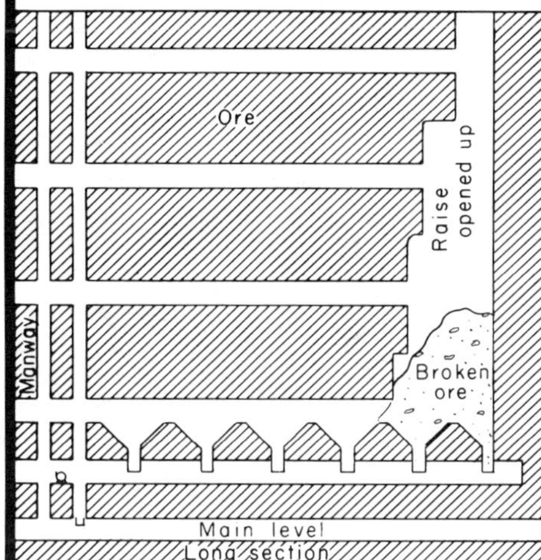

D. Mining in the stope progresses from opening raise toward manway

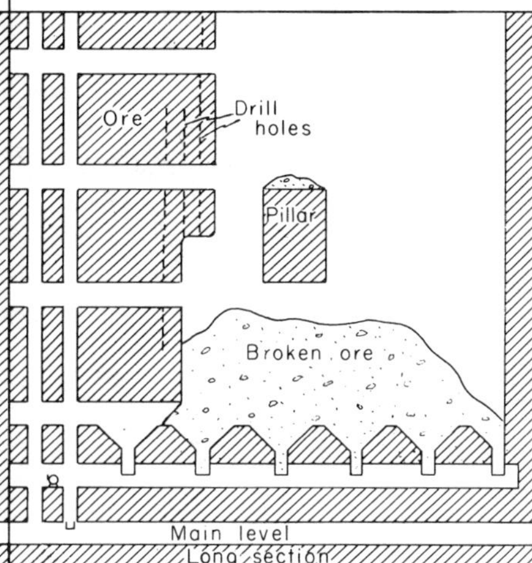

E. Stope being mined. Pillar is left as support for hanging wall

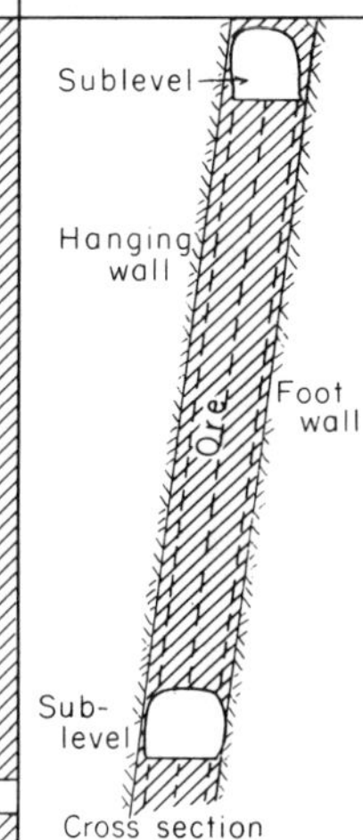

F. Long drill-hole pattern in narrow stope

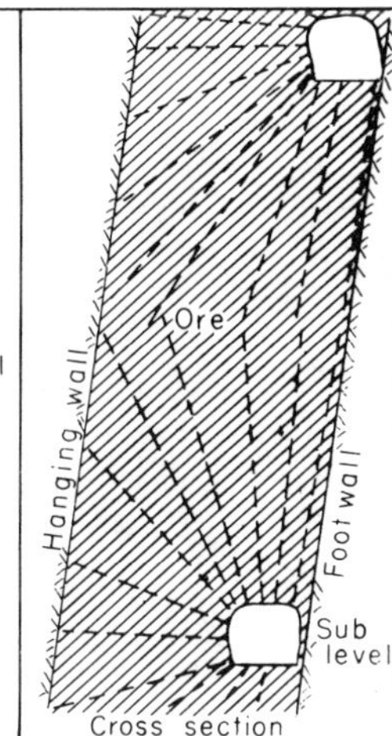

G. Ring drill-hole pattern in wide stope

H. Long-hole drill in a buslevel stope

I. Model of a sublevel stope

J. Model of a sublevel stope showing ore by dark color

DRILLING IN SUBLEVEL STOPES

Some ore bodies are pipes (Plate 2-ID). In the model (Plate 8-8J) of a sublevel stope designed for an ore body of this type, the dark color represents the ore and the white horizontal lines are sublevels. Because the ore body is so wide, it is necessary to use two sets of sublevel drifts.

Special long-hole drills A are used to drill the deep holes necessary for this system. Designed to drill with sectional steel, this drill can drill a hole 4 feet deep as it travels up the track. When the drill reaches the upper end of the track, the coupling is unscrewed and another 4-foot length of steel is inserted in the machine. Alternatively, the 4-foot length is unscrewed and an 8-foot length inserted. These machines can drill 100 feet or more.

RING DRILLING

Ring drilling is shown in B. The finished holes are plugged. The distance between the rings of holes is called the burden on the holes; it is the distance between the explosive and the free face. The machine is of slightly different construction than the one shown in A.

Sublevel mining, where applicable, has the advantage of producing ore at low cost, and the safety features of the system are good. Drilled holes can be loaded and blasted as the ore is needed.

A. A special long-hole drill used for ring drilling. Note sectional steel connectors. **B.** Ring drilling in a sublevel stope. The drilled holes are plugged to keep them clean.

PLATE 8-9 DRILLING IN SUBLEVEL STOPES

OPEN-STOPE MINING IN GENTLY DIPPING VEINS

Some bedded ore does not dip steeply enough to permit mining by sublevel methods because the broken ore will not run down the footwall. Therefore, it is necessary to scrape the ore down the stope.

The haulage level may be driven in the footwall just below the ore D. Chute cutouts are made at intervals A. After the chutes are installed, the short raises are widened to full stoping width B. The stopes advance in stairstep fashion C to reduce the effects of ground pressure in the stope faces.

As these headings are blind, i. e., not open to the level above, it is necessary to break into the neighboring stope for ventilation C. Air is circulated through the stopes by fans in these connections.

The system shown is actually a combination of timber and pillar methods. Narrow pillars separate the stopes, which would be called rooms if the deposit were horizontal, but two or three rows of timber stulls in the stope help to support the center until the stope is mined B, and C. These stope faces can be 70 feet or more in width, depending on the strength of the ground.

ANGLE OF REPOSE

A question may arise as to how steep the stope must be for the ore to run by itself. Theoretically, ore will run if the slope is steeper than the angle of repose of the broken rock. The angle of repose is the angle (measured from the horizontal) that the sides of a pile of the material assume naturally. For broken rock this angle is close to 40°. Therefore, if the ore is hard, it should run of its own accord at angles exceeding 40°, but if it contains much clay, it may require a steeper angle.

Production rates per man shift compare favorably with those of other underground mining systems; this is regarded as a low-cost underground method. The basic requirement is firm strong walls, which will stand unsupported over large spans.

PLATE 8-10 OPEN-STOPE MINING IN GENTLY DIPPING VEINS

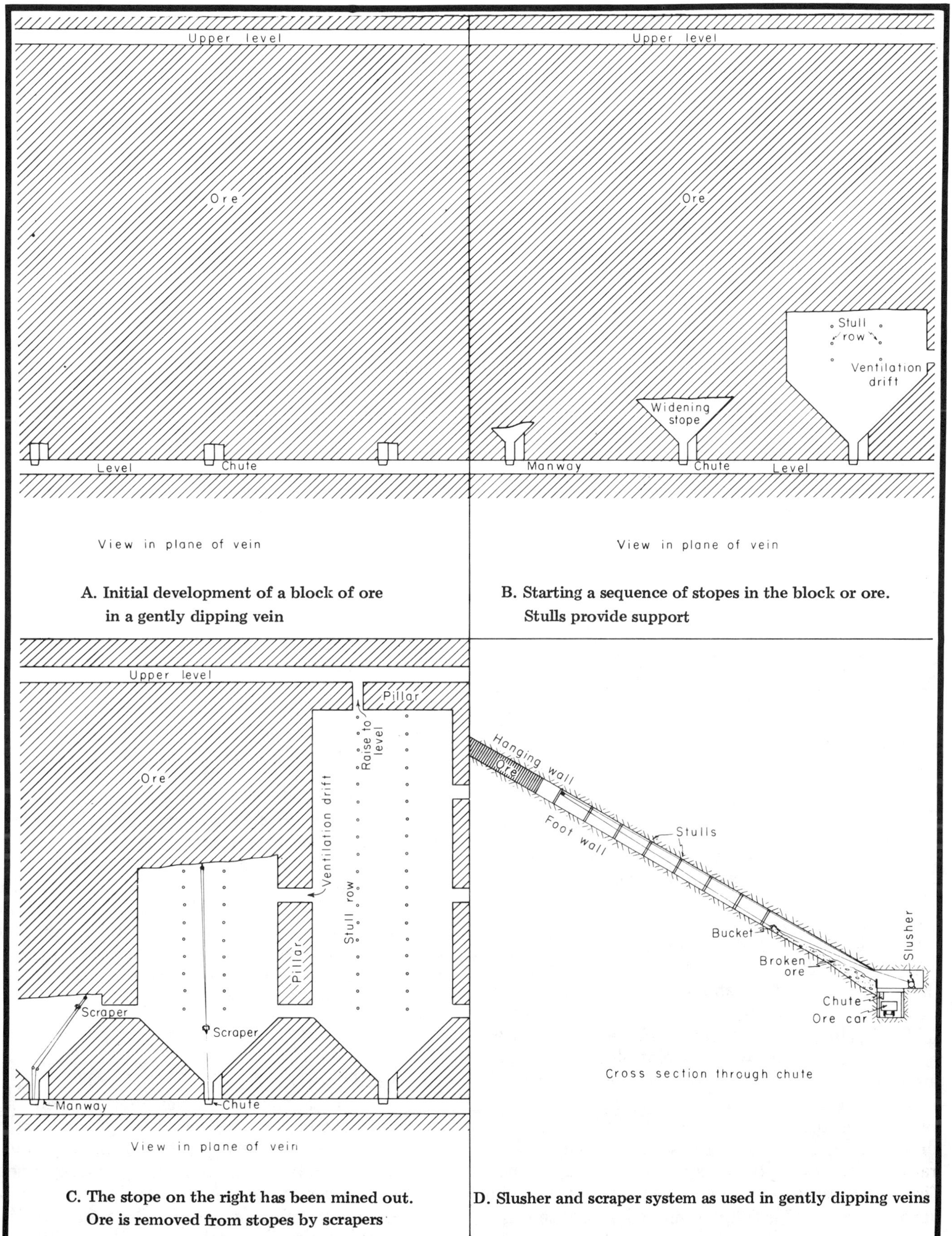

A. Initial development of a block of ore in a gently dipping vein

B. Starting a sequence of stopes in the block or ore. Stulls provide support

C. The stope on the right has been mined out. Ore is removed from stopes by scrapers

D. Slusher and scraper system as used in gently dipping veins

SHRINKAGE STOPES

The shrinkage method can be employed in steeply dipping veins if the ore and walls are strong and self supporting. Broken ore is left in the stope to provide a working platform for the miners. Because breaking the ore increases its volume by 40 percent or more, some ore must be pulled out of the stope as mining progresses, otherwise the broken ore would choke the stope.

ACCESS TO STOPE

At each end of the planned stope, raises are driven through to the level above A, and small drifts are driven at about 20-foot intervals to provide access to the stope as mining progresses upward. This stope is being developed with chutes at the mouth of draw holes. Not all shrink stopes are developed in this fashion. Timbered or cribbed raises at each end of the stope may be used instead of unsupported raises, and the short drifts are not driven. To gain access to the stope, the sides of the raise are removed. If raises are not driven through first, they must be carried up with the stope. The stope shown in B is fully developed except that the chute raises are not yet coned or funneled out.

After the chute raises are funneled out, mining is started C and progresses back and forth across the stope D. Air-leg drills are commonly used in these stopes, and a bench as much as 15 feet in length can be drilled and blasted in one shift. The series of benches across the stope makes one vertical cut in the stope. In pulling broken ore away from the solid ore face to provide working room, caution must be exercised not to pull too much, or the miners cannot reach the working face. When the stope is completely mined out E, it is full of broken ore, which is then removed F. Commonly a pillar is left near each level so that the drift will not be destroyed. The stope may be filled with waste at some later time. The pillars may or may not be recovered.

HIGH PRODUCTION PER MAN SHIFT

The shrinkage method gives a large production per man shift because scaling, drilling, and blasting are the main operations. The ore in the stope must be carefully drawn to avoid losing time by under pulling or over pulling. A big disadvantage is that a large amount of mined ore must remain in the stope until mining is completed.

PLATE 8-11 SHRINKAGE STOPES

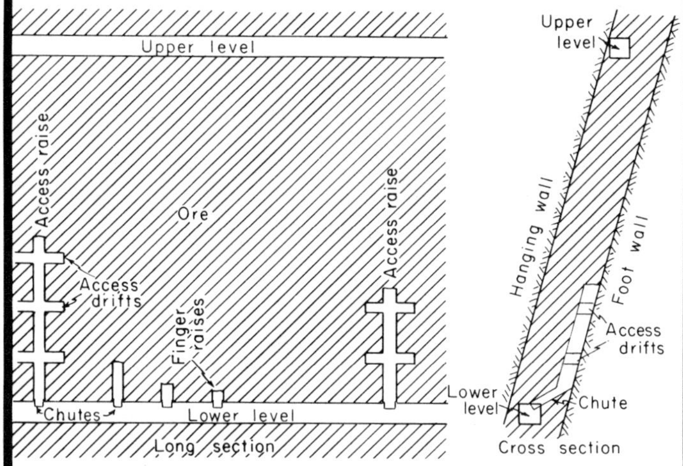

A. Driving raises and fingers to prepare a shrink stope for mining

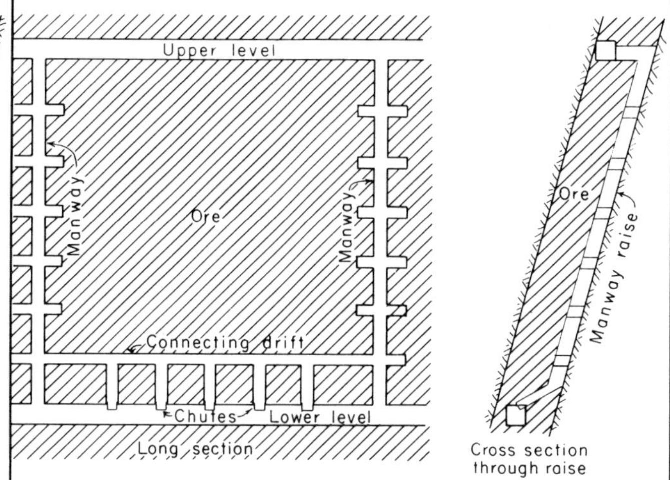

B. Stope developed and ready to be mined

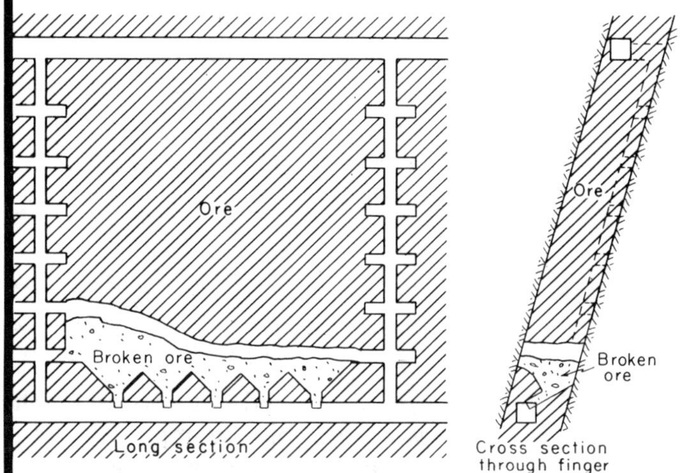

C. Finger raises are belled out and mining has started

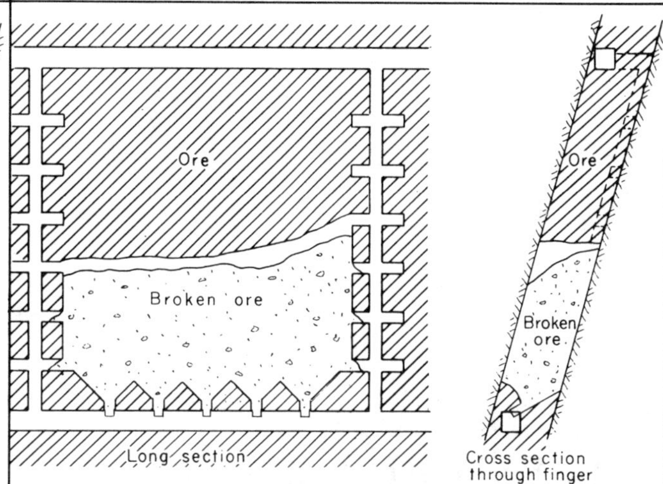

D. Mining progresses in the shrink stope

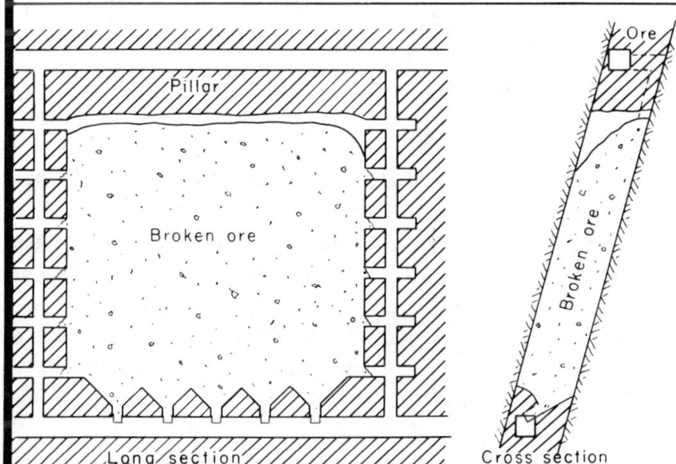

E. Mining is completed but the stope is still full of broken ore

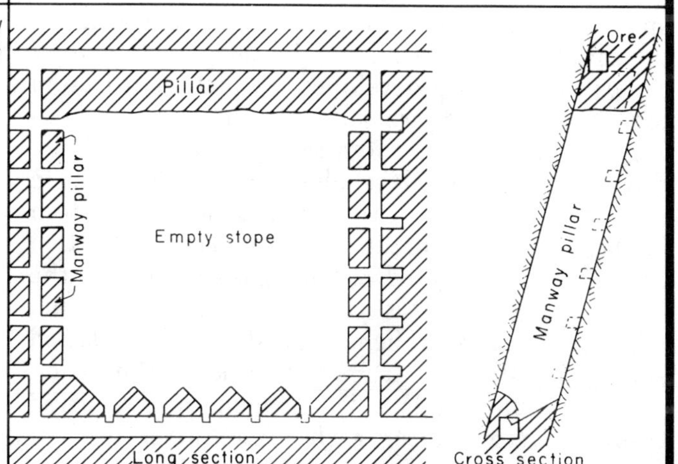

F. All broken ore pulled from stope; mining of this stope is completed

HORIZONTAL CUT-AND-FILL STOPES

In a cut-and-fill stope, a cut of ore is mined and waste is brought in to support the walls of the mined-out portion of the stope. Plate 8-12 shows this method developed by a lateral driven in the footwall and crosscuts driven into the ore zone at intervals A. From the crosscuts, raises are driven in ore to the level above. A drift is driven in the ore to connect two crosscuts at the bottom haulage level. When the drift connecting the two crosscuts is completed B, mining is started, and the mined-out portion is filled with waste or mill tailings. A series of benches is drilled and blasted across the stope C from one raise to the other, which completes a cut. The broken ore is removed from the stope after each blast. After the cut across the stope has been completed, the remaining broken ore is scraped out of the stope to the ore pass on the left side of the sketch D. After the ore has been scraped out, waste rock is dumped down the raise on the right from the level above. The waste is scraped into the mined-out portion by the slusher E. The waste is not piled completely to the back, as room is needed for the expansion of blasted ore. Mining of the stope progresses upward F.

The cut-and-fill method can be used only if the ore is fairly firm and the walls will stand unsupported until waste fill is brought in.

SAND FILL BY SLURRY

Sand or mill tailings are now commonly used instead of waste for support in cut-and-fill stopes. Sand can be brought into the stope as a water slurry in a pipeline. This reduces the scraping required in the stope. The sand fills all voids and forms a tight compact support after the water has drained away. Stopes of this type are shown in Plate 8-14 C, and D.

As a general rule, a floor is not built on top of the waste fill to keep the ore and waste separated. Repeated building and removing of a floor would cost more than the value of any ore lost in the waste fill.

PLATE 8-12 HORIZONTAL CUT-AND-FILL STOPES

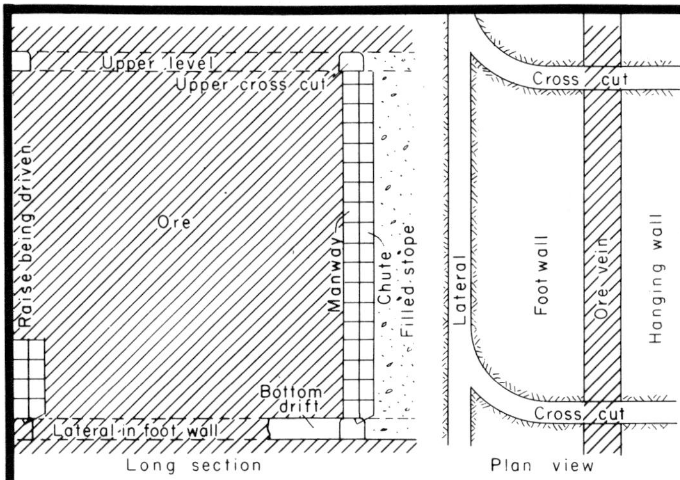

A. Initial development of a cut-and-waste-fill stope involved driving 6-post raise and bottom drift of stope

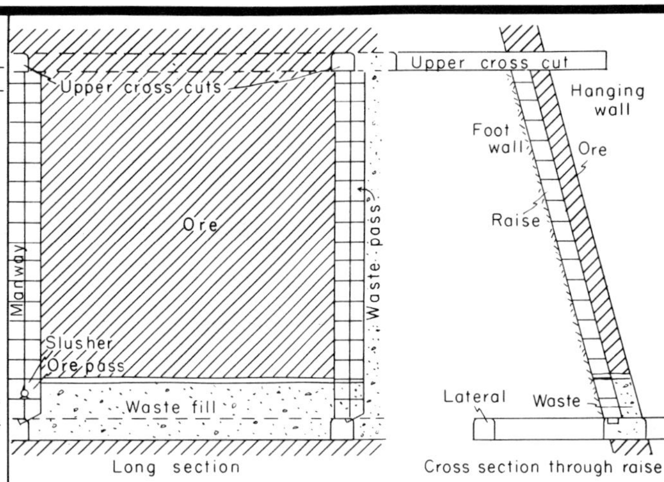

B. Stope during mining cycle. Sill and bottom two floors of stope are filled with waste

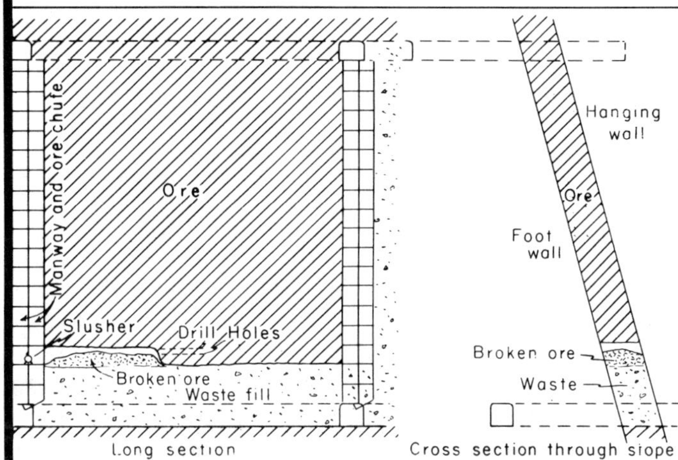

C. Breaking another cut across the scope. Only enough ore is removed to give working space for drilling

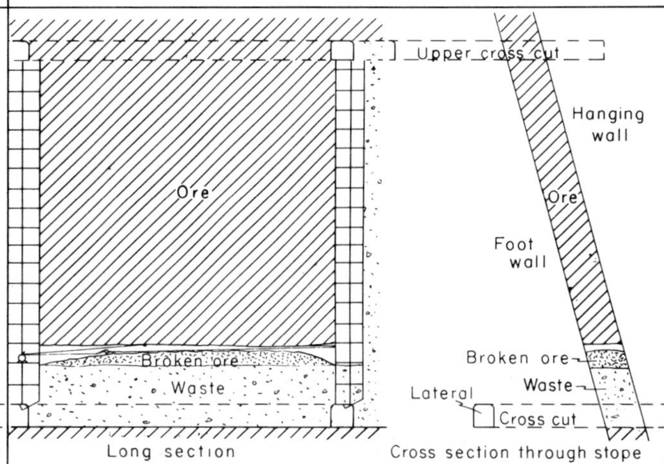

D. Breaking cut completed across stope. All of the broken ore is next removed.

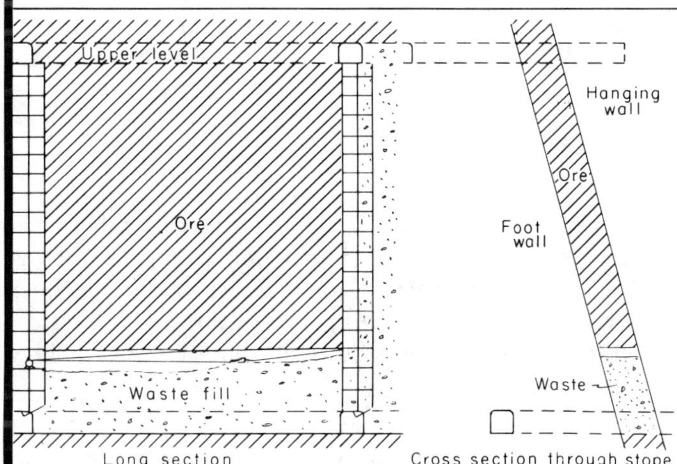

E. After all broken ore has been removed from the cut, waste rock is slushed in to fill the open space

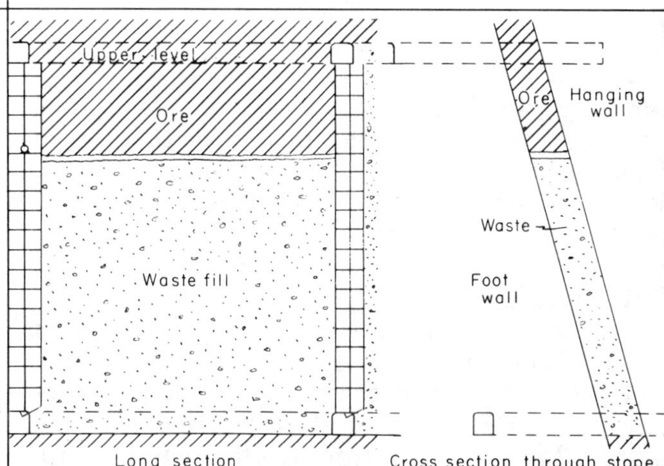

F. Stope cut filled with waste and ready for next cut. This view shows work much higher in the stope than E; several cuts have been made

RILL OR INCLINED CUT-AND-FILL STOPES

Efficiency in a horizontal cut-and-fill stope requires a slusher and scraper or other type of ore mover. Before the widespread use of slushers and scrapers, the rill stope was devised to use gravity to move the ore and to emplace the supporting waste fill. This sytem is seldom used today, but in special situations in a small mine this system could be used efficiently. Mining equipment may be at a premium in a small operation, and slushers and scrapers may not be available.

THE RILL STOPE

The rill stope is developed either by driving a raise at each end or by utilizing the raise of the previous stope A and driving a raise at the other end. The lower corners of the stope are mined, and waste is brought down from the level above to fill the corners B. The waste flows in and stops at its angle of repose. When the ore is broken, it slides down the top of the waste pile into the drift or into a chute F.

In this system, a center 8-post raise is required, but it is usually brought up as the stope is mined C. A cut is started in ore and is taken up from the center raise D. The ore is held in by timber stops placed against the timbers of the raise. The miners work from the pile of broken ore. After the cut is completed, left side D, the ore stops in the center raise are pulled out and the ore slides into the ore pass. When all of the ore has been removed, the stops are permanently placed against the center raise and waste is run into the stope from the level above E.

Common practice is to mine in one side of the stope while the other side is being emptied of ore and filled with waste D, and E. After one has been filled with waste, mining will start on that side, while the other side is being drawn of ore and filled with waste.

The ore must be strong enough to support itself over the long opening. The walls should stand over the unsupported height until fill can be placed. The ore should run well at the angle of repose and should be free of sticky clay.

PLATE 8-13 RILL OR INCLINED CUT-AND-FILL STOPES

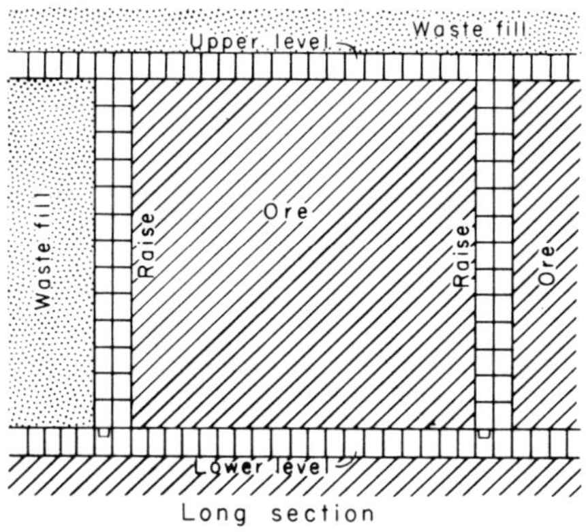

A. Rill-type stope developed with raises at each end

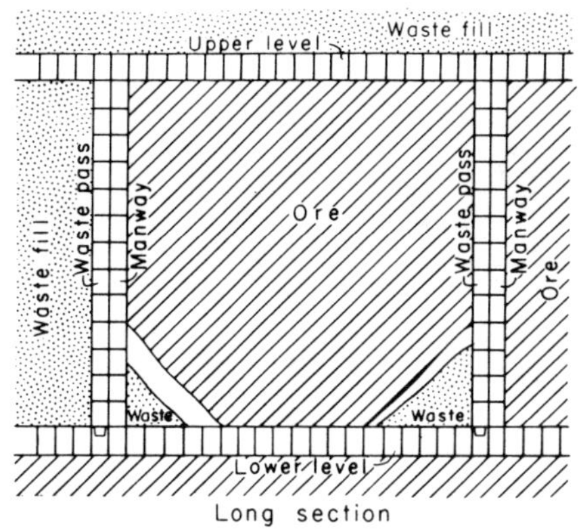

B. End cuts are being made and the mined portion filled with waste

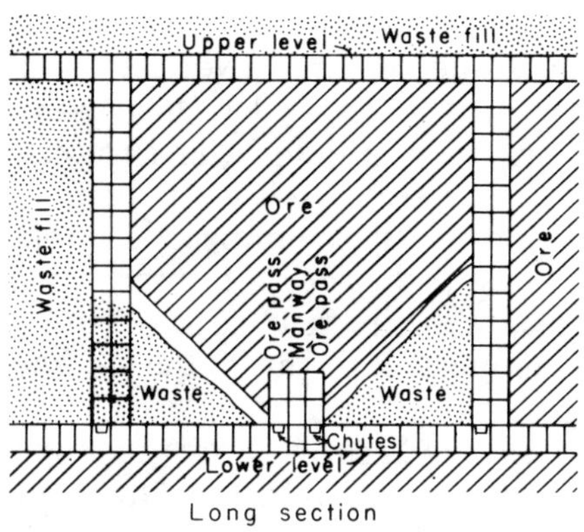

C. End cuts have progressed far enough that the 8-post center raise is started

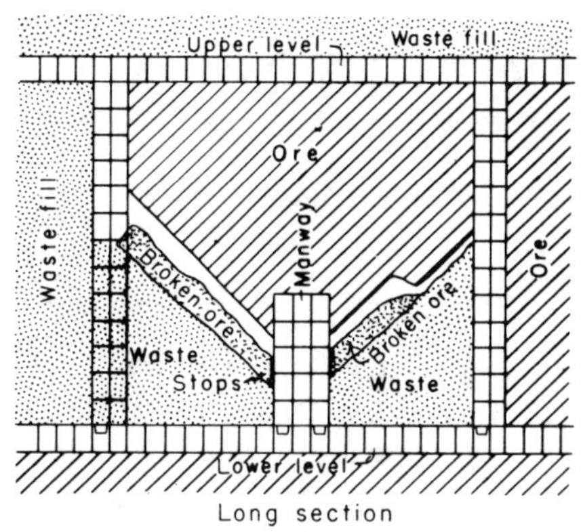

D. Stope in mining cycle showing the ore-breaking process

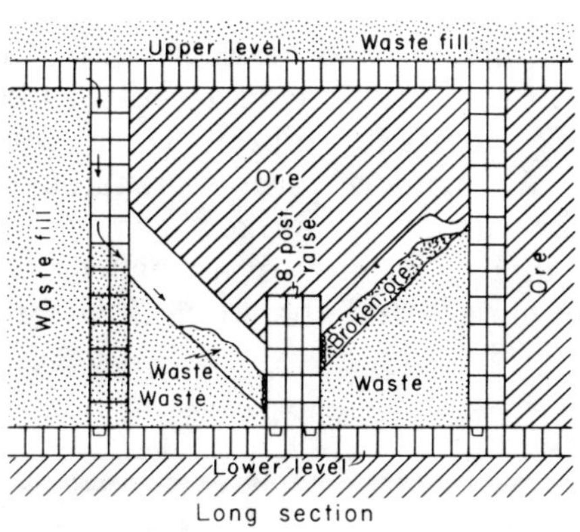

E. In cut on left the ore has been removed and waste fill is being introduced

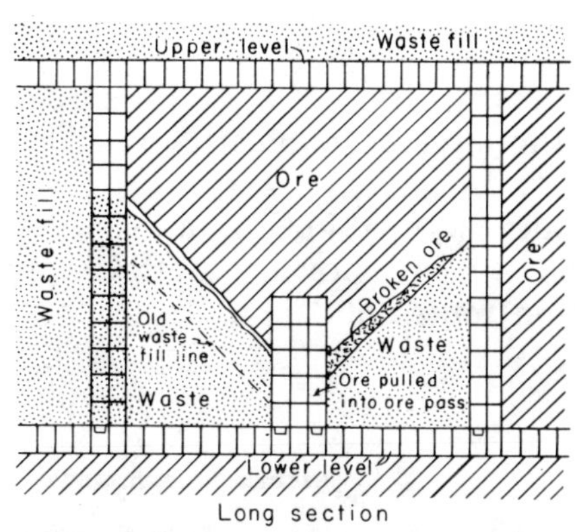

F. Cut on left has been filled with waste, and ore is being removed from cut on right

MINE MODELS

Picturing a mining method in two dimensions may be difficult unless a model of the system is made.

The models pictured in A show a very wide cut-and-fill stope. Ore passes at certain intervals in the stope are shown on the left side as light-colored semi-circular openings. The fill is below the horizontal opening or cut, and the ore is above this opening. Waste pass raises, one of which is shown, are driven to the level above. The model on the right in A shows a timbered system, which is used when the stope approaches the upper level. The ore must be very strong if this system is to be successful.

The model pictured in B is a rill stope system, which is the system explained in Plate 8-13.

The model pictured in C is a cut-and-hydraulic-emplaced-fill system showing several stopes. The white portion represents fill and the dark part is ore. The stope on the lower left is ready for fill; the stope in the upper left has been filled and is again ready for mining. The first cut in the preparation of a stope is shown in the lower right. When the sand fill is emplaced, it sets almost like concrete, and therefore the ore is mined horizontally in the first cut toward and to the waste already emplaced in the stope on the lower left.

The picture D shows how the stope looks when all of the ore is removed and the stope is prepared for filling. Note the rock bolts and headboards supporting the ore back and the hanging wall. The manner in which the raise is prepared in order to stop the sand from flowing down the raise is shown in Plate 5-12F, as this cut is being filled.

SLUSHER DRIFT

A slusher drift to draw ore out of the stope is shown in E. The white represents broken ore, and the dark represents unbroken ore, which is to be mined. A shrinkage stope prepared with chutes through which the broken ore is drawn out is shown in F. The chutes are belled out, the raises are completed, and the stope is ready for mining.

A. Models showing cut-and-fill stope method used in wide veins. Model on right shows conversion to timber. B. Model of rill stope. C. Model showing hydraulic stope fill; stopes in different stages of operation. D. Cut-and-hydraulic-filled stope in Butte. *(Courtesy Anaconda Co.)* E. Model of shrink stope. As mining progresses, ore is removed from this stope through slusher drift. E. Model of shrink stope. As mining progresses ore is removed from this stope through slusher drift. F. Model of shrink stope just as it is ready for mining.

PLATE 8-14 MINE MODELS

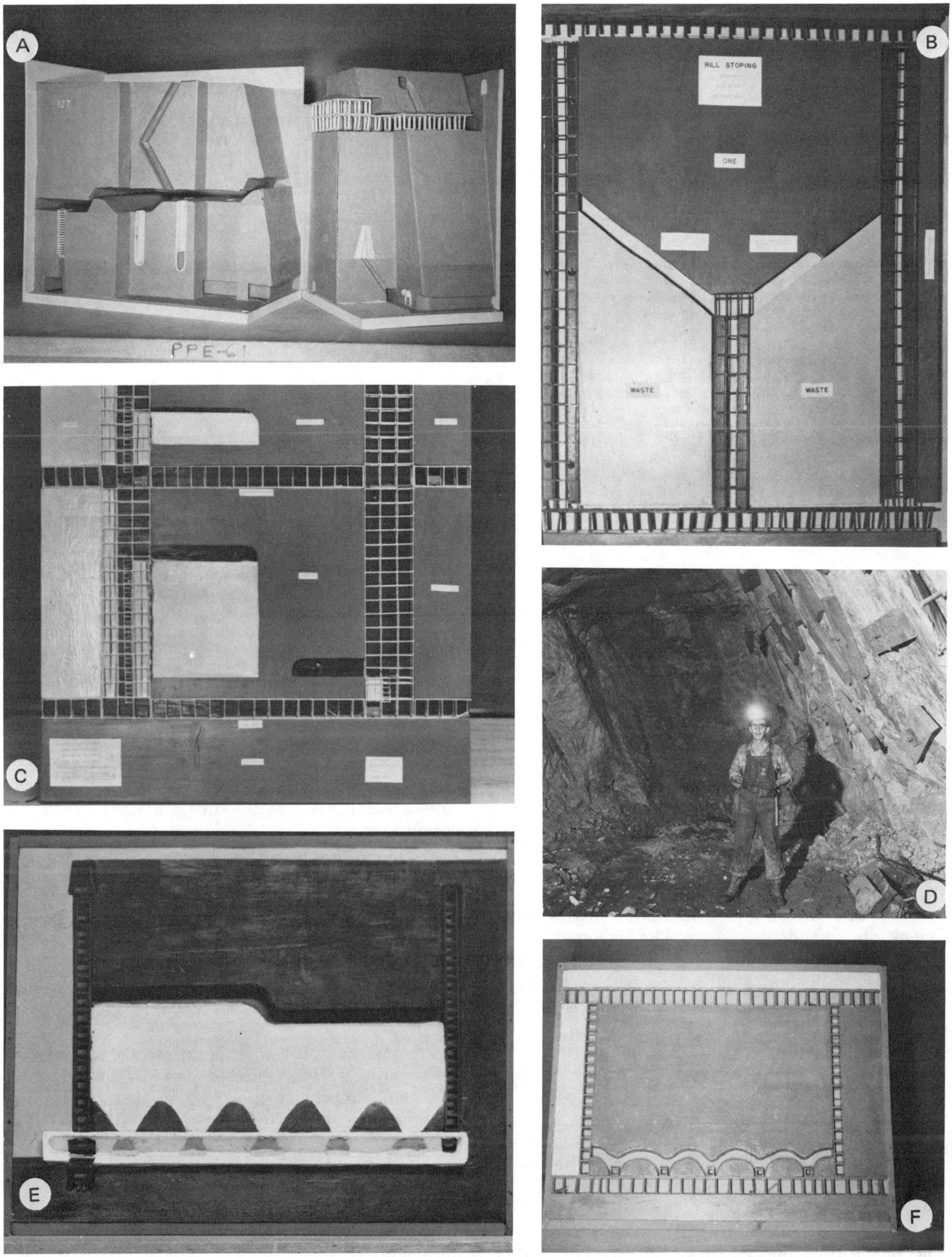

RESUING METHOD

Much high-grade ore occurs in narrow veins, but to mine the ore it is necessary to have sufficient mining width and room. The high-grade ore is not rich enough to pay for hauling waste to the smelter or mill, and hand sorting the waste from the ore is ineffective, tedious, and expensive. The resuing method is an attempt to keep the ore and waste separated during mining.

SEQUENCE OF OPERATIONS

The sequence of operations is shown in Plate 8-15. When the stope has been started A, the ore and waste are drilled (cross section). The waste is loaded and blasted first (cross section, B). The waste is leveled off and if the ore is high grade, a wooden floor may be laid on the waste. The ore is then blasted down and scraped out of the stope C. After the ore is blasted down, the stope is ready to be drilled again (cross section, D).

PRODUCTION PER MAN SHIFT LOW BUT LITTLE ORE WASTE

Production per man shift by this method is low, but the ore is not diluted with waste. The walls and the ore must stand very well, or stulls may be required. Keeping the ore from breaking with the waste may present a problem.

Sometimes the ore is blasted and removed before the waste is blasted down. Usually a trial-and-error approach must be used to determine whether the ore or waste should be blasted first.

If the waste is insufficient to fill the mined-out portion, a raise may be driven in the hanging wall to provide additional waste for fill. Another plan of a resuing stope is shown in the picture of a model, Plate 8-22 E.

PLATE 8-15 RESUING METHOD

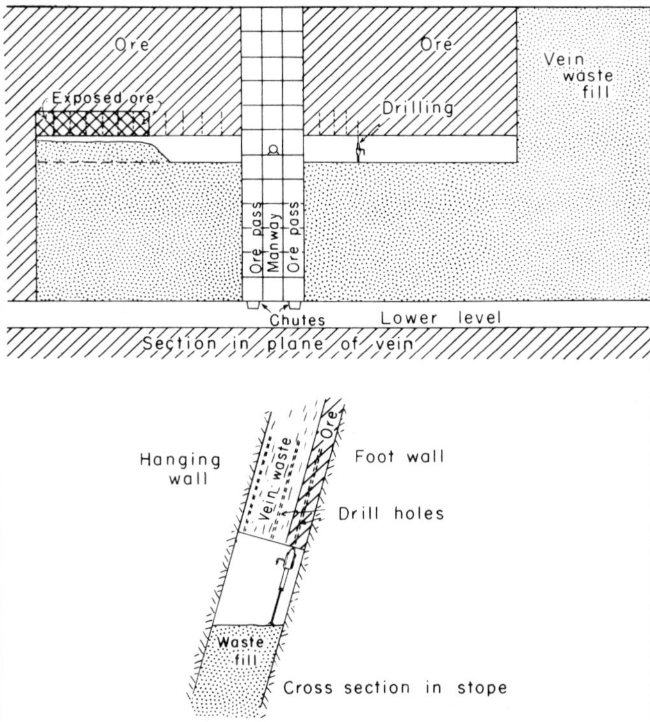

A. View of resuing type stope showing drilling operation

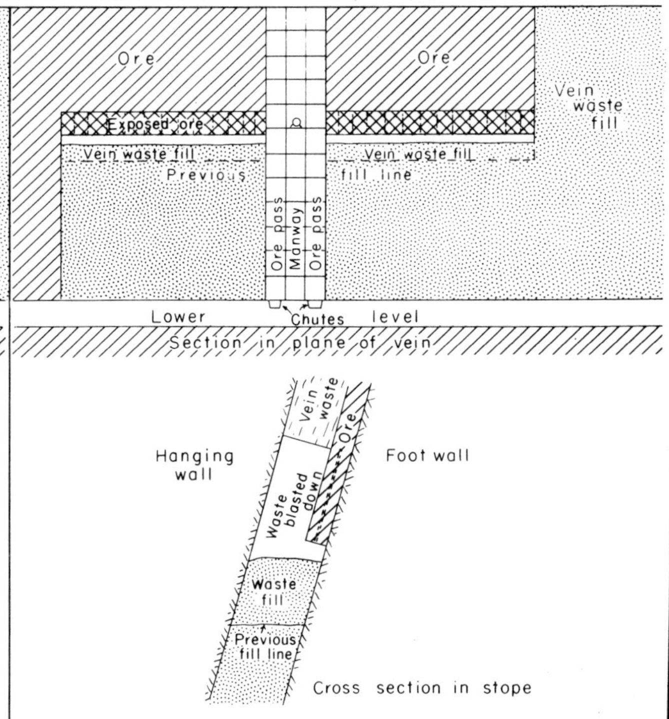

B. View of same stope after vein waste is blasted into slope and levelled

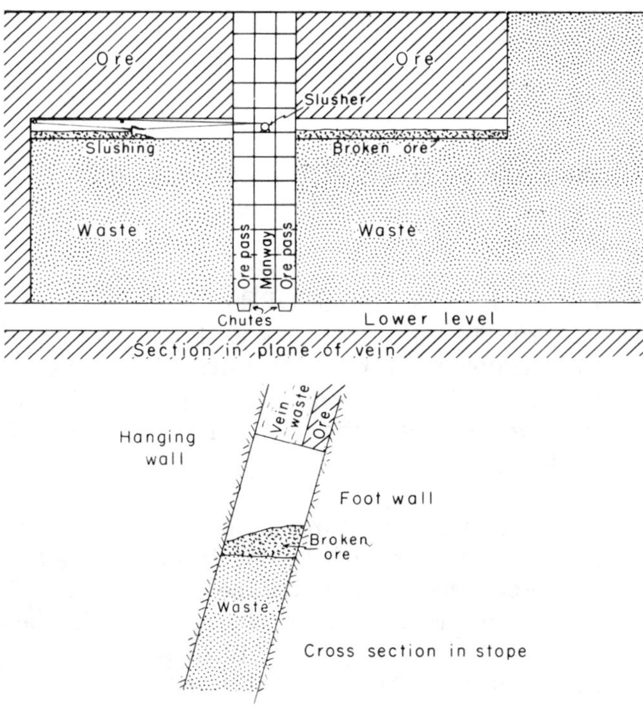

C. View of same stope after ore is blasted down on waste filling

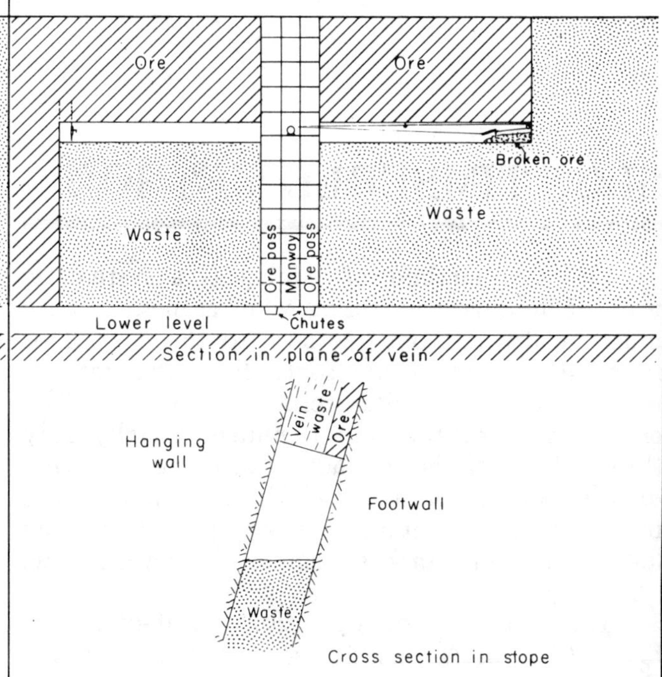

D. Ore scraped out; drilling cycle can start again

SLOT TIMBERED METHOD

The slot method of mining is a systematic way of mining and filling wide stopes. A raise must be driven from one level to the next. This raise extends from the footwall to the hanging wall of the vein and is termed a slot raise. The entire width of the vein is mined. Usually only three sets of timber are advanced along the strike of the vein in each stope. During mining, it is necessary to maintain a waste pass from the level above and an ore pass to the level below.

The sequence of operations is depicted in Plate 8-16. In A, one slot stope has been mined through and two more stopes are being driven from this initial mined-out and filled stope.

In the stope on the left side, the slides direct the broken ore to the ore pass. After the slides are raised, waste is dropped down the waste pass from the level above to fill in under the slides. Mining in the stope on the right is progressing upward. After two or three floors are mined out, the slides must be raised.

PROGRESS IN THE STOPES

Progress in the stopes is shown in C, and D. This sequence is practiced until both stopes reach the upper level. Then new stopes are started in the ore next to the bottom of the mined-out stopes.

A cross section through an ore pass is shown in B. Ore breakers of heavy timber on alternate timbered sets tend to reduce the velocity of the broken ore as it falls, thereby reducing the damage to timber in the raise. Alternatively, one compartment may be lined with lagging on all four sides to provide a narrow ore pass.

Timber methods are used in weak ground that requires continual support. These methods are slow and costly, therefore the ore must be reasonably high grade.

PLATE 8-16 SLOT TIMBERED METHOD

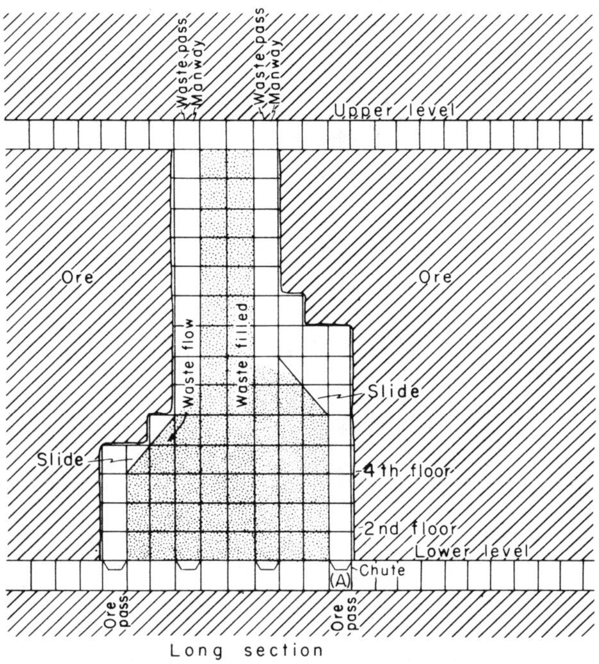

A. Timbered slot type stopes progressing in both directions

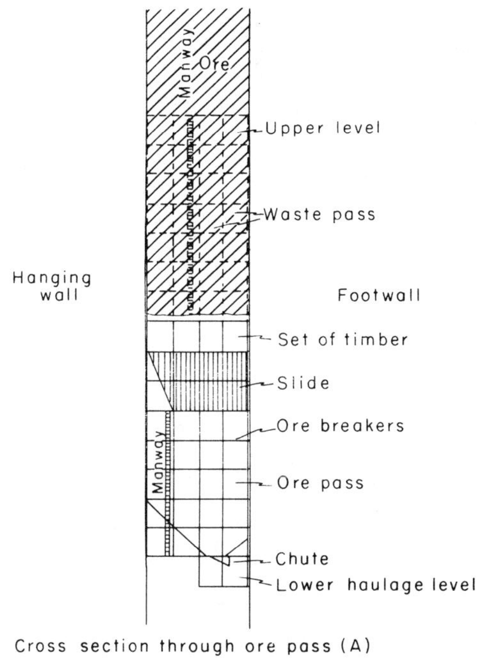

B. Cross-section of a slot slope four sets wide

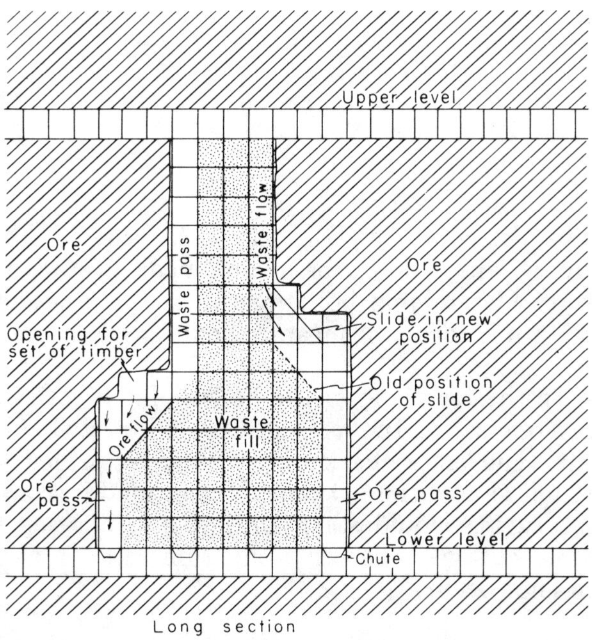

C. Another view of the stope shown in A but each cycle more advanced. Mining is proceeding in the left side; the slide is being moved on the right

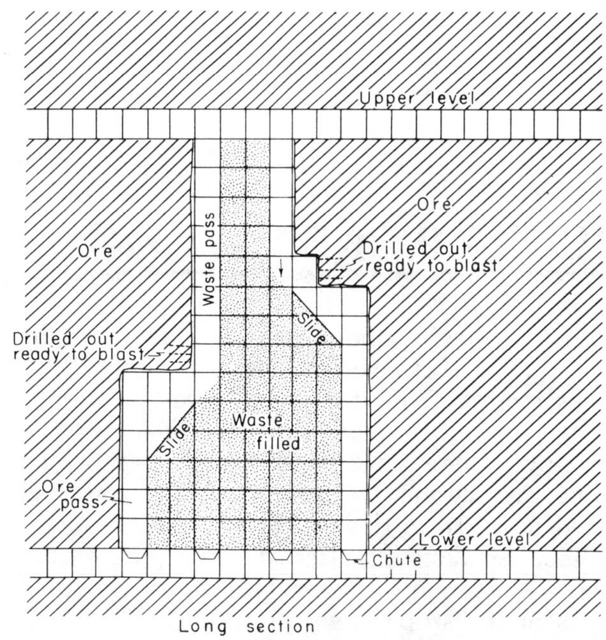

D. One more set of timber is to be installed on left side before raising slides; several more sets are to be installed on right side before raising slides

SLICE TIMBERED STOPE

A narrow vein, about one set wide, can be mined by the slice system (cross section, B). A stope is started from the lower level by mining three or four floors up and dropping the ore to the level below, where it can be loaded with a mucking machine. A slide may be installed A and waste fill dropped down from the level above to fill in under the slide. Breakers in alternate sets protect the timber in the stope from breakage by falling rock. Progress in the stope is shown in views B, and C.

When the stope has progressed to a certain point C, the slides must be raised because of heavy ground. Waste fill is introduced below the slide, and mining continues in the stope D. Note the manway, on the right side in these stopes, which must be maintained for the next stope. Maintaining both manways as the stope is mined provides two means of exit from the stope.

In the benching method of drilling B, space for only one set of timber is drilled and blasted at one time. The entire back or cut may be drilled and blasted at one time D unless the ground is so weak that caving would start. Drilling and blasting the whole cut gives better production rates if caving does not occur.

If the ground is not too weak, the slide may not have to be installed until the stope is completely mined. Waste is not introduced until the stope is completed. A model illustrating such a method is shown in Plate 8-18D.

PLATE 8-17 SLICE TIMBERED STOPE

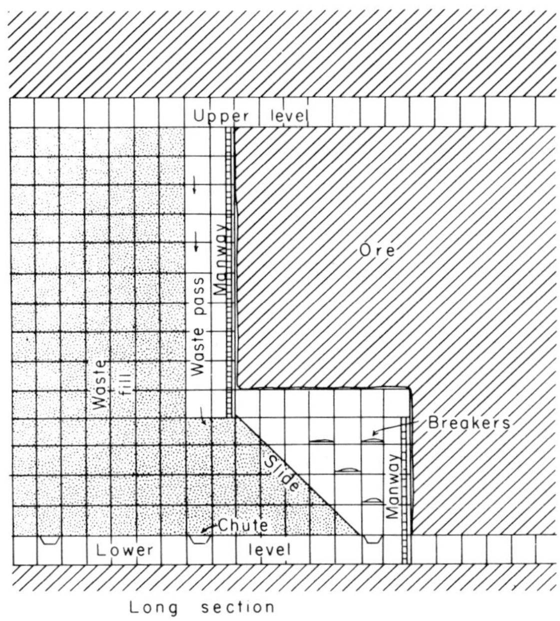

A. Slice type of timbered stope. Ore flow is interrupted by breakers in stope

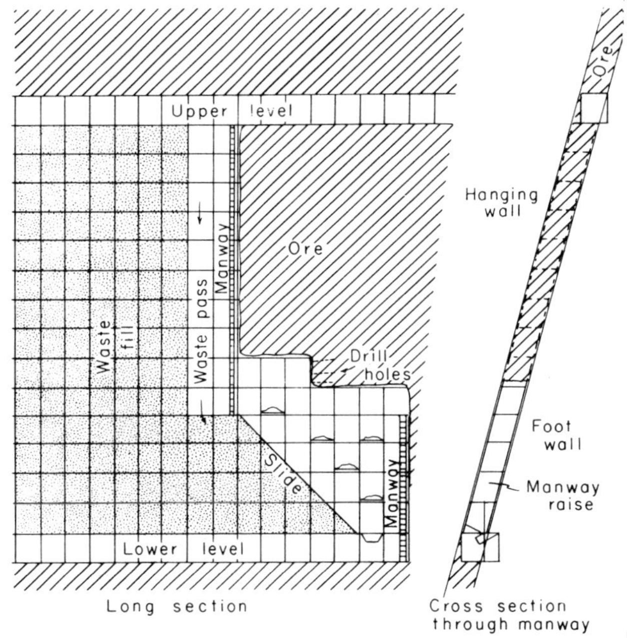

B. Another floor is being mined, one set of timber at a time

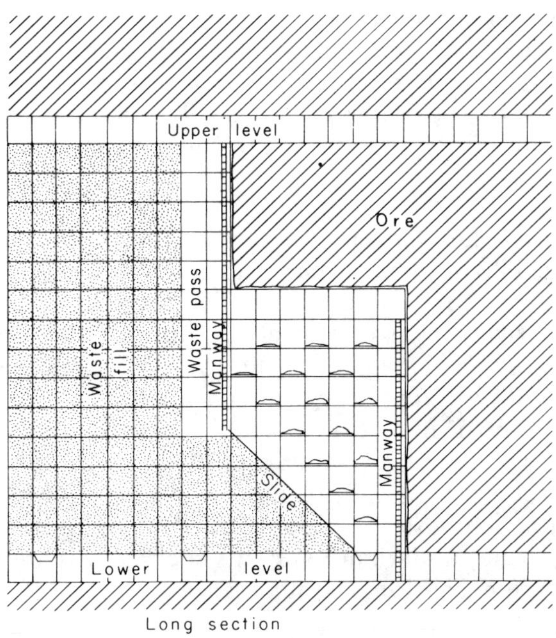

C. Stope after several floors have been mined out. It is necessary to clean the breakers and raise the slides

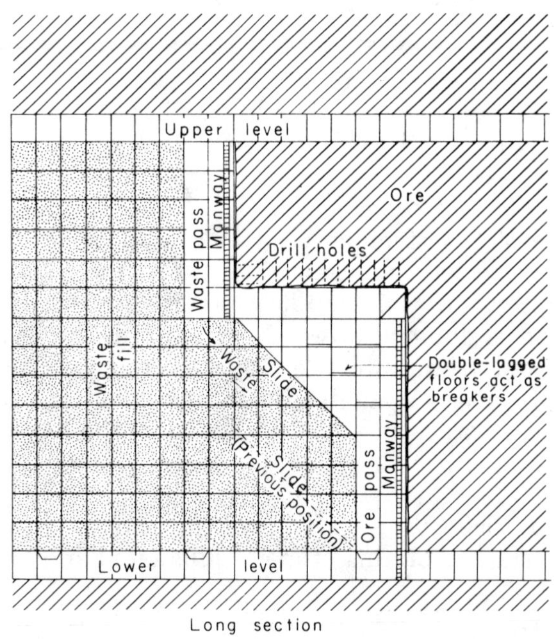

D. After the slide has been raised, waste can be introduced below the slide. Entire back of stope has been drilled

MODELS OF TIMBERED STOPES

Two views of a model A, and E illustrate a slot stope. The ore is at the left in A. That part of the model is removed in E to give a view of the ore pass.

The slot system as used in a dipping vein C requires special arrangement of the timber. When sand instead of mined-out waste is used for fill, the sand is dumped down the raise from the level above and flows into the stope by gravity. Sand not only makes a tighter fill and gives better support than waste rock but also reduces the loss of ventilation air.

The slice stope D is used in narrow veins. The operation of this stope is illustrated in Plate 8-17.

Fill is not used in some applications of the square-set timbered method, or if fill is used, it may not be introduced as systematically as in the slot or slice method. The model pictured in B shows a square-set stope with the raise in the center of the stope. Ore is usually blasted down on a timber floor and scraped to the ore pass.

Another variation of the timber method is the Mitchell top-slice stope F. Two slot raises (extending from foot to hanging wall) 15 to 20 feet apart are driven from one level to the next. The pillar between is mined from the top down. As the ore is removed, long heavy timbers are used to connect the timber in the raises to keep them from collapsing. As mining progresses downward, waste fill is brought in to fill the mined-out area from above, and to support the walls of the mined-out portion. As a rule, underhand systems such as this are used only in exceptional circumstances.

A. Side of slot-stope model. Block of ore is at left. **B.** Model showing square-set stope. Waste fill need not be introduced until the stope is empty. **C.** Model showing slot stope adapted for dipping vein. Sand is used for fill. **D.** Model of slice stope in vertical vein. **E.** End view of model shown in A with ore removed to show breakers and slides, which direct flow of broken ore. **F.** Model showing Mitchell slot-stope method. Ore is mined underhand, progressing downward.

PLATE 8-18 MODELS OF TIMBERED STOPES

BLOCK CAVING

Some ore caves readily, and if the ore body contains enough tonnage, the block-caving method may be used. After the stope is developed, the ore breaks of its own accord; it does not have to be drilled and blasted. Caving is a large-production low-cost method. If an opening is large enough, it will eventually cave, even in the firmest and strongest rock, but a caving system of mining requires that the ore or rock will cave over a small unsupported area.

SLUSHER DRIFTS

A common method of developing a block of ore for mining consists of first driving a slusher drift in the ore A, and B. Slusher drifts are spaced at suitable intervals in the block to produce efficient caving above the fingers and are usually spaced so two or three cars in the ore train can be loaded at the same time. There may be as many as five slusher drifts under each block. Slusher loading of cars is shown in Plate 5-19 C, G.

From the slusher drift, finger raises are driven to the undercut level C, and D. The tops of the finger raises are connected by crosscuts and drifts E, and F. The crosshatched portion in F represents the supporting pillars, which keep the overlying ore from caving. The tops of the finger raises are drilled and blasted to a funnel shape G, and H. The supporting pillars are either drilled and blasted when the raises are enlarged or immediately afterward in sequence H, 1 to 11. As blasting of the raises and pillars progresses, the stope begins to cave, cross sections, I, and J.

CAVING ACTION

When the broken ore is pulled from the back of the stope by drawing ore from the raises, cracks form and the ore still in place tends to break by its own weight and fall to the pile of broken ore. Because the ore increases in volume when broken, the broken ore will soon fill up to the back, which in turn gives support to the back and thus stops the caving. The more rapid the rate of draw the more rapid the caving action. Too rapid a draw of one finger may cause overlying waste to come through the stope and into a finger. Therefore, all of the fingers must by carefully drawn to insure even caving action and to prevent overlying waste from coming through the fingers before all of the ore is pulled out.

PLATE 8-19 BLOCK CAVING

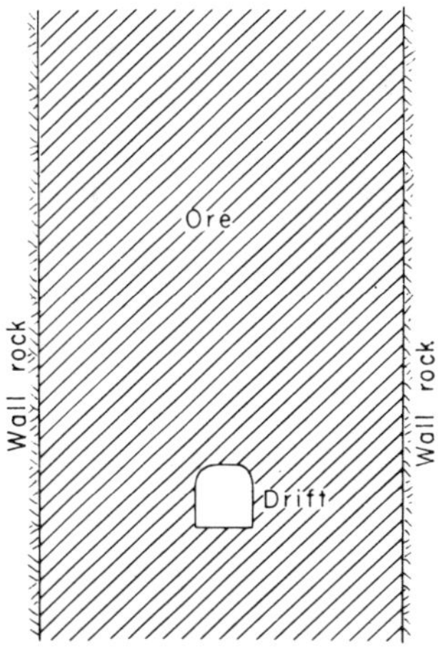

A. Cross-section of slusher drift in ore body to prepare for caving

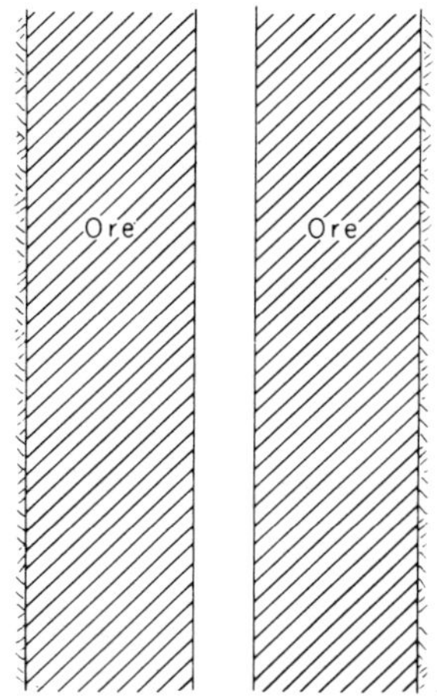

B. Plan of slusher drift in ore body to prepare for caving

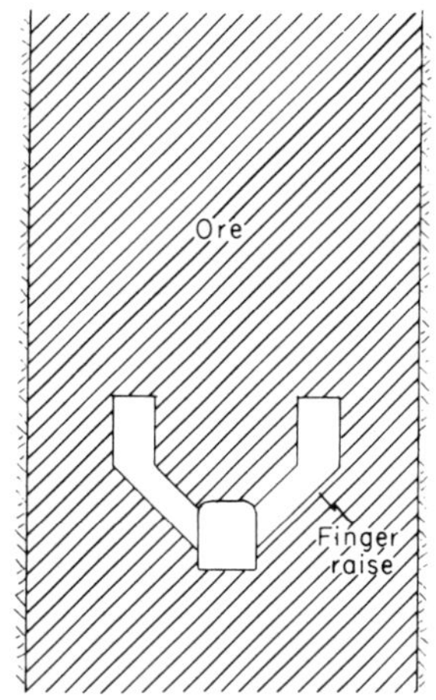

C. Cross-section of finger raises from slusher drift

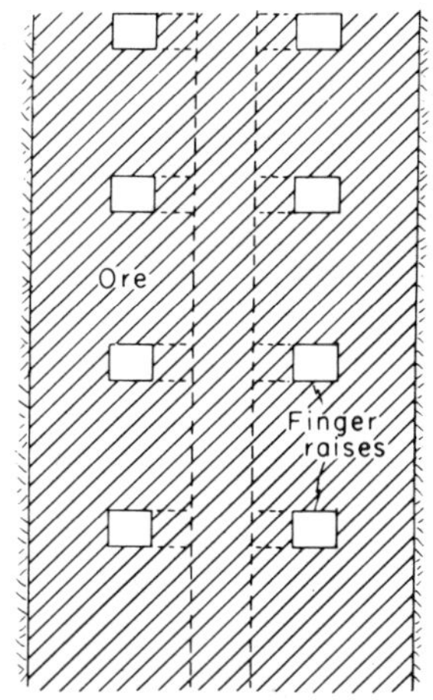

D. Plan of finger raises from slusher drift (dotted lines outline lower workings)

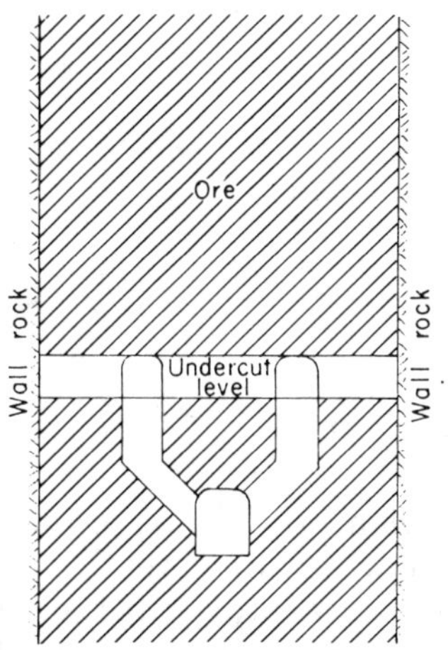

E. Cross-section of undercut level to prepare for caving

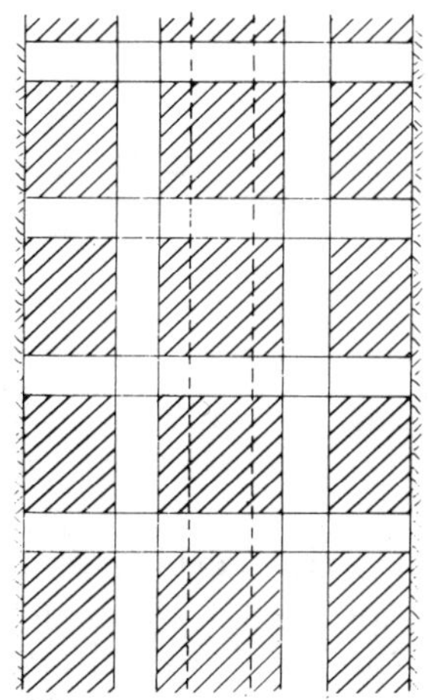

F. Plan of undercut level

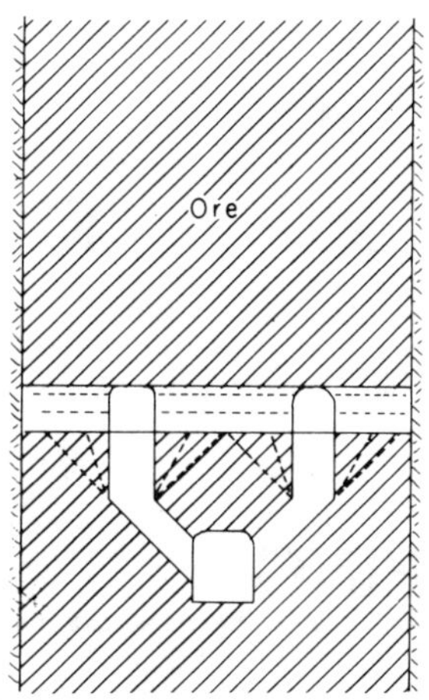

G. Cross-section showing drill pattern to enlarge fingers and start caving

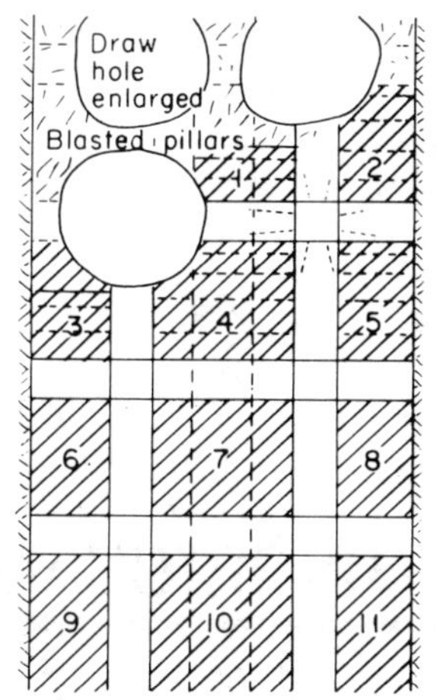

H. Plan view of undercut level showing pillar-blasting sequence

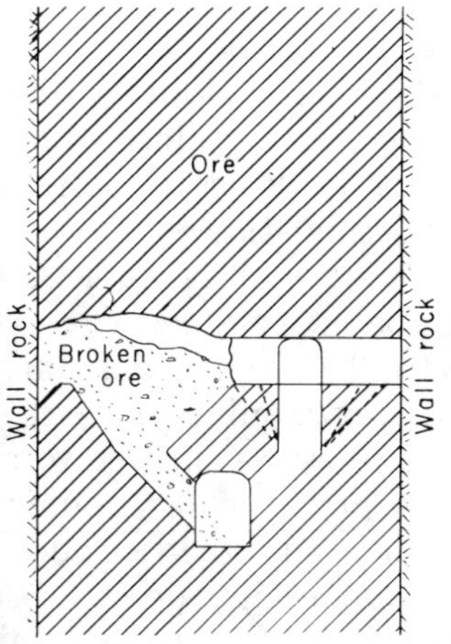

I. Cross-section of undercut level after part of ore has started to cave

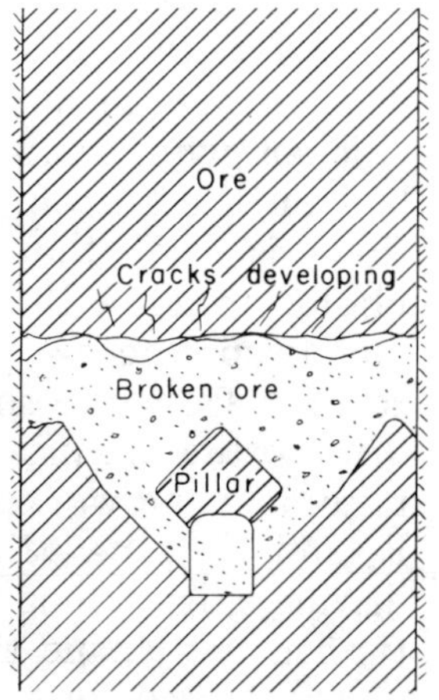

J. Cross-section of cave stope after undercut level has been blasted

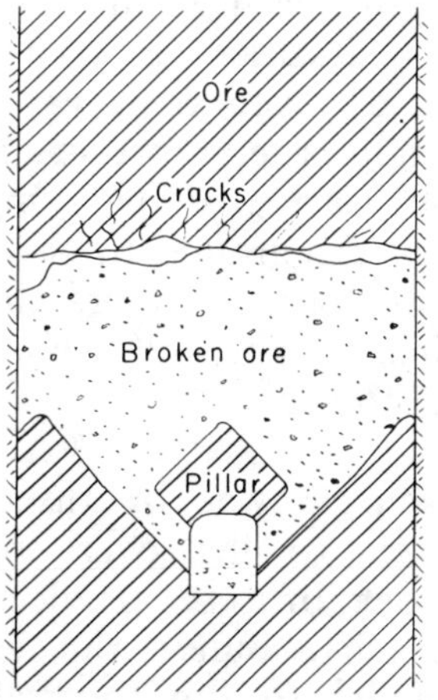

K. Cross-section showing development of cave

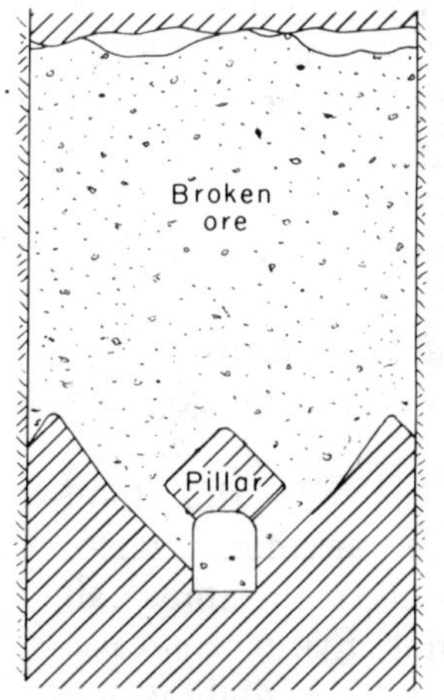

L. Cross-section showing progress of cave

TOP-SLICE STOPING

The top-slice method yields greater production per man shift than the square-set timbered method. The supporting timber in a top slice is only temporary, therefore a cheaper grade can be used, and only that for the first stage of development need be framed. This method is classified as a cave system because the overhead waste rock caves downward. The ore does not cave, however; it must be drilled and blasted. Caving of overlying rock will cause the surface of the ground to cave, therefore this method cannot be used where the surface property is valuable. Almost all caving methods destroy the surface, however.

From the haulage level, a two- or three-compartment raise is driven to the overlying rock, sometimes called cap rock, or to the previous mat. From this raise a drift is driven under the cap rock or mat near and parallel to the footwall (long sections, A, B, and plan, B). The drift extends to each end of the block. From the ends of the drift, crosscuts are driven to the hanging wall (cross section, A, and plan, B). These crosscuts are then mined out to a width of several sets of timber, and mining progresses toward the raise (plan, B). The small crosshatched areas in the long sections on the mining floor represent slices that have been mined out but not yet caved.

SEPARATING ORE AND ROCK BY MATS

After the slice has become wide enough, usually three or more sets wide, a floor of timber, chicken wire, scrap lumber, etc., called a mat, is laid on the bottom of the workings C. After the floor is laid, the supporting timber posts are blasted to allow the overlying rock to cave D. The mat keeps the ore and rock separated. The ore is scraped from the slice to the drift and from the drift to the ore pass. The ore is mined toward the raise, and the worked-out areas are allowed to cave E, and F. A new floor is prepared for mining by driving a drift along the footwall directly under the timber mat F, and the sequence is repeated. Stulls or props used for support of the mat while mining, and which are blasted as mining progresses, become part of the mat that descends to form the next mining floor.

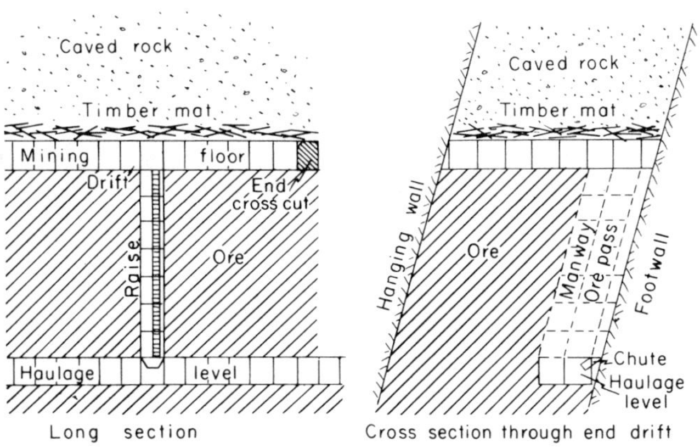

A. Long and cross-sections of top-slice caving stope. Initial development of mining floor

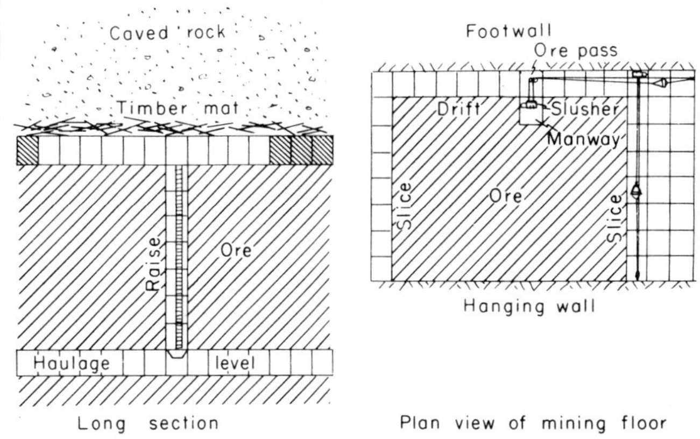

B. Long-section and plan view of top-slice stope as mining progresses

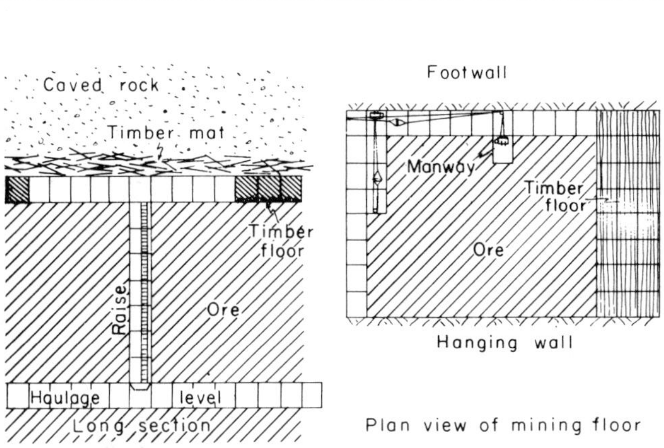

C. View of stope as part of mining floor is prepared for caving

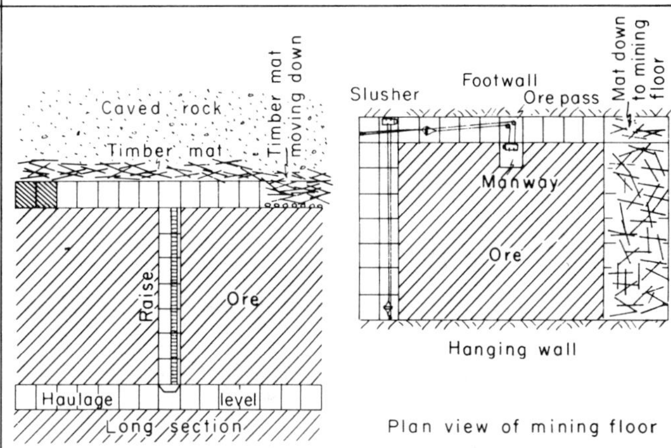

D. View of stope after supporting timber has been blasted and overlying mat and rock allowed to cave

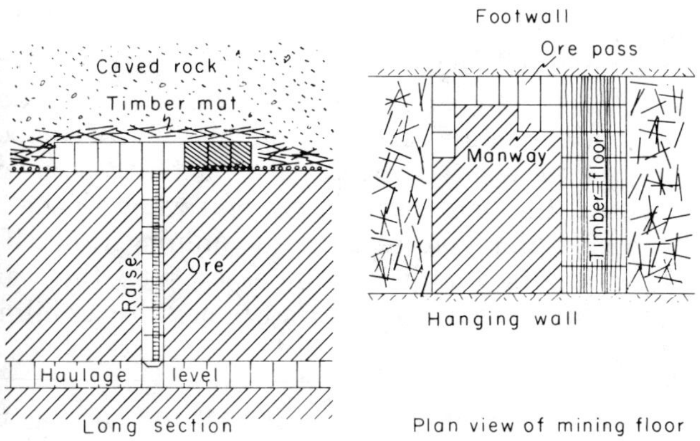

E. View of stope as mining progresses on both sides of the raise

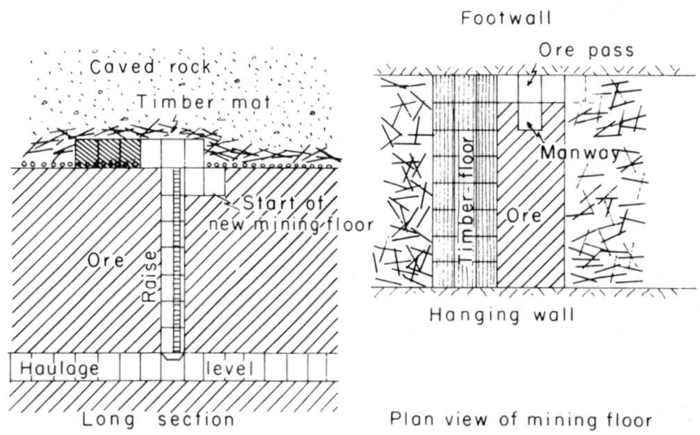

F. View of stope as most of floor is mined out and new mining floor is started

SUBLEVEL CAVING

Sublevel caving resembles top slicing except that the mining floors are spaced farther apart in the sublevel system, and not only the overlying cap rock but part of the ore is also caved. The cap rock and ore must be weak enough to cave readily. Production per man shift in sublevel caving is greater than in top slice, and the cost of explosives and timber is likely to be less.

Sublevel caving should not be confused with sublevel longhole stoping. The sublevel longhole system requires strong rock and is not a caving method.

INITIAL STAGE

The initial stage of a sublevel cave stope is similar to the top slice, i. e., a raise is driven up to the cap rock or the overlying mat. A drift is driven along the footwall as in a top-slice stope except that some ore is between the mat and the drift B. The crosscuts or sublevel slices are spaced a certain distance apart A, and C. Sublevel caving normally requires framed timber for ordinary drift sets in the slices.

After the slices are driven A, and C, the lagging on the side of the slices is removed and the ore is allowed to cave into the sides of the slice. The ore is scraped through the slice to the drift and from there it is scraped to the manway. If the ore does not cave readily, drilling and blasting of the ore may be necessary to start caving action. As the ore caves C, the mat caves down, and when all the ore is mined the mat reaches the mining floor. The slices are floored with timber to keep the waste from mixing with the ore on the next lower mining floor.

On the next mining floor D, notice that the slices are not directly below the previous slices but are offset between the upper slices. The upper slice positions in D are indicated by round timbered ends. By this arrangement the mat will eventually be evenly distributed between the ore and waste.

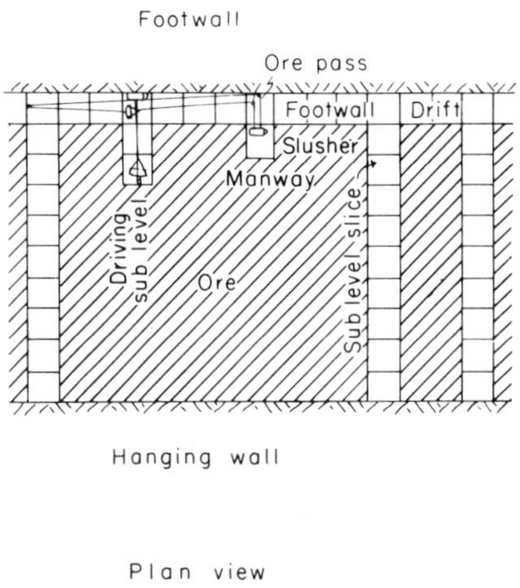

A. Plan view of sublevel cave stope

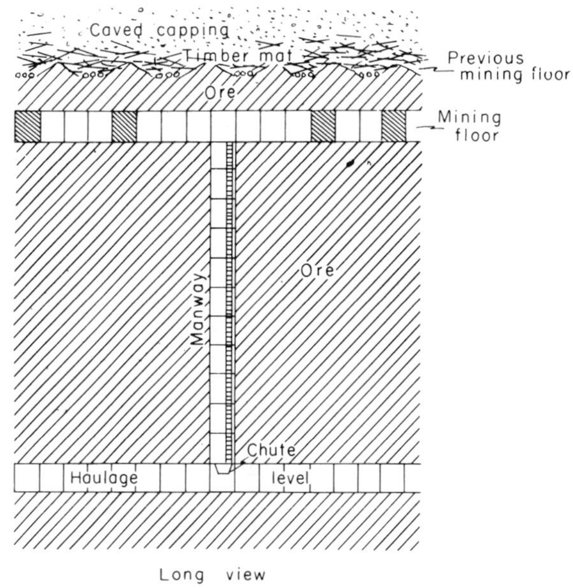

B. Long view of sublevel cave stope showing overlying mat and caved capping

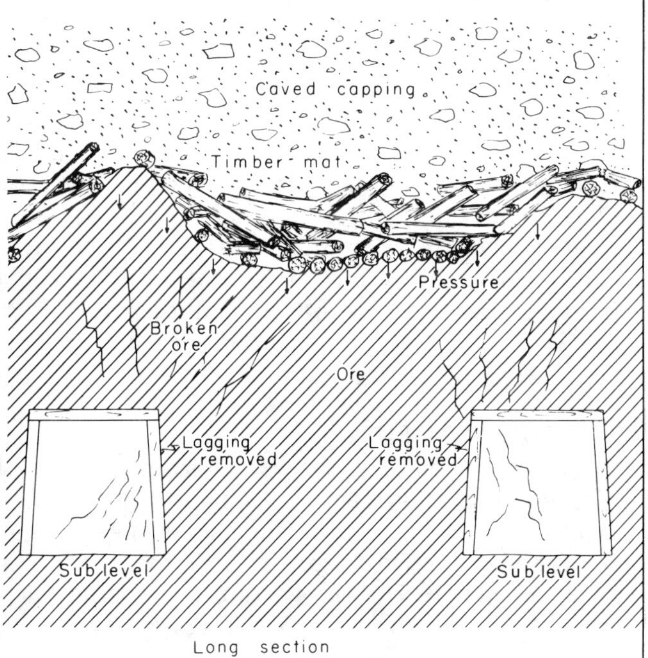

C. Close view of caving action around sublevels

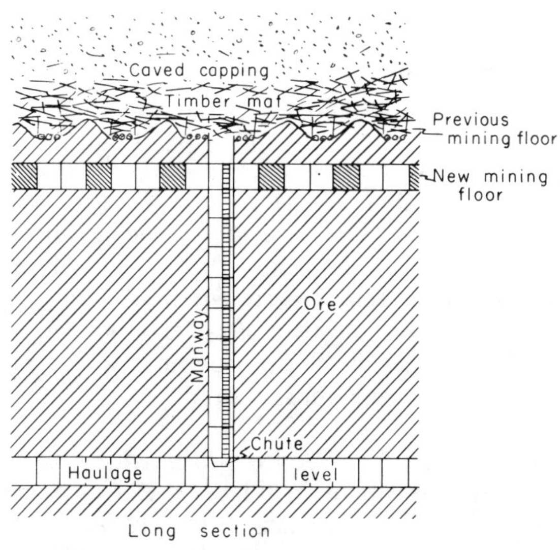

D. View of sublevel cave stope showing next lower development level

MINE MODELS (CAVING)

To better illustrate the block-caving method, models of the system are pictured. In the early caving system used in Butte A, and B, raises were driven from the lower haulage drifts to the grizzly level. From the grizzly level, finger raises were driven to the undercut level. The undercut level was developed as explained in Plate 8-19. The grizzly level was used to break large boulders and to pull old timber out of the ore. Parts of the area caved in Butte had previously been mined by square-set timber methods. Two styles of models of the same system are shown in A and B. In A the solid represents the actual openings underground, whereas in B the openings are shown as they actually occur underground.

The system has been modified in Butte to the operation described in Plate 8-19. Ore can be scraped directly into cars from slusher drifts C. The fingers are driven as explained in Plate 8-19. The undercut level is much more complicated in this model because the system is designed to leave larger pillars over the slusher drifts and haulage levels. Pillars over the development headings are shown in Plate 8-19. The system depicted in the model C is included to present a modification of the block-caving system.

SUBLEVEL CAVE SYSTEM

The model shown in D depicts a sublevel cave system. The mining floor can be seen as the top of the model, and the mat overlies the model. The small side sets on the side of the raise show where the other mining floors will be driven. The drift in this model is in the center of the stope rather than on the footwall as shown in Plate 8-21. Any system can be modified to suit the peculiar conditions at a mine.

RESUING STOPE

The model in E shows a resuing stope rather than a caving system. Resuing was described in Plate 8-15. This model shows a system designed to take advantage of gravity in moving ore, somewhat similar to the rill system, rather than using a slusher and scraper as shown in Plate 8-15.

A. Model of block-caving stope in which all openings are represented by solid material. **B.** Model of block-caving stope depicting the same system as A, but in this model the openings correspond to actual openings in the ground. **C.** Model of block-caving stope in which a sloping type of undercut level is used. **D.** Model of sublevel cave stope showing arrangement of raises, drifts, and caving crosscuts. **E.** Model of resuing stope.

PLATE 8-22 MINE MODELS (CAVING)

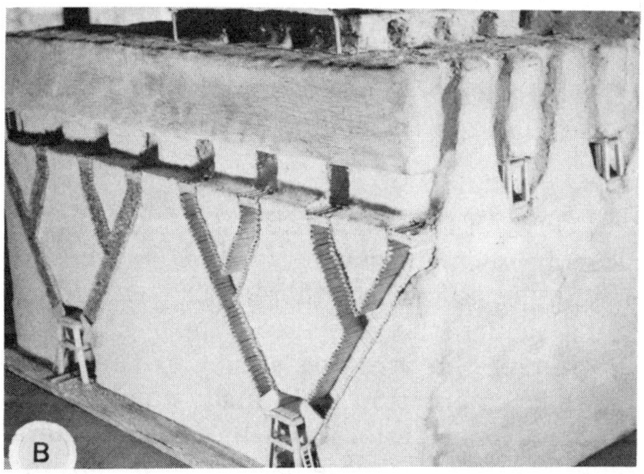

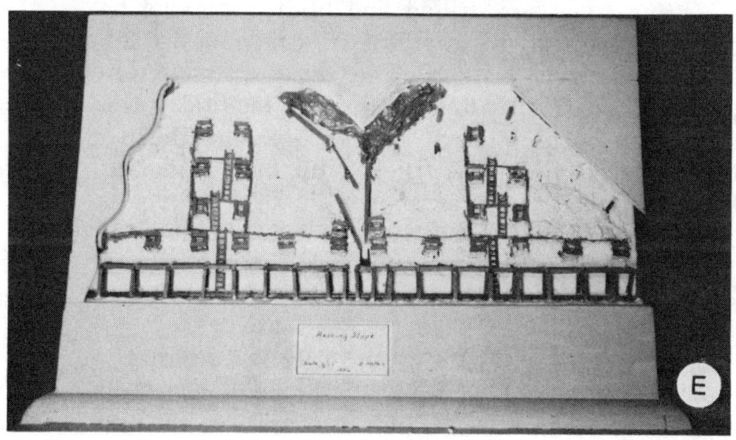

SUBLEVEL CAVING TECHNIQUES IN A STRONG ORE BODY

A relatively new mining method has been devised for sublevel caving strong ore bodies that have a weak capping and hanging wall. Mobile drilling, loading and hauling equipment have made this system very successful. No timber mat is used as in the conventional sublevel caving system, and the ore must be drilled and blasted.

A typical ore body that may be mined by this method is shown in A. It is necessary to sink a decline or shaft in the footwall and develop the ore body with a series of sublevel drifts, as shown in B. The elevation difference between each horizon of sublevel is commonly about 10 meters (about 30 to 35 feet). Also, the sublevels are not driven directly under each other, but in staggered positions as shown in C.

The usual procedure is to drive the upper crosscuts to the hanging wall on the same level. Then raises may be driven along the hanging wall from the ends of the crosscut to the top of the ore body or to upper crosscuts. These raises are often opened up to form an open slot between the hanging wall and the ore body.

Long holes are drilled and blasted parallel to the hanging wall so the slot of ore next to the hanging wall is broken. The raises provide expansion for the blasted ore which facilitates the formation of the slot of broken ore. This falls to the sublevel where it is picked up by a loader or load-dump-haul unit as shown in D. Mucking is continued until the overlying waste alluvium or broken hanging wall appears in the muck pile in significant quantities.

LONG HOLE DRILLING PATTERNS

Long hole drills, usually jumbo mounted, drill a series of fan holes in a pattern similar to the one shown in C. Several of these patterns may be drilled at one time, and each pattern may be five to nine feet apart along the length of the crosscut, as shown in D.

Usually the ore deposit is worked from the top down, and each level is worked so that the cave is uniform over the ore body, as shown in B. While the upper crosscuts are being mined, the lower crosscuts are being developed.

CROSSCUTS CONNECTED BY DECLINES

The various levels of the crosscuts are frequently connected by declines to enable the mobile equipment to move from one location to another. A haulage level to the shaft is located at some lower elevation in the mine and raises serving as ore passes are driven to the crosscut levels. In this way, the loader or load-haul-dump unit does not have to travel long distances.

NUMBER OF PATTERNS BLASTED

As can be seen in viewing C, and D, when a pattern of holes is blasted, there is broken ore or waste lying against the free face. Apparently this does not hinder the breaking action of the blast because ore fragmentation is normally good. Some mines blast only one pattern at a time, while others have found that two or more patterns, with appropriate delays between each pattern, can be blasted and mucked at one time.

The pattern of drill holes may be vertical or may tilt as much as 70° to the horizontal. This angle is governed by the way the ore draws down over the waste. There is always some waste dilution by this method, so it is used in large ore bodies where waste can be tolerated.

The advantages of this system are that it can be used in a wide variety of ore bodies; it can be highly mechanized; and it is a flexible method because it can be modified for unusual conditions or changed to some other method. One disadvantage is that there is always some dilution with this method, and some ore can be lost.

PLATE 8-23 SUBLEVEL CAVING TECHNIQUES IN A STRONG ORE BODY

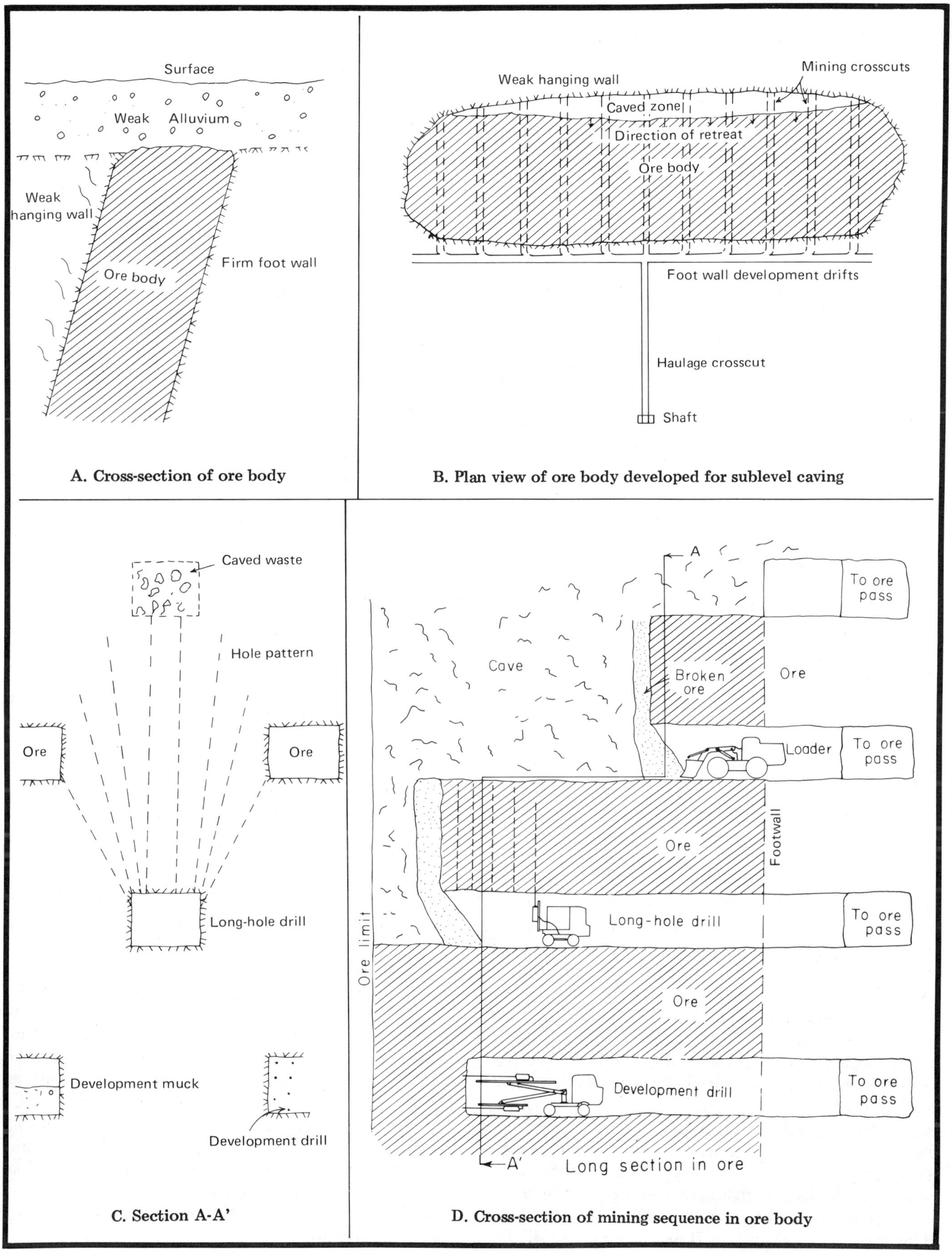

A. Cross-section of ore body

B. Plan view of ore body developed for sublevel caving

C. Section A-A'

D. Cross-section of mining sequence in ore body

UNDERHAND STOPING METHODS

The majority of mining methods either mine upward on the ore body called overhand stoping, or horizontally, sometimes called breast stoping. In the underhand system mining is carried on from the top down as shown in A. The reason for doing this is to have some kind of artificial back or roof support of known strength characteristics so personnel will not be exposed to overhead weak ore and wall rock. The top slice method explained in Plate 8-20 is an example of underhand stoping.

In some mining situations the ore and wall rock become so fractured and loose they cave. These caves sometimes are difficult to control, as well as being costly and time consuming. The caves at times are so uncontrollable that mining is abandoned in the area. This, unfortunately, leaves a valuable resource in the ground, with little chance for recovery.

Pillar recovery often is a problem when a major part of the ore has been removed by some other method. Frequently the pillar becomes broken and unstable, and overhand methods haven't always been successful; even with timbered methods.

CEMENTED SANDFILL

To cut down on the amount of timber required, and to provide a better overhead structure in recent years, cemented sandfill has often been used. For narrow veins, up to twelve feet wide, the system shown in B has been successfully used. Actually, it is similar to the horizontal cut-and-fill technique, except that after the cut has been completed, heavy caps or stulls are placed on the floor of the cleaned out cut. They are wedged and pinned firmly in place. Commonly a foot of unbroken ore is left on top of the cut as shown in B to protect the timber when the solid ore below is blasted. These may be from three to eight feet apart, but are usually 5 feet. Lagging or wooden planks are installed on top of the heavy caps or stulls, and the zone is prepared for filling. The rock bolts shown in B are usually installed as the cut is being made to keep the hanging wall in place.

The bottom three feet or so of the sandfill is mixed with cement as the zone is being filled. This tends to put a firm hard material over the timber, and the rest of the zone is then filled with ordinary sandfill to the overlying cap and filled zone.

The method requires much more timber preparation than the ordinary untimbered cut-and-fill. But when ground gets unstable, timber frequently must be used with conventional cut-and-fill for support. Even with timber, caves occur; consequently some operators feel the underhand system has merit because caves are usually eliminated or reduced.

WIDER UNDERHAND STOPES

When the pillar or ore zone is wider as shown in C, and D, the caps or stulls required are too long for support. A special method for wider underhand stopes was developed by the International Nickel Company of Canada (INCO) and their system is shown in C, and D. In this process the stope is started from the top, and as the ore is mined out the space above is filled with cemented sandfill material. A scissor set, developed by INCO, is shown in D. In this method, hardened fill instead of a loose ore zone is overhead. Safety in these stopes has been good. When underhand mining is resumed under a scissors mat, posts are not required for support. The absence of posts increases the efficiency of mucking equipment.

COMBINATION TIMBER AND SANDFILL

The Magma Copper Company in Arizona reports efficient mining of very weak ore and wall rock by a combination of timber and sandfill method. The mining plan is similar to a top slice stope, but instead of a timber mat, cemented sandfill is used to fill in the void. The plan is to catch the overhead timber posts and stringers with posts on the mining floor to keep them in place. After a mining cut has been completed, it is filled with cemented sand about three feet thick, and the rest of the cut is filled with ordinary hydraulic sandfill.

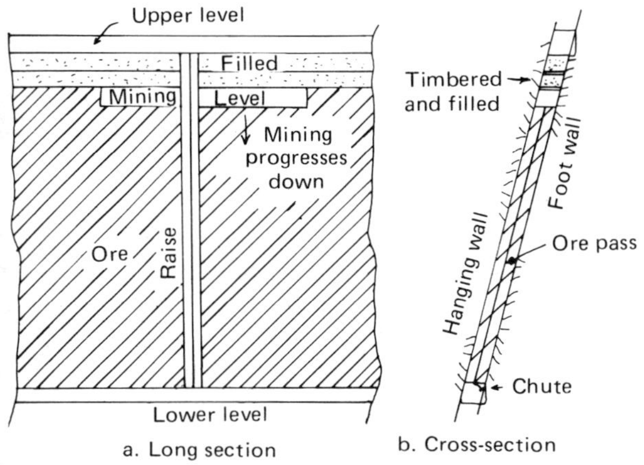

A. Mining narrow vein by underhand method

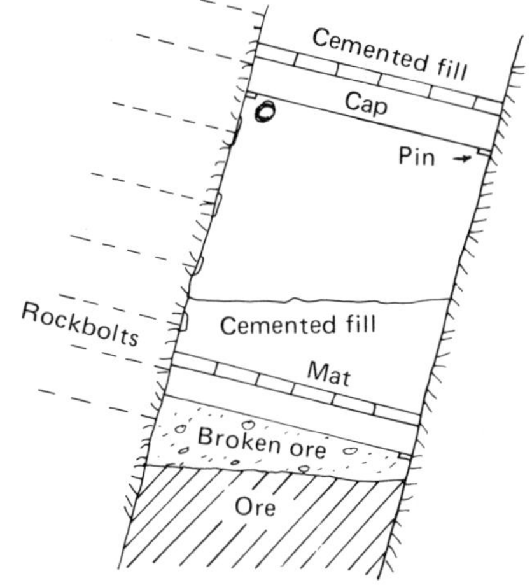

B. Preparation for fill in underhand system

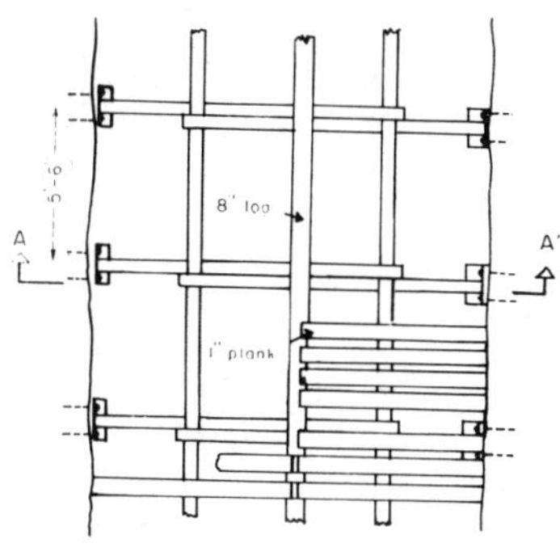

C. Stope preparation in wider vein

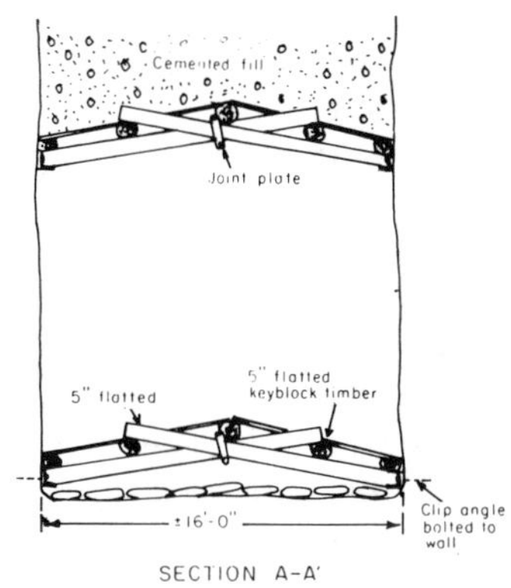

D. Cross-section of view in C

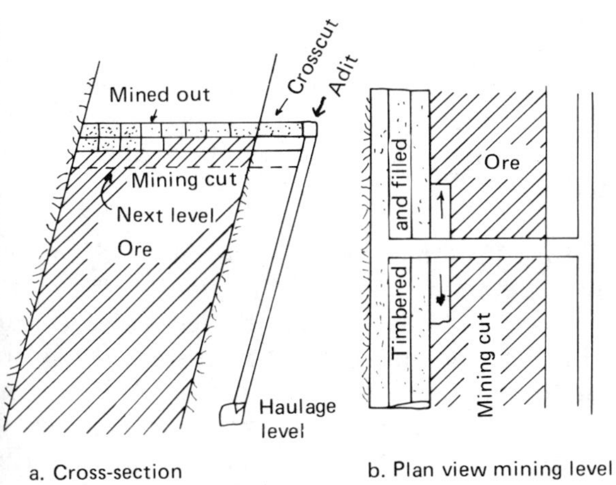

E. Underhand mining in wide veins

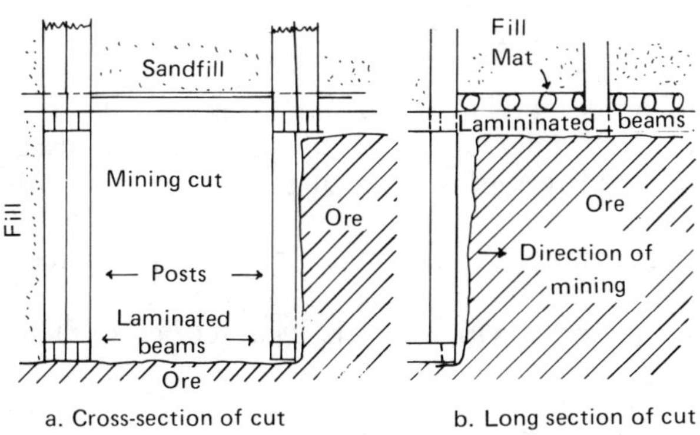

F. Mining sequence in wide stope

WIDE VEINS AND PILLAR RECOVERY

Ore bodies sometimes occur in wide deposits that do not necessarily lend themselves to caving systems of mining. The ore and rock may be relatively strong so that it does not cave readily. Another factor may be that dilution of ore with waste is not desirable because of a particular milling or marketing situation.

CRATER RETREAT

A typical wide vein situation is shown in A, and B where the ore is mined in stopes by various means. A rather popular method is cut-and-fill, but shrinkage, long-hole sub-level stoping or other types of open stoping are also often used. An innovation has been the crater retreat method as shown in F. Although F shows the system retrieving pillars, it is also used in mining the initial stopes. Large diameter, nearly vertical holes, are drilled from the level above to the bottom of the stope in a pre-determined pattern. It is necessary to have an opening over the width of the stope so the blasted ore will have a place to fall and occupy. Explosive charges are placed and blasted in the holes at the calculated optimum distance above the undercut. The same holes are used again for lowering and placing explosives at the calculated distance above the previously blasted zone. The hole pattern and explosive placement location is based on the spherical charge and cratering theory which is briefly described in Plate 5-2.

NUMEROUS ADVANTAGES

Although this is a new development in mining, users of this system claim that it greatly saves development work, reduces labor, gives better fragmentation, reduces blasting damage to pillars and increases mining efficiency.

RECOVERY OF PILLARS

Pillars are required to support the walls and back and act as a dividing line between the initial stopes. The pillars contain considerable ore, therefore, recovery of the pillars is often justified. Before pillar recovery can be started it is necessary to fill the mined out stopes with waste rock or tailings so the filled space will support the walls. If a cut-and-fill system was used to mine the initial stopes the fill is in place, however, if the stope was mined by some system of open stoping it is necessary to fill the voids as shown in C. Mill tailings, if available, are used, but surface gravels, sands and even crushed rock may also be used. Mixtures of coarse and fine material are often used. To give stronger support to the walls, cement is sometimes added to the fill material.

Recovering the pillars is a tedious task because they usually take some weight from the walls or back which breaks and sometimes distorts the remaining ore. This shattered and displaced ore is seldom very strong. One overhand system is shown in D where the pillar is mined in a series of drift type cuts. After one cut is mined the length of the pillar, the area is filled with cemented sandfill. The next cut is made alongside the filled cut as shown in D a, and b. When this cut is made it too is filled with cemented sand, and the next cut is mined and filled. As shown in D c, the first level is below the crosscut. When this level is mined, the down ramp is filled and the second level is mined as described earlier. Level three then is mined. Another lateral is driven three levels above the one shown, and the whole sequence is repeated.

Cut-and-fill methods are often used in pillar recovery, but the back is frequently difficult to hold. Some operators have drilled holes through the pillar as shown in E and reinforced the broken ore with cemented rods and cables. As the stopes are mined the supporting rods or cables are cut off. The vertical crater retreat method described earlier has been used with success to recover pillars. However, the fill on either side of the pillar must be competent so it doesn't flow into the open slot.

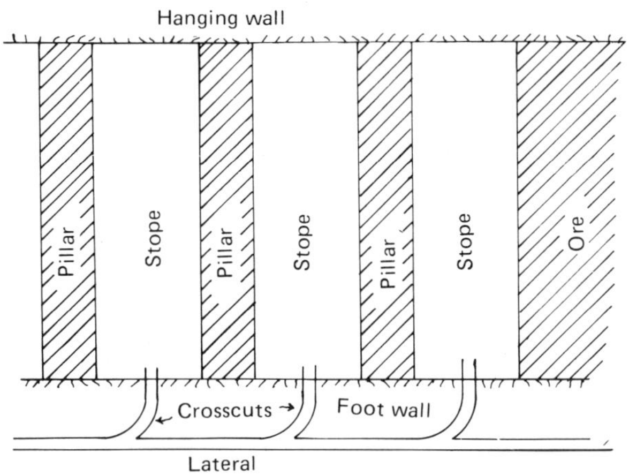

A. Plan of a wider vein stoping system with pillars

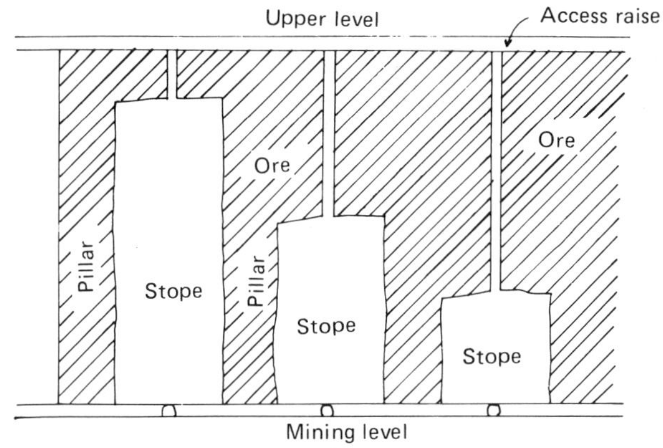

B. Long view of a wide vein stoping system with pillars

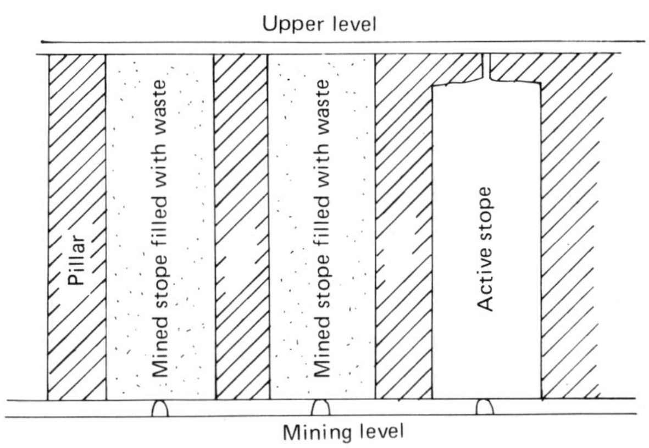

C. Mined stopes filled with backfill material

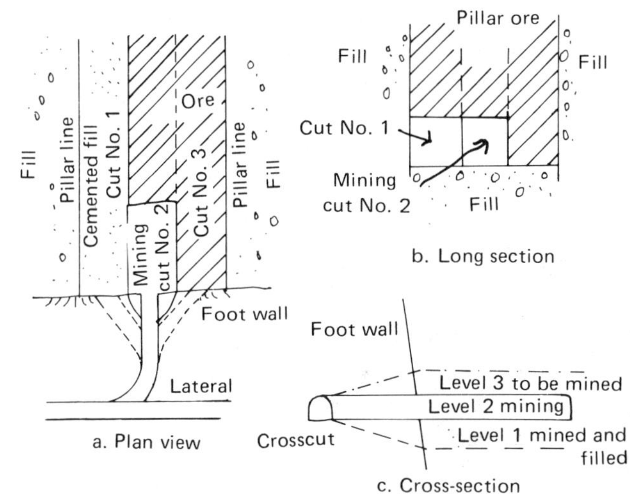

D. Overhand and fill method of recovering pillar

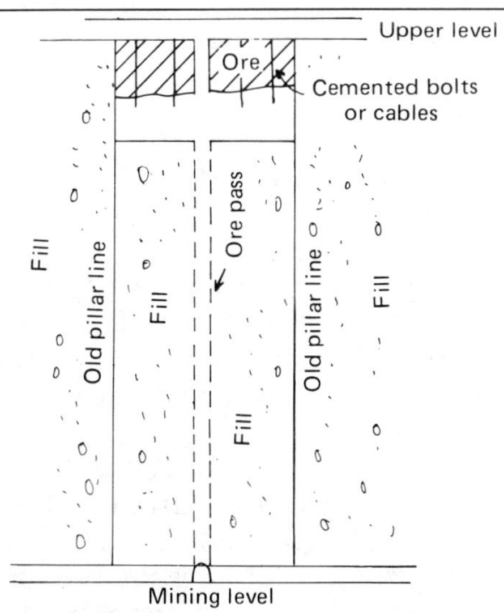

E. Mining pillar with overhand cut-and-fill

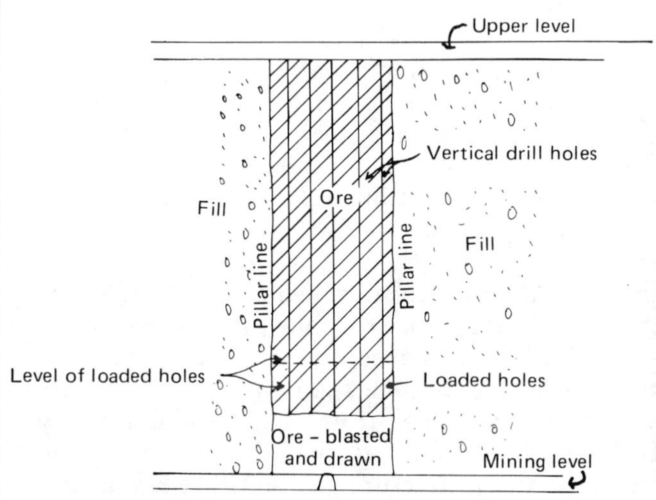

F. Vertical crater retreat method of recovering pillars

SOLUTION MINING AND LEACHING

Solution mining is an old system used for producing soluble minerals such as salt, or minerals that can be melted by hot water, such as sulfur. One system for producing salt from an underground formation is shown in A. This is called the bottom injection method. Water injected through the center pipe dissolves the salt. The salt carrying solution flows up between the inner pipe and the outer casing. The arrows in A show the direction of the solution. Reversing the flow is called the top injection method. Other injection techniques are used to improve recovery, but eventually the efficiency of the well decreases because the size of the cavity and the collapsed roof may clog the well. When this occurs it is necessary to drill another well.

FRASCH PROCESS

Sulfur production by the Frasch process is shown in B. Hot water is introduced into the hole melting the sulfur. The molten sulfur collects at the bottom of the hole and starts to flow up the inner pipe. Compressed air is introduced which helps raise the liquid sulfur and excess hot water up the pipe.

IN-SITU LEACH MINING

In-situ leach mining has been used in recent years, espcially in recovering low grade uranium ores. A series of holes is drilled into the ore zone, the solvent is pumped down injection wells and recovered in production wells. A two-well system is shown in C. The solvent is pumped down the injection well into the mineral formation where it dissolves the mineral, and then flows to and up the production well. The ore zone must be permeable enough so that the solvent can flow from the injection zones through the mineral zone, dissolve the minerals and then flow to the production well. Techniques used in the oil industry to increase production by fracturing the oil bearing soils around oil wells are also used for in-situ leaching. One of the problems encountered in this system is the formation of a channel between the injection and production well so the solvent will flow directly between the two wells, and will not permeate into the mineral zone and dissolve the minerals. As more experience is gained with this system recovery of minerals should improve since many of the operational problems are being solved. One more drawback to the system is the possible contamination of the natural ground water with the solvent. Care must be taken to control the flow of solvent and any wash or cleaning solution to the production wells. The dissolved minerals are recovered from the solution by various recovery systems when it comes up the production well.

After a mine has been worked out it may be possible to flood it with solvent and dissolve much of the remaining mineral. This practice has been done in copper mines as shown in D.

IN-SITU LEACHING AFTER ATOMIC DETONATION

Some low-grade ore bodies have been found that cannot be mined by conventional methods. Plans have been made to detonate an atomic bomb at the bottom of the ore body to melt and fracture the ore body so in-situ leaching can be done. The location of the explosive is shown in E,a, and the resulting mass is shown in E,b. After the area has cooled, production and injection wells will be drilled into the mass. It is believed that there will be sufficient fragmentation for good recovery by solvent mining.

HEAP LEACHING

Heap leaching for recovery of copper, gold and silver has increased in recent years. An impermeable pad is prepared as shown in F,b. The broken ore is placed on this pad as shown in F, and the solvent is sprayed over the top of the flattened piles. The solvent percolates down through the pile, and the mineral bearing solution is collected on the impermeable pad and run through a refinery to collect the dissolved mineral. Sulfuric Acid is commonly used to recover copper, while a cyanide solution is used to recover gold and silver. Again, control of the solvent flow is necessary so the ground water will not be contaminated.

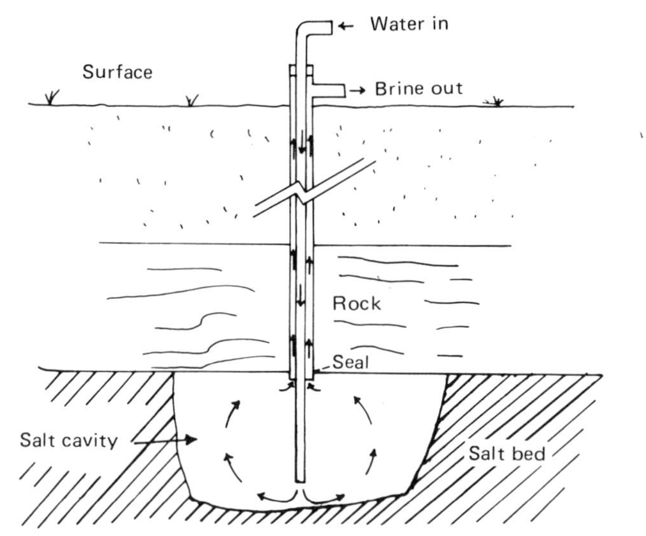

A. Bottom injection system for mining solubles

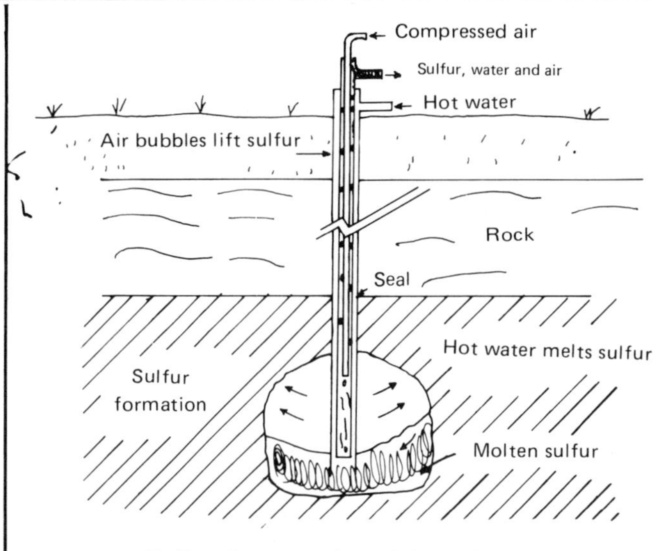

B. Frasch system for mining sulfur

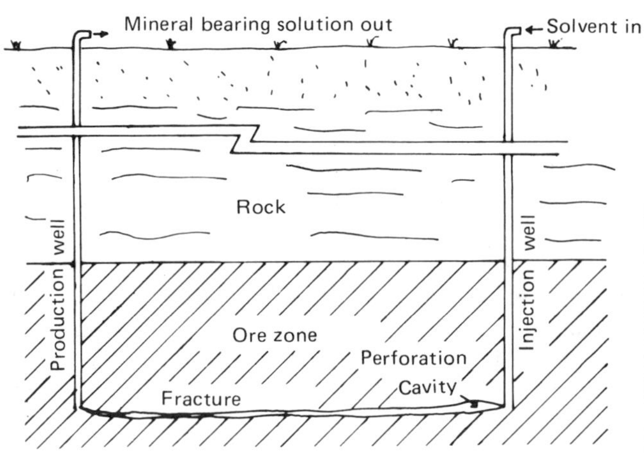

C. Multiple well system for in-situ leach mining

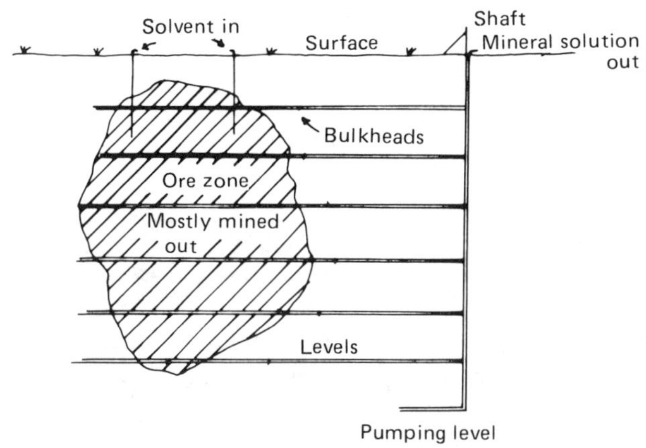

D. Flooding and leaching a mine

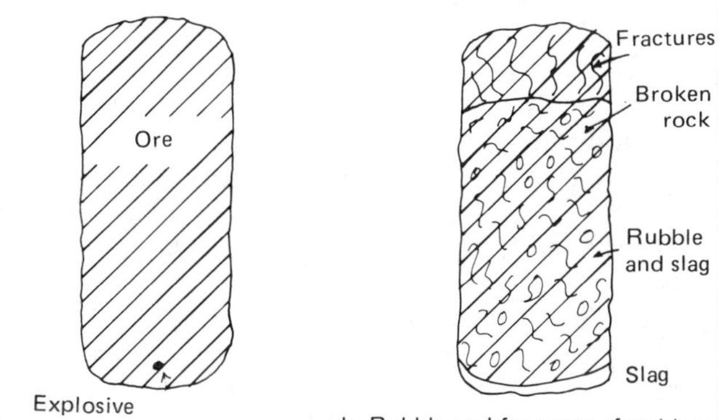

a. Explosive in position

b. Rubble and fractures after blast

E. Breaking ore for leaching process

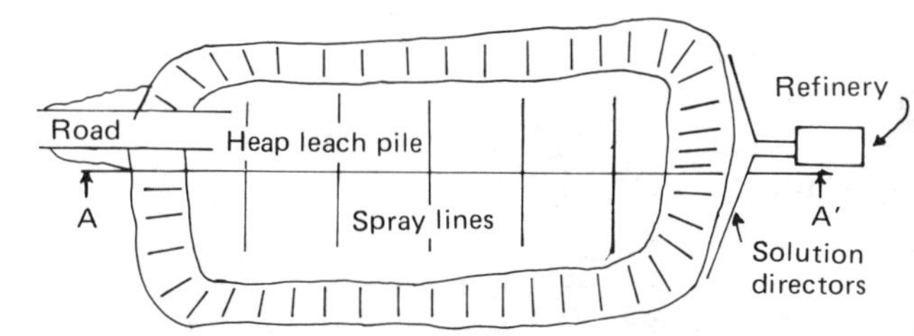

a. Plan view of heap leach pile

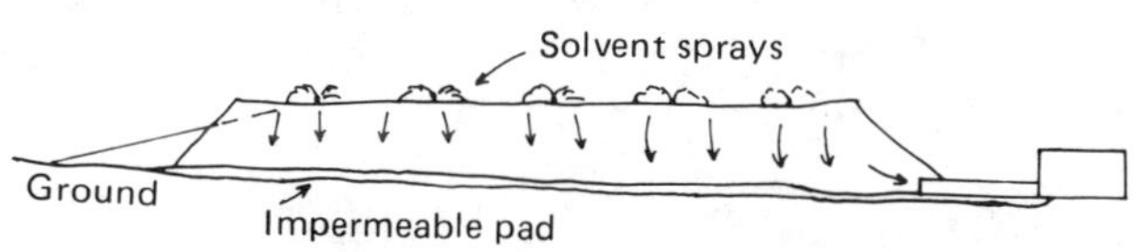

b. Section A-A' of heap leach pile

F. Heap leaching for low-grade ores

PART 8
CONCLUDING REMARKS

Various methods of underground mining were discussed in this section. Underground miners are either seeking ways to improve their present methods, or they are looking for new methods. For example, in room and pillar mining much of the ore or coal, from 30 to 60 percent, is left underground. With the energy crisis or high value of minerals, methods to extract all of the minerals in first mining are examined with interest. But one of the deterring factors to complete mineral extraction is the surface subsidence problem. All factors must be weighed when choosing or changing a mining method.

Developments in the past two or three decades that have changed the underground mining picture have been the widespread use of rock or roof bolts, the use of the diesel powered equipment underground, and the development of continuous mining machines. These factors certainly have had a marked effect on underground mining.

SIZE AND QUALITY OF EQUIPMENT

As in surface mining, underground equipment is getting bigger, more versatile, and more productive. Too, the machines are becoming more durable and dependable. Special machines are made to relieve much of the laborious activity of mining. The development of the jumbo drill, operated from one position by an operator, is an example of this. Versatile (LHD) units are replacing track-bound and slusher equipment in many operations. Roof bolting machines have replaced the tedious timber post or prop placing in coal mines.

Underground mining commonly takes place in a hostile environment. The problems of gas formation, concentration of dust, and roof or back falls are ever present in underground mines. Except for dust, these problems are practically nonexistant in surface mines. The health and safety of the workmen are paramount in all mine decisions. Falls of ground are still one of the major hazards in underground mining; therefore, special attention is given to support and ways to protect personnel from falling debris. Adequate ventilation is necessary to reduce the gas and dust hazard.

SAFETY PROVISIONS

A few of the other safety provisions consist of protective canopies on machines to protect the workmen from overhead rock falls, noise reduction devices on drills, better communication and warning devices, self-rescue units for miners, and alternate ways to reach safety underground or on the surface in case of an emergency.

RECLAMATION PRACTICES

Until the past decade or two there were few requirements for reclamation of mined land. The surface of the ground where mining occurred was generally considered expendable in the mining operation. The value of the mineral was usually considered to be far in excess of the worth of other surface uses for the land.

Actually, the total land area disturbed by mining activities is fairly small when compared to other surface uses. The American Mining Congress recently reported that less than 0.3 of one per cent of the surface land in the United States has been disturbed by all types of mining, including strip mining and sand and gravel pits. About one half of this has been reclaimed. The total area of disturbed land is about equal to the amount of land used for airports, equal to the land used by railroads and equal to about one-sixth of the land used for highways.

GOVERNMENTAL REGULATION

However, in the early 1970's regulations were enacted to insure that surface lands would not be destroyed by mining. If an area was to be mined, it had to be reclaimed to its original condition or to as near that as possible, or to some beneficial use. Most states and government agencies now have reclamation requirements with which mining companies must comply.

As reclamation is a relatively new aspect of mining, considerable experimentation must take place in land restoration and revegetation. In most cases the mining company must submit mining and reclamation plans to a state or federal agency, or to both, for approval before mining can commence. The company must post a bond to insure that the reclamation work will be done. It takes a great deal of time to comply with these requirements, and often mine development is delayed for years. Many companies resent the delays and uncertainties in interpretation of the laws rather than the actual requirements for reclamation. All of this causes a considerable loss in time and money.

Although relatively new, reclamation practices have generally been successful in the arid western part of the United States. Vegetation failures are studied and new methods are tried to improve the success ratio. The mine planner must now consider reclamation requirements in addition to the mining sequence. He must see that the overburden material isn't handled more than is absolutely necessary. Any rehandling reduces efficiency and increases costs.

RECLAMATION PRACTICES IN STRIP MINES

Practically all states and government agencies in the United States require that land disturbed by surface mining must be reclaimed. The agencies may require coal mine operators to restore land to its original form or as near to that as possible. In practically all cases the slopes of spoil piles must be reduced, soil must be put back on the disturbed ground, and vegetation must be planted.

In the arid western part of the United States the soil cover generally is thin. Hence the available soil is stripped from the mining areas and spoil pile areas, as shown in A, and stockpiled for future use. The next phase of mining is the removal of the overburden which is piled on top of the stripped overburden, as shown in B.

As shown in C, after the exposed coal has been mined, part of the spoil pile is pushed into the exposed cut by bulldozers. Enough room is left so the overburden in the next mining cut can be piled into this cut, as shown in D.

REDUCING SPOIL PILE SLOPES

The slope of the original spoil pile is reduced by bulldozers or other equipment. It is usually not required that the spoil pile be leveled completely. A small hill or mound is permitted, as shown in E. After the spoil pile has been formed, the stockpiled top soil is spread back over the overburden as shown in E.

As the mining sequence progresses, leveling and forming the spoil pile surface and removing and replacing the top soil becomes an integrated part of the mining cycle. The distance the soil has to be transported is kept to a minimum. After the mine has been in production for some time, the reclamation work progresses, and the mined out portion is seeded to produce vegetation again. In most strip mines unless the area is a permanent road or structure, it seldom will be without vegetation for more than a few years.

RECLAIMING THE LAST CUT

When the strip mine is finished, the last cut does not have much spoil material available for filling. This creates a problem. Sometimes the overburden bank is pushed in to make a more gradual decline in the contour of the land. In other cases if the ground water level is high enough, the cut will fill with water and a rather long narrow lake results. In some cases these lakes have been developed into recreational areas, and receive use by the surrounding population.

Although grasses and shrubs usually grow in one season, it takes much longer for trees to grow. However, trees have recorded good growth where they have been planted and nutured. So for the most part reclaimed land is as productive if not more productive than it was originally.

PLATE 9-1 RECLAMATION PRACTICES IN STRIP MINES

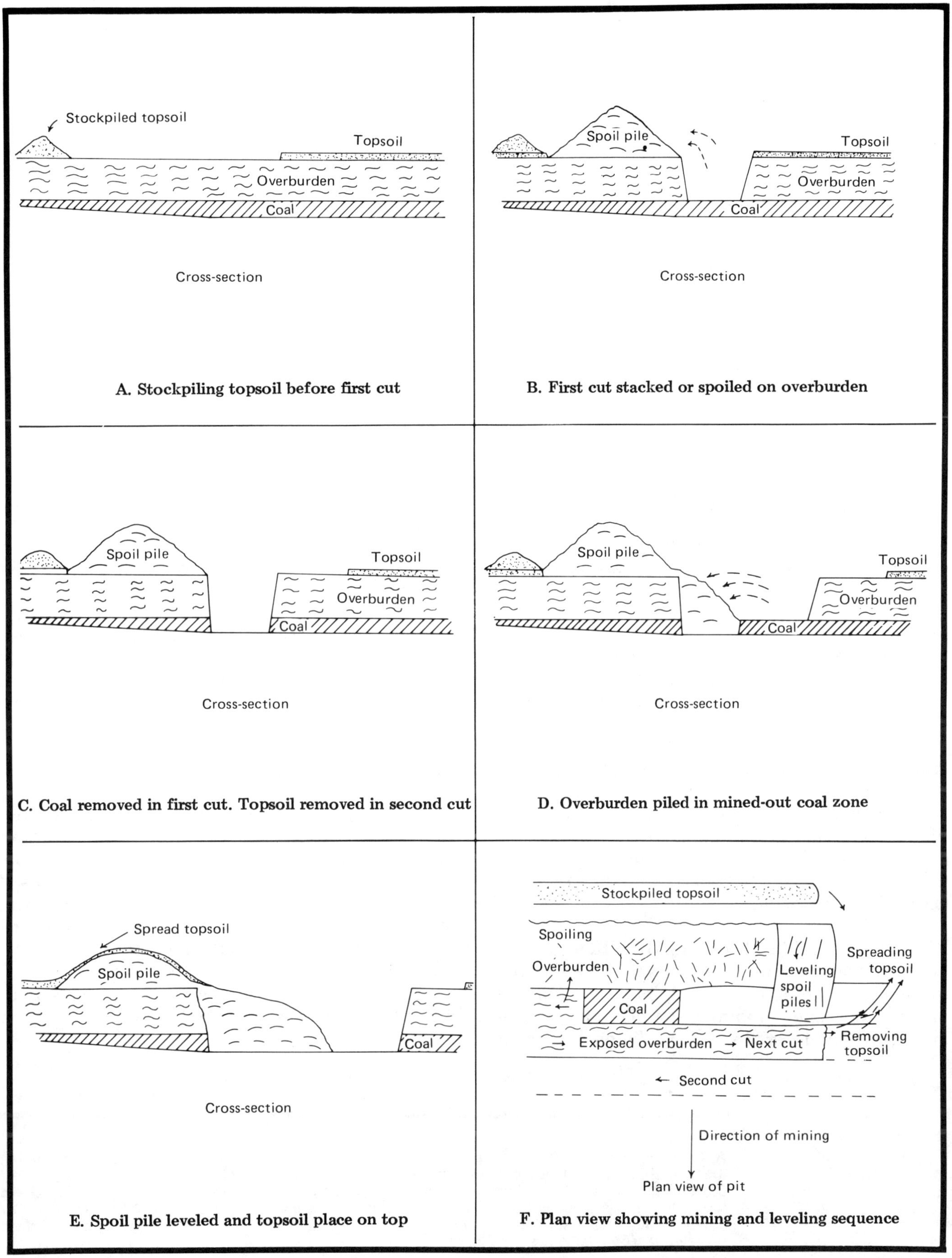

A. Stockpiling topsoil before first cut

B. First cut stacked or spoiled on overburden

C. Coal removed in first cut. Topsoil removed in second cut

D. Overburden piled in mined-out coal zone

E. Spoil pile leveled and topsoil place on top

F. Plan view showing mining and leveling sequence

RECLAMATION OF COAL STRIP MINES

As explained in Part 7, on surface mining, the overburden is stripped from the coal and piled to the side. When a cut of coal has been mined, the overburden is piled into the mined-out cut. The draglines shown in A and B are casting the overburden into the mined-out cut. Because the dragline sits in one position to dig and dump, it is difficult not to form piles or stacks of the overburden material. Also the dragline cannot fill the cut completely unless there is considerable moving. Generally, it is more economical to level the spoil piles with bulldozers or other types of excavating equipment.

BULLDOZERS SPREAD OVERBURDEN

Bulldozers are most often used to spread the stacked overburden material. A bulldozer working on spreading and smoothing the piles can be seen in B. Two bulldozers smoothing the piles are seen in C. Note how rough the spoil pile looks in that picture.

Bulldozers have a limit of how far they can push material efficiently. The machine shown in D is an experimental model using two crawler tractors pushing a 48 foot angle blade. It was developed by the United States Bureau of Mines and the Pittsburg and Midway Coal Mining Co. This machine requires only one operator. The second tractor is controlled remotely.

SPREADING TOP SOIL

After the spoil piles have been smoothed and contoured to the final design, the stockpiled top soil is spread over the spoil. A bulldozer is shown doing this in E. Often scrapers, used in the construction industry, are used for spreading top soil because they can make more uniform layers. These machines also are more efficient for carrying material greater distances than bulldozers.

After the top soil is in place, it is necessary to prepare the surface for planting which is shown being done in F.

A. Coal loading and stripping dragline filling cut. *(Courtesy Western Energy Co.)* **B.** Stripping dragline filling mined out cut and dozers leveling spoil piles. *(Courtesy Western Energy Co.)* **C.** Conventional spoil pile leveling technique. *(Courtesy U.S. Bureau of Mines, SMRC)* **D.** 48 foot angle blade leveling spoil piles. *(Courtesy U.S. Bureau of Mines, SMRC)* **E.** Close up of bulldozer leveling a spoil pile. *(Courtesy Western Energy Co.)* **F.** Preparing reclaimed surface for seeding. *(Courtesy Western Energy Co.)*

PLATE 9-2 RECLAMATION OF COAL STRIP MINES

RECLAMATION OF COAL STRIP MINES

The intent of reclamation practices is either to put the disturbed land back into its natural state, or to prepare the land for an equal or more beneficial use. Many conservationists have questioned whether disturbed land can be restored to its natural state, or even if vegetation would grow on the land. This is especially true in the arid regions of the Western part of the United States.

Intensive research was done at Colstrip, Montana on ways to plant crops, types of crops to plant, and techniques to get the original kind of vegetation to grow again. Reclamation was, and still is being, monitored by private companies, state and federal agencies. The pictures in this section show success was attained in growing grain and grazing crops on the reclaimed land.

The prepared land was fertilized and seeded. Progress of crop growth is noted in A, B, C, and D. As the grain matured, it was harvested, as shown in E. Note the power plant in the background and the dragline boom in the right background.

The general procedure is to plant grain the first year or so to prevent erosion of the land. Later the native grasses and vegetation, which take longer to become established are planted. Another factor which worried environmentalists and conservationists was whether the revegetated land would hold up for livestock grazing. Cattle are shown grazing on this reclaimed land in F. Preliminary observations show that reclaimed land is at least as productive, and in many cases, more productive than the original land.

A. Checking early plant growth on reclaimed land. **B.** Measuring the rate of growth on reclaimed land. **C.** Inspecting the quality of the vegetation on reclaimed land. **D.** Grain ready for harvesting on reclaimed land. **E.** Harvesting grain on reclaimed land. (Note power plant in background) **F.** Cattle grazing on reclaimed land. *All pictures courtesy of Western Energy Co.*

PLATE 9-3 RECLAMATION OF COAL STRIP MINES

RECLAMATION OF OPEN-PIT MINES

Open-pits are normally much deeper than coal strip mines so reclamation presents a different set of problems. A typical open-pit mine is shown in A. The pit slopes are commonly formed in rock with very little, if any, soil available to cover the slopes. Too, the rock slopes are often very steep, such as ½ : 1, 1 : 1 or 2 : 1. These steep slopes are not conducive to reclamation. Because they are composed of rock, and much drilling and blasting would be required to lower the slopes, it becomes impractical. In some instances the pit may be backfilled if a supply of waste rock is readily available.

Ordinarily the waste rock from an open-pit is deposited a considerable distance from the pit, and the cost of hauling it back for filling the pit is uneconomical. Often the waste dumps are located in mountainous areas where the toes or bottom of the dumps are several hundreds or even thousands of feet below the top level of the open pit.

In this situation open pits are usually not reclaimed because of the costs and problems encountered. The land in the pit may be dedicated to mining and no artificial attempt to reclaim it is made. The dumps, however, are often reclaimed and vegetation will grow on them and reduce the hazards of erosion.

REHABILITATING DUMPS

Dumps are often made with an upgrade which the trucks can negotiate, and the material is deposited over the end of the dump. This disposed of material flows down the end of the dump at about the angle of repose, as shown in B.

The Forest Service, a branch of the U.S. Department of Agriculture, shows the probability of vegetation on different slopes. As can be seen in C, the steeper the slope, the less likely the success of revegetation by ordinary planting and seeding means.

HYDRO-MULCHING AND HYDRO-SEEDING

In rehabilitating dumps, the slopes are usually lowered and roads which act as water barriers are built down the slope every so often, as shown in E, and F. Because the slopes in dumps are frequently devoid of soil or fine material, some sort of organic material must be supplied to the surface which will allow seeds to have some bed material for germination. Highway cuts and fills have the same problem, therefore hydro-mulching and hydro-seeding machines are used to form an organic base to enhance vegetation.

Hay, bark or other types of organic material are chopped and pulverized in a machine and then mixed with some sort of binding material and water to form a sticky organic mass. After the chopping and pulverizing process, seeds are mixed in the mass for even distribution throughout. This mass is blown on steep slopes and will usually adhere long enough for vegetation to start. These machines are called hydro-mulchers and or hydro-seeders. Once the seeds start to germinate, the roots will hold the mulch in place and form a protective vegetative layer to minimize water erosion.

Natural slopes of 3 : 1 and 2 : 1 are often vegetated by nature. If the hydro seeding process can be accomplished successfully on the steeper artificial slopes, the slopes are generally stabilized because natural growth will tend to follow the intitial planting.

Successful slope vegetation is shown in F. Often it is necessary to make more than one attempt with hydro seeding before a high success rate is reached.

PLATE 9-4 RECLAMATION OF OPEN-PIT MINES

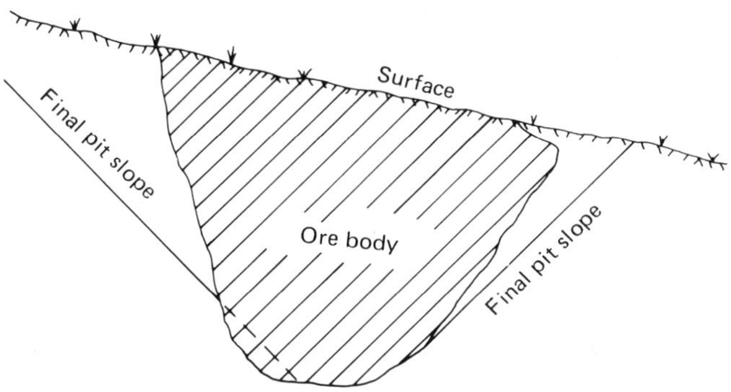

A. Cross-section of a surface ore body

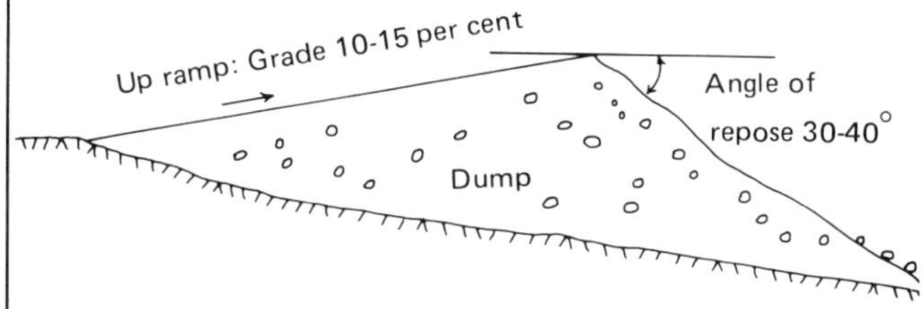

B. Typical dump from open-pit overburden

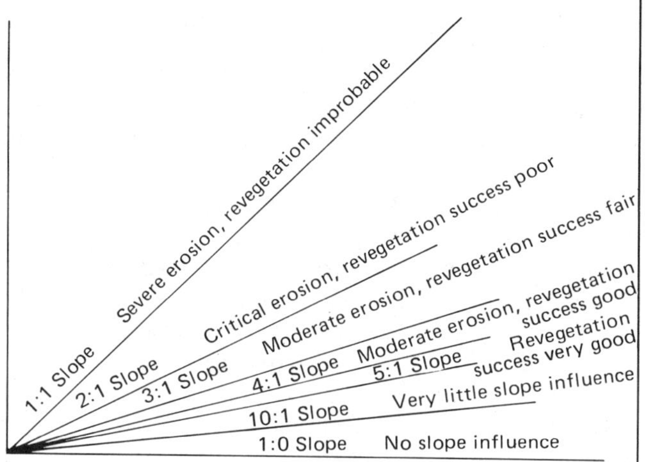

C. Influence of slope on revegetation. U.S.D.A. Forest Service

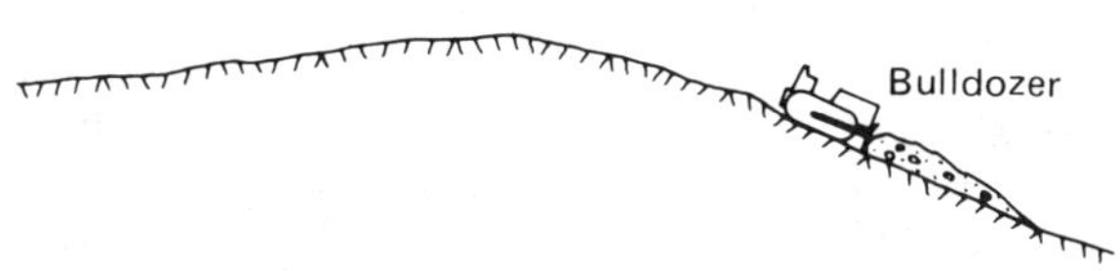

D. Lowering slope angle with bulldozers

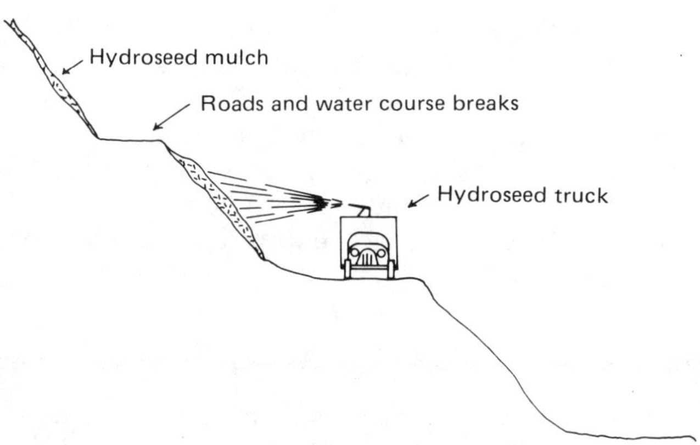

E. Hydroseeding

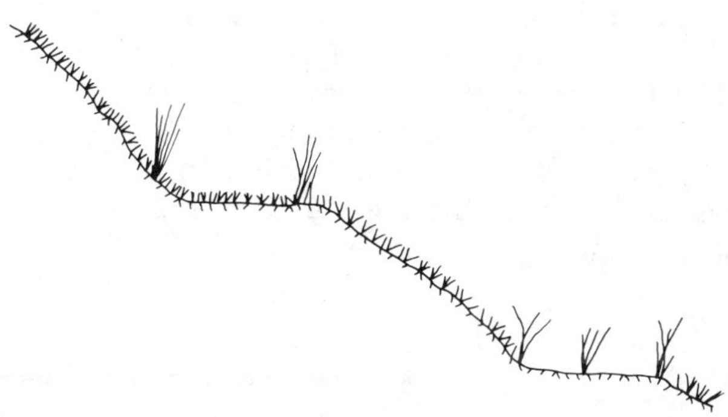

F. Vegetation on sides of dump

RECLAMATION OF OPEN PIT DUMPS

The problems of reclaiming open pits was discussed in Plate 9-4. The dumps are often reclaimed, and procedures developed and used in stabilizing highway cuts and fills are applied to mine dumps.

Nature provides protection and nourishment to the soil through the mulching process, i.e. decaying organic and mineral matter enrich the earth. Also, nature reseeds the land with help from wind, rain, animals, and birds. However, it takes time to start this process on newly formed slopes.

To speed up the process, hydro-seeding was developed. This method applies seed, fertilizer, lime, and mulch by a spray system to steep slopes.

HYDRO-SEEDING MACHINES

Hydro-seeding machines are shown in A, and B. These machines are powerful enough to spray up to 200 feet from the nozzles, and hoses can be attached to get at hard-to-reach areas. The seeds, mulch, nutrients, and binder stick to the steep slopes until germination begins and root growth is established.

Various mixtures of seed, nutrients, and mulches can be mixed and sprayed in the hydro-seeder. Sometimes just the seed and nutrients are placed with the hydro-seeder, and straw mulch is placed with the mulch spreaders as shown in C. Liquid adhesive can be mixed with the mulch to enable it to stick on steep slopes. This machine also has hose attachments for better maneuverability.

THE KRIMPER MACHINE

Protecting the mulch blanket against weather extremes can be a problem until the root system is established. The machine shown in D is called a Krimper. This machine forces some of the mulch vertically into the soil giving it much more support to stay in place. This machine can work on fairly steep slopes.

The pictures in E, and F show partially reclaimed mine dumps and tailings piles. As mentioned previously, it may take several applications of the seeding process before a good vegetation cover is established.

The last cut in a coal strip mine is difficult to reclaim. There is no soil or spoil available to fill the cut. Such a situation is shown in G. If the ground water table is high enough, these cuts fill with water and make lakes. Fish will often thrive in these ponds. Even boating and water recreation activites are enjoyed on these ponds.

A. Hydroseeding a mine dump. *(Courtesy Finn Equipment Co.)* **B.** Hydro-seeding a highway cut. *(Courtesy Finn Equipment Co.)* **C.** Placing mulch on a dump. *(Courtesy Finn Equipment Co.)* **D.** Krimping a seeded and mulched area. *(Courtesy Finn Equipment Co.)* **E.** Reclaiming a mine dump in Western Montana. **F.** Reclaiming a tailings dam in Western Montana. **G.** Remaining high wall in a coal strip mine. *(Courtesy Montana Bureau of Mines and Geology.)*

PLATE 9-5 RECLAMATION OF OPEN PIT DUMPS

PARTS 1-9 CONCLUDING REMARKS

We have looked at past and present techniques of mining in the preceding sections. It is probable that more advances have been made in production per shift, and in safety improvements in the past two decades than in the entire history of mining. Underground coal mining suffered a gradual loss of production in the 1970's, and its safety record did not change dramatically. But as more surface coal mines came into production in this period, overall output and safety rates did improve.

Equipment and techniques in mining have improved. Energy problems have created an increased demand for coal and uranium. An improving standard of living for much of the world has created a demand for more energy and non-energy minerals. There is no reason to believe this demand is going to slacken.

The challenge of mining will always be present. As surface mines are depleted more emphasis will be placed on underground techniques. Lower-grade deposits will have to be mined. Fortunately the total mineral content in a low-grade ore body is often many times the mineral content in a high-grade ore body. Thus the challenge of producing minerals at lower costs will be greater in future decades. Most of the high-grade bonanza ore bodies are gone.

Legislation designed to protect surface lands, to keep air and water pollution at a minimum, and to restore mined areas to beneficial use will not diminish in the future. A new dimension, protecting the environment, has been introduced in the past decade. This is an added burden to mining a mineral deposit successfully.

Mining people are resourceful and will face each new challenge as it arises. The advances in techniques since this book was first published in 1967 are amazing. Yet, on close examination, the basic principles of mining are still utilized today. New machines, new techniques and new applications cause changes which affect the industry.

The future of mining is encouraging. If this or a similar book is brought up to date twelve years from now the advances will undoubtedly be extensive. Although new mining methods may evolve, improvements of existing systems will be paramount. Basic concepts that work well don't change much, but improvements can be made on all existing systems.

PART 10

DEFINITIONS AND RECOMMENDED READING

This section defines commonly used mining terms, gathered in one place for handy reference. In many cases plates or previous material are referred to for more complete explanations of the terms.

This book is designed as an introduction to the fundamental concepts of mining. There are many books available explaining in much more detail topics covered in this book. A partial list of those books is to be found in the second part of this section.

DEFINITIONS OF MINING TERMS

Adit—a horizontal gallery or opening driven from the surface which gives access to the ore body. The term "tunnel " is frequently used in place of adit, but, technically, a tunnel is open to the surface at both ends. (Pl. 4-1)

Air-leg—a pneumatic-operated cylinder attached to a rock drill. The cylinder, with a small bore, and a piston, with a long stroke, supports the drill when it is used in drilling operations. (Pls. 5-4, 5-5)

Air-shaft—a shaft used for ventilating mines, downcast when transferring fresh air from the surface to underground workings, upcast when discharging exhaust air to the surface. (Pls. 1-4, 5-23)

Alluvial—generally pertains to loose gravel, soil, or mud which has been deposited by water.

Altered rock or mineral—a rock which has undergone chemical change since its deposition or emplacement.

Anisotropic—the physical properties of the material are not identical in all directions within a body.

Amalgam—an alloy of mercury with another metal or metals. Amalgamation is the process of recovering gold and silver with mercury.

Ampere--a unit of electrical current flow.

Analysis—the determination of one or more constituents of a substance either as to kind or amount.

Angle brace—a wooden brace used to prevent mine timbers from moving or leaning.

Apex (vein)--this term is used in mining law to denote the outcrop of a vein which is exposed on the surface, or the upper-most limit of a blind vein which does not outcrop.

Assay—the determination of valuable minerals in a sample. A wet assay is one determined with both chemicals and fire. Gold and silver are usually assayed by fire.

Attitude—is the direction and degree of strike and dip of a a vein or bed.

Axial—acting parallel to and along an axis.

Azimuth of a line--the angle that the line makes with the north direction, and it is always measured clockwise from the north.

Back—a term having different meanings underground according to the mining district. Usually means the overlying roof of a stope or drift.

Back-sight—a term used in surveying for any sight taken in a backward direction.

Bar mining—a long steel bar (usually 5 feet) with a sharp point on one end and a chisel point on the other end. Used to pull down loose rocks. (Pl. 6-6)

Bar placer—a bank of gold bearing sand or gravel usually found in the slack or quieter part of a stream.

Beam theory—treatment of a stope wall or roof as though it acts as a loaded beam.

Bearing of a line—always measured in a horizontal plane and defined as the acute angle (less than 90°) that the line makes with a north-south line.

Bedrock—any solid rock underlying gravels or other detrital material.

Benches. placer—step or terrace-like deposits of rock or gravel.
Open pit—step or terrace-like working areas on the sides of a pit to keep the sides from becoming too high.

Biaxial forces—stresses applied in two directions.

Bit—the part of a drill that comes into contact with and actually cuts or chips rock. Bits may be formed directly on the drill steel or they may be detatachable. (Pl. 5-1)

Black sand—grains of heavy, dark minerals, magnetite, ilmenite, chromite, etc., found in streams and commonly collect in sluice boxes. It may carry gold and platinum.

Blast—an explosion or violent detonation caused by the discharge of dynamite or other explosive material. (Pl. 7-6)

Blasting theory—discussed in plate 5-2.

Breast—the vertical end surface of a working heading; the same as a face.

Brunton compass—a special compass used in surveying with an attached clinometer and sighting device. (Pl. 3-6)

Brittle—breaking under load after minimum deformation.

Bulkhead—a timber or concrete dam to hold air, tailings or water in place.

Burst—a sudden violent release of energy contained in a stressed rock.

Canopy—the roof and steel section of a hydraulic roof support. (Pl. 8-5)

Cantilever beam—a beam fixed and supported at one end only.

Cap (explosive)—a small shell, closed at one end, and loaded with a charge of extremely sensitive explosives. A cap is used to detonate a dynamite charge. Avoid subjecting caps to shock or heat.

Cap (timber)—the uppermost heavy timber in a set, usually at right angles to the line of the drift and at right angles to the foot and hanging walls. (Pl. 5-12)

Capillarity—the attractive force between two unlike molecules, an example is the rising or the drawing-up of water in small interstices such as those between the grains of rock or soil.

Capillary water—that portion of soil water that is held by cohesion as a continuous film around the particle and in the capillary spaces. (Glossary of Geology)

Cave-in—the partial or complete collapse of a mine working.

Caving systems—stoping systems designed to take advantage of the natural tendency of some ores to cave. (Pls. 8-19 to 8-23)

Cemented soil—soil in which the grains are held together by some cementing agent such as silica, carbonates, iron minerals, or gypsum.

Centroid—the center axis of a beam where one side is the physical duplicate of the other side.

Chocks—a hydraulically actuated roof support. (Pl. 5-13)

Chute—an opening in the ground through which ore can pass from one level to another. It is also the structure built to load cars from a stope or ore-pass. (Pl. 5-19)

Claim—a portion of the earth's surface claimed by a prospector and marked out by stakes. Maximum size of a placer claim for a single person is 20 acres.

Clastic zone—the fractured, broken, or loose zone around an underground opening.

Clay—a natural material having plastic properties; it is composed of very fine particles, most of which are fragments of crystals of hydrous aluminum silicates or hydrous magnesium silicates. (Glossary of Geology)

Clean Mining—a system of mining where the ore is not diluted by waste rock.

Clinometer—an instrument used to measure angles in a vertical plane.

Cohesion (c)—the ability of a rock or soil to stick or stay together when force is applied to it.

Cohesive—to stick together. Cohesive strength resists pulling the body apart by a tensile force.

Collar—the surface at the top of a shaft, or at the start of a drill hole.

Compressive forces (stresses)—forces (stresses) acting on opposite sides of a body, which tend to compress or push it together.

Concentrate—the valuable material produced from an ore by a separating or concentrating process.

Concurrent—all forces, if moved along their lines of action, would pass through one common point in a plane.

Consolidometer—a device used to measure compressibility of a soil.

Continuous beam—a beam continuous over several supports.

Contour lines—lines connecting points of equal elevation. (Pl. 2-3)

Conveyor—a moving belt or scraper chain assembly that moves material in a near horizontal plane. (Pls. 7-11, 8-6)

Coordinates of a point—its position relative to some origin, generally designated 0, and the axes going through this origin.

Coplanar—all forces are acting in one continuous plane.

Couple—two forces equal in magnitude but opposite in direction and separated by a distance.

Creep—the slow transportation of soil by gravity alone. In mechanics creep is defined as the permanent deformation of a material subjected to a load for a long time, even though the material is not loaded to the elastic limit.

Crib—a system of timbering in which the members are laid one upon another to form a rectangular opening in the center or to build a solid timber support.

Cross-cut—a horizontal level driven at a large angle to the strike of a vein. (Pl. 4-1)

Cut—that part of a drill round that must break out before the rest of the holes can break. (Pl. 5-4)

Deformation—change of the shape of a body by the action of a force or stress.

Degree of slope—slope expressed in degrees. The tangent of the slope angle is equal to the vertical rise or fall per 100 feet of horizontal distance.

Density—the mass of a substance expressed in units of weight per unit of volume; ex. pounds per cubic foot.

Diamond drill—a machine-driven rotary drill with a diamond set bit which may or may not cut a core. The abrading action of the hard diamonds wears away the rock. (Pls. 3-7, 5-1)

Dip—the maximum angle of inclination downward that a vein or bed makes with a horizontal plane. (Pl. 1-2)

Double-jack—8 or 12 pound hammer used in hand drilling.

Drift—a horizontal underground opening driven along the course of a vein. It does not necessarily have to be driven in the vein.

Drifter—a percussion-type air drill used in drifting and long-hole drilling operations. It has special mountings which makes it suited for this type of work. (Pl. 5-5)

Drill-hole nomenclature:

Cuts—holes drilled to make the initial opening in the ground.
Rakers—holes drilled to clean out and enlarge the cut.
Easers—holes drilled to enlarge the cut.
Relievers—same as easers.
Enlargers—holes drilled to enlarge the opening.
Trim—holes drilled to outline the final opening.
Back—holes drilled at the top of the round to outline the opening.
Lifters—holes drilled at the bottom of the opening to bring the heading to grade.

Driving—the process of advancing a mine working, especially a drift or raise.

Elasticity—the ability of a material to regain its original shape and dimensions upon the removal of a force.

Elastic limit—the point where the stress-strain curve deviates from a straight line.

End lines—the end boundary lines of a mining claim which cross the vein. They should be parallel.

Entry—a term commonly used in coal mining for a horizontal, or nearly so, development opening. (Pl. 8-1)

Equilibrium—a state of balance between forces or actions so that the algebraic sum of all forces and all moments acting in and on a body equals zero.

Erosion—the weathering of rocks. May produce both physical and chemical changes in the material.

Exploration—the work involved in looking for ore. Prospecting. (Part 3)

Exposure—any part of a vein or rock outcrop that can be easily seen.

Extensometer—a device used to record movement of a pit slope or an underground opening with reference to some stable point.

Extralateral right—the right of a mine owner to follow a vein down beyond his side lines under a neighbor's ground if he has the apex on his claim, and his end lines are parallel.

Face—the working area or end of a development heading.

Failure angle (α) of a slope—angle between plane of failure and a horizontal plane; can be estimated from Mohr's circle of failure.

Failure envelope—a plot of normal and shear strength of rocks and soil to show where loading failure can be expected.

Fault—a fracture in the earth, one side of which is displaced with respect to the other. (Pl. 2-1)

Fee angle (ϕ)—the angle of internal friction as measured by direct shear or triaxial tests.

Fence—a bulkhead or dam constructed in a stope to contain tailings.

Finite element method—a method of solving complex structural problems; it is facilitated by using an electronic computer.

Firing (blasting)—the detonation of a charge of dynamite or other explosive. (Pl. 7-6)

Fissure—an opening or crack in rocks. A fissure vein is a fissure that has been filled with vein minerals.

Float—the loose and scattered pieces of ore which have broken from an outcrop. (Pl. 3-1)

Floor—the bottom or lowermost part of an underground working.

Footwall—the bottom or lower enclosing wall of a vein. (Pl. 3-5, 4-1)

Force (F)—load exerted by one body on another or the cause of acceleration of movement.

Friction—the force resisting relative motion between two bodies in contact. Its direction always opposes motion.

Friction angle (σ)—defined as cos σ = coefficient of friction (μ).

Gallows-frame—another name for a headframe. (Pls. 5-20, 5-21)

Geology—a science dealing with the formation of the earth and ore deposits.

Girt. Timbering—the upper horizontal member of a timber set, parallel to the direction of the drift or strike of the vein, which holds the posts and caps in position. (Pl. 5-11, 5-12)

Grade—amount of vertical rise or fall in 100 ft. of horizontal distance expressed in per cent. Also, value of the ore.

Gravel—small stones and pebbles or a mixture of sand and small stones; more specifically, fragments of rock larger and coarser than sand, worn by the action of air or water. (U.S.G.S.)

Grizzly—rails spaced at equal intervals over raises and hoppers to keep over size material from flowing into the openings. (Pl. 5-9)

Grout—a filling agent inserted by pressure into loose and broken rock or ground. It can be a fine cement slurry or certain types of chemicals.

Gunite—an air-blown mortar consisting of one part cement and three to five parts sand.

Hanging wall—the top or upper enclosing wall of a vein. (Pls. 3-5, 4-1)

Head—the pressure on a unit area due to the height of the fluid surface above the unit area. Usually it is expressed as feet of head of column or in pounds per square inch.

Headframe—a structure erected over a shaft to support the sheave wheel for hoisting purposes. (Pls. 5-20, 5-21)

Headgate—the conveyer system is located in this entry as shown (Pl. 8-5), also called maingate.

Heading—any part of the mine where work is under way. Usually confined to development workings only.

Highgrade ore—an ore that yields a large margin of profit per ton.

Hitch—a depression or hole cut into the rock to hold timbers in place. Stulls are commonly put in a hitch so that they will stay in position.

Homogeneous—of uniform consistency throughout, in every direction.

Hooke's law—the constant ratio between linear stress and the strain produced, within the elastic limit.

Horse—a mass of waste rock in the ore or in a vein.

Humidity—moisture or dampness contained in the air or atmosphere. See also relative humidity.

Hydraulic gradient—a quanity determined by dividing the amount by the length of soil through which the water is moving.

Hydrostatic—the forces imposed on a body immersed in a fluid.

Igneous rock—rock that has solidified from a molten state.

Impermeable—not allowing fluids to move readily through the rock or soil.

Impervious—same as impermeable.

Inclined—at an angle with the horizontal.

Inclinometer—a device used to measure the dip or inclination of a hole.

Intermediate level (sublevel)—a level driven from a raise or manway instead of the working shaft. (Pl. 1-4)

In-situ—in place-as used in mining, it means that the mineral zone is recovered by techniques used in solution mining.

Isotropic—the physical properties of the material are identical in all directions within the body.

Jackhammer—a trade name in common usage for a sinker-type, percussion air-drill.

Jumbo—a carriage containing drills which can be positioned with ease. Plates 5-5 and 5-6.

Lagging—planks or split-round timber placed around timbered sets to keep material from falling into the opening. Also used for flooring.

Lateral—a horizontal mine working. A drift in the footwall of a vein is often called a lateral. (Pl. 8-25)

Leaching—using chemicals or solvents to win the valuable material from ore-ore can be in heaps or left in place. (Plate 8-26)

Lease—"A contract by which one conveys real estate for life, for a term of years, or at will, usually for a specified rent; also, the act of such conveyance, or the term for which it is made."

Ledge—a horizontal layer, but often a vein or lode is referred to as a "ledge" of ore.

Lessee—a person who obtains a lease on mining land.

Lessor—the grantor of a lease.

Level—all the connected horizontal mine openings at a certain elevation. Generally levels are 100 to 200 feet apart, and are designated by their distance from the collar of a shaft or some other point of reference. (Pl. 1-4)

Load-haul-dump (LHD)—is a machine that can load and carry a large quantity of material to a dump site. (Pls. 5-16)

Longwall mining system—used especially in coal mines, is the system whereby most of the coal is mined and the roof over the worked-out area is allowed to cave. (Pl's. 8-4 through 8-7)

Manway—underground passage, vertical or inclined, for the accomodation of ladders, pipes, and timber slides.(Pl. 8-8)

Milling ore—ore that must be concentrated at or near the mine before it is shipped.

Mill tailings—the crushed and ground rock reject material from an ore concentrating plant-often used as hydraulically placed fills in mines.

Mineral—a homogeneous substance of fairly definite chemical composition and physical properties found in nature and not directly a product of life or the decay of a living thing.

Mineralogy—the science of minerals and their identification.

Miners inch—water measure: In Montana, 2½-cubic feet of water flowing per second is one hundred miners inches.

Misfire—failure of a dynamite charge to explode.

Modulus of elasticity—the stress divided by the strain at the elastic limit.

Muck—a common term for any broken ore or waste underground.

Mucker—a shoveler, a person or machine who handles muck.

Mudflow—a flow of heterogeneous debris when lubricated with ample water. Mudflows tend to follow stream channels.

Nonhomogeneous—not of uniform consistency throughout.

Normal force—the component of a force acting 90° (perpendicular) to a plane.

Normal strain—the linear deformation of a stressed body divided by the original length before being stressed.

Normal stress—a stress acting 90° (perpendicular) to a plane. Mathematically $\sigma = F/A$ where σ = stress in psi, F = Force in lb., A = Area in sq. in.

Option—"The power or right of choice." In the case of a mining property, it is the right to purchase at a stated price.

Ore—a mineral aggregate of sufficient value, both in quantity and quality, which can be mined at a profit.

Ore body—the part of a vein that carries ore. Generally all parts of a vein are not ore.

Ore shoot—the valuable parts of a vein. Same as ore body.

Orientation—the trend or position of a feature in spatial relation to its environment or to other features.

Outcrop—the outcrop is the edge or surface of a mineral deposit or sedimentary bed which appears upon the surface.

Overburden—the valueless dirt and material overlying the pay zone in a placer deposit or the valueless solid outcrop of an orebody.

Oxidation—combination with oxygen. Burning is one form of oxidation, although it can take place without flame.

Oxide—a compound of a metal and oxygen.

Oxidized zone—the upper part of an orebody that has been altered by weathering and as a result contains oxygen-bearing minerals.

Pack—an artificial support system made from shaped rocks, concrete or pumice blocks.

Patent—a written title to land granted by the government after filling certain obligations. A mining claim can be patented after $500.00 worth of work has been done. The owner is then relieved of annual assessment work, but the land is taxed.

Percussion drill. Mining:—a compressed-air-actuated hammer-type rock drill. (Pl's. 5-1, 5-5, 6-11)

Permeable—soil or rock permits fluids to flow through its pores or openings.

Perpendicular—at right angles (90°) to a given line or surface.

Pervious—same as permeable, the property of a soil or rock that permits fluids to flow through its pores or openings.

Piezometer—a tube used to measure the static water level at a specific point.

Piezometric level—height to which water will rise in a tightly sealed standpipe, casing, or tube with respect to a fixed datum plane.

Pillar—ore or waste left in place underground to give support to the wall or roof (back). (Pl's. 5-11, 8-25)

Piping—the action of underground water to form a channel or to produce a "quick" condition.

Pitch—the same as the rake of ore.

Pit failure—collapse or slump of pit wall; some ways are shown in Plate 4-4.

Placer—any deposit of minerals not in place (gold, platinum, or tin occuring in gravel is called a placer deposit). (Pl's. 7-1 through 7-3)

Plane—a flat even surface. A horizontal plane is a flat, even, and level surface.

Plastic—material which will deform under load and not regain its original shape.

Poisson's ratio—the lateral strain divided by the axial strain.

Porosity—the void space in rock or soil, usually expressed as the ratio of the volume of voids to the total volume of the material.

Porous—describes a body of rock or soil that contains tiny spaces in or between particles of solid material. These spaces may be filled with air, water, or other fluids, or a mixture of fluids.

Post—an upright supporting member in a set of timber.

Powder—the miner's term for dynamite.

Pressure—the force per unit area.

Pressure dome theory—another name for the high-stress and clastic zone phenomena around an underground opening.

Principle stress (strain)—stress acting on a plane in a stressed body where the shearing stress is zero. It can be a maximum or a minimum value of stress.

Promontory—a sharp-angled pillar or rock projecting into a mined-out stope.

Prop—a post (timber is commonly used) to support loose rocks, slabs, or the roof. This terminology is common in coal mines and compares to a stull in a metal mine.

Protractor—an instrument for laying out and measuring angles.

Public domain—areas of mineral land owned by the United States Government open for location of mining claims.

Qualitative. Ore testing:—what kind of valuable minerals or elements are present in a sample of ore.

Quantitative. Ore testing:—how much valuable mineral or element is present in a sample of ore.

Quick—condition that results when a soil loses all shear strength and bearing capacity and in effect becomes a liquid.

Radioactive—"The emission of radiant energy; the property possessed by certain elements (chiefly radium, urnium, thorium, and their products), of spontaneously emitting alpha or beta rays, and also gamma rays, by the disintegration of the nuclei of atoms." These elements can be detected by radiation counters.

Raise—an exacavation of restricted cross section driven upward, either vertically or at an angle, from a level in the mine. (Pl's. 1-3, 1-4)

Rake—the trend of the ore body within the vein.

Relative humidity—the ratio of the amount of water vapor actually present to the greatest amount possible at the given temperature; usually expressed in per cent. Air that is completely saturated has a relative humidity of 100 per cent.

Riffle—grooves, channels, slats, or wire screens in a sluice box or rocker to catch gold or other valuable minerals. (Pl. 7-3)

Right angle—an angle of 90°.

Rock bolts—steel bolts inserted in holes drilled in the rock and anchored to help support the rock in place. (Pl's. 5-11 through 5-13)

Rock burst—a sudden violent release of energy contained in a stressed rock. Most bursts occur in brittle rocks and can have devasting consequences.

Roof—the overhead part of a mine working. Same as back.

Roof sewing—a method of inserting and grouting steel cable into drill holes to support the opening.

Room and pillar—a system of mining in which part of the coal or ore is mined out and the rest is left in place as pillars for support. (Pl. 8-1)

Rotational failure—the common way in which soil walls fail, usually producing a failure surface that has a spoon or bowl-shaped appearance. For example, in pits.

Round—the group of drill holes in a face to ensure ccm-plete breakage of the ore or rock when the holes are blasted. (Pl. 5-4)

Sand—detrital material ranging from 1/16 to 2 mm in diameter. (Glossary of Geology) Very coarse 1-2 mm; course 1/2-1 mm; medium 1/4-1/2 mm; fine 1/8-1/4 mm; very fine 1/16-1/8 mm.

Saturated—thoroughly soaked, all pore spaces filled with water (or other liquid).

Scale—a graduated rule used to measure or plot distances on a map or chart.

Sensitive clay—clay that may, if remolded, suddenly sof-tens, and lose much of its strength.

Shaft (inclined)—a shaft that is not vertical but follows the dip of the vein or coal bed. (Pl. 6-5)

Shaft (vertical)—a vertical excavation, commonly made from the surface; compared to its depth, it has a small cross-section area. A shaft may be used for finding and mining ore, for draining water, for ventilation, or for lowering and hoisting men and materials. (Pl's. 6-1 thru 6-4)

Shear force—the component of a force acting parallel to a plane.

Shearer—machine used to shear coal in the longwall mining system. (Pl. 8-6)

Shearing modulus of elasticity—the shear stress divided by the shear strain at the elastic limit. It is seldom, if ever, equal to E.

Shear-strain—the angular deformation of a stressed body, measured as deformation, perpendicular to the length, divided by the original length. It is expressed as in./in. or radians.

Shear stress—stress acting parallel to a plane. Mathematically $\tau = F/A$ when τ = stress in psi, F = Force in lb., A = Area in sq. in.

Shortwall—a new mining development using continuous miners and hydraulic roof supports. (Pl. 8-7)

Shotcrete—the name usually given to mixes containing coarse aggregate, sand, and cement, mixed and applied under hydraulic pressure (sprayed).

Shield—a recent hydraulic actuated roof support of special design-becoming popular in longwall systems. (Pl. 5-13)

Shrinkage compensated cement—cement which contains enough expansive cement so that shrinkage is com-pletely compensated by the expansion.

Silicosis—lung trouble caused by silica or quartz dust.

Sill—the floor or bottom of a passage in a mine, the bottom part of a timbered slope: a piece of timber laid across a drift or crosscut to hold the posts in a drift set.

Single-jack—a light hammer, commonly 4-lb., used in hand drilling.

Sinking—the driving or excavating of a shaft or winze.

Sinker—a hand-held drill primarily designed to drill down holes.

Slip—a small fault.

Slope—the horizontal distance traversed for vertical rise or drop 2:1 slope; for every 2 ft. of horizontal distance a rise or drop of 1 ft. occurs.

Sluice box—a trough paved with riffles through which gravel and wash from placer mining operations is passed so the gold and other valuable minerals will be caught and saved. (Pl's. 7-1, 7-3)

Sluicing—washing gravel through a sluice-box. (Pl's. 7-1, 7-3)

Slusher—a small double-drum hoist used to pull a bucket in underground openings to move ore or waste. (Pl. 5-17)

Soil—as used in this publication, is any unconsolidated material composed of individual solid particles; gases, liquids, or both fill the spaces between solid particles.

Soil sampler—a device used to collect an undisturbed soil sample.

Specific gravity—the ratio of the mass of a body to the mass of an equal volume of water at a specified tem-perature.

Speed—the time rate of motion, measured as the distance moved per unit of time.

Steel change—different lengths of steel needed when drilling a hole. The common steel-length difference when using a stoper drill is 18 in.

Stemming (tamping)—material tamped into a drill hole, after the dynamite has been placed, to increase the effect of the blast.

Stope—any excavation underground to remove the ore, other than the development work. The outlines of a stope are determined either by the limits of the ore body or by raises. (Pl. 1-4)

Stoper—a type of drill used in stoping and raising operations. (Pl. 5-5)

Stowing—a mining method in which broken waste rock is placed in a mined-out zone.

Strain—the total deformation of a stressed body divided by the original length before being stressed.

Strain gauge—an electric wire designed so that it is distorted, a different electrical resistance will be displayed. The gauge is securely fastened to an unloaded member and its electric resistance is noted. After loading, the resistance is read again; by proper calibration, the strain can be read.

Strain rosette—strain gauges mounted in a precise relationship to one another to measure strains in certain directions, from which strains in all directions can be calculated.

Stress—the total force divided by the area. It is expressed either as normal stress σ (sigma) or shear stress τ (tau).

Stress riser—a configuration such as a sharp notch, a sharp junction, or a crack, which will cause stress in a member to be above average next to the configuration.

Stress-strain curves—plots of strains produced by stresses.

Strike—the bearing of a horizontal line in the plane of a vein, bed, or fault with respect to the cardinal points of the compass.

Stripping—the removal of overburden from a placer deposit, coal bed, or ore deposit. (Pl. 7-7)

Stull—a timber used to support loose rock, slabs, or the hanging wall. It may also be used to support a platform in a working place. (Pl. 5-11)

Subsidence—the sinking or settling of material, especially over an underground mining operation. (Pl. 4-4)

Sump—a hole or excavation used to collect water on a level or at the bottom of a shaft. (Pl. 5-22)

Tailgate—the passage way in a longwall coal mine that the tail end of the longwall face conveyor occupies. (Pl. 8-5)

Tamping—the gentle forcing and compaction of dynamite and stemming into a drill hole.

Tenor—the average amount or percentage of valuable constituents in an ore, concentrate or tailing.

Tensile forces (stresses)—acting on opposite sides of a body, which tend to pull it apart.

Tonnage—the number of tons (2,000 lbs. = short ton; 2,240 lbs. = long ton) in an ore body.

Torque—effect tending to produce rotation in a shaft; the magnitude of torque is equal to force times distance.

Transported (referring to solid particles)—material carried by wind, glaciers, or water from a former location.

Trend—the general direction or bearing of a vein, ore shoot, fault, or rock outcrop.

Triaxial forces (stresses)—forces that act on all sides of a body.

Triaxial test—one in which the sample undergoes lateral or side stress.

Triaxial test cell—a device to test soils for shear strength under triaxial states of stress.

Trommel—a revolving screen used in placer mining. (Pl. 7-1)

Tugger hoist—a small air-operated hoist often used underground for hoisting timber and other mine supplies.

Turbelent flow—flow of a fluid in which the paths of the individual particles are irregular and haphazard.

Unconfined compression test—one in which the lateral or side stress is zero.

Uniaxial forces (stresses)—forces (stresses) acting in one direction only, but these forces (stresses) act on a body at 180° to one another.

Unpatented claims—a claim held by location, which must have $100.00 worth of improvement made on it each year for the owner to retain undisputed possession. When a claim is patented, the assessment work is not required.

Vane-shear tester—a device used to test shear strength in soils, especially in clays.

Vector—a physical quantity that has both magnitude and direction.

Vein—a well-defined tabular mineralized zone, which may or may not contain ore bodies.

Velocity—the time rate of motion in a given direction; it is expressed as distance traveled per unit of time, e.g. miles per hour.

Void—a general term for openings or pore spaces in rock or soil.

Void ratio—the volume of voids divided by the volume of solids in a soil.

Wall—the waste or country rock on each side of a vein. The upper or overlying wall of an inclined vein is the hanging wall; the underlying or lower wall, the foot-wall.

Wash—barren rock or mineralized material that does not have enough value to be classified as ore.

Weathering—the alteration of rock by such near-surface processes as the chemical action of air, water, plants, and bacteria, and the mechanical action of temperature change. Rock exposed to prolonged weathering changes in character, decays, and finally crumbles into soil.

Winze—a shaft-like opening sunk from a level in a mine to develop, dewater, or ventilate the ore zones. (Pl. 1-3)

Work—the product of force times distance, but the force must move through an actual distance.

Working face—any part of the mine where work is under way, such as the face of a drift or the face of a raise.

Yield—to give way to overpowering force.

PARTIAL LIST OF FURTHER READING IN MINING

1. **Anatomy of a Mine**, U.S. Forest Service SEAM, U.S. Department of Agriculture.

2. **An Introduction to Mining**, Thomas, 1977, Halsted Press. (ISBN 0-470-99220-4, LC 73-14857)

3. **Coal Age Operating Handbook of Underground Mining**, 1977, McGraw Hill. (ISBN 0-011457-9, LC 77-85375)

4. **Elements of Mining**, Lewis and Clark, 3rd Edition, Wiley & Son.

5. **E/MJ Operating Handbook of Mineral Surface Mining and Exploration**, Volumes 1 and 2, 1978, McGraw Hill. (ISBN 07-019518-8, LC 78-4501)

6. **E/MJ Operating Handbook of Mineral Underground Mining**, 1978, McGraw Hill. (ISBN 07-019521-8, LC 78-13250)

7. **Energy and Transport: Recent Advances in Mining and Processing of Low-Grade Submarginal Mineral Deposits**, 1977, Pergamon. (ISBN 0-08-021051-1, LC 76-11771)

8. **Exploration and Mining Geology**, Peters, 1978, Wiley. (ISBN 0-471-68261-6, LC 77-14006)

9. **Gold and Silver Cyanidation Plant Practice**, Monograph 1975, Soc. Mining Eng., AIME. (ISBN 0-89520-027-9)

10. **Handbook for Small Mining Enterprises**, Earll, Bull. 99, 1976, Montana Bureau of Mines and Geology, Butte, MT 59701.

11. **Introduction to Mining: Exploration, Feasibility, Extraction, Rock Mechanics**, Thomas, 1973, Halsted Press. (ISBN 0-470-85951-2, LC 73-14857)

12. **Mineral Industry Costs**, Hoskins, Northwest Mining Assoc., West 1020 Riverside Ave., Spokane, Washington 99201.

13. **Mining Engineers Handbook** Volume 2, Peele, 1941, Wiley-Interscience. (ISBN 0-471-67716-7)

14. **SME Mining Engineering Handbook**, 1973, Soc. of Min. Eng., AIME. (ISBN 0-89520-021-X, LC 72-86922)

15. **Soil and Rock Mechanics Illustrated**, Stout, Bull. 96, 1975, Montana Bureau of Mines and Geology, Butte, MT 59701.

16. **Surface Mining**, 1968, Soc. of Min. Eng., AIME. (ISBN 0-89520-002-3, LC 68-24169)

17. **Symposium on Solution Mining**, 1974, Soc. of Min. Eng., AIME. (ISBN 0-89520-025-2, LC 73-94005)

ISBN – International Serials Book Number
LC – Library of Congress